Additional Praise for
Breaking the Spell: Religion as a Natural Phenomenon

"*Breaking the Spell* continues to spark controversy with its Darwinian take on the all-too-human urge to believe. . . . Dawkins, Dennett, and Harris argue that religious beliefs, particularly those derived literally and selectively from religious texts, can lead to behavior that is dubiously moral according to more universal principles of right and wrong. The killing of innocents in the name of holy war is only the most obvious instance. . . . Dennett extends a conciliatory hand to believers so long as they are willing to subject any purportedly God-given moral edict 'to the full light of reason, using all the evidence at our command.'"
—*U.S. News & World Report*

"Rich and rewarding . . . the main business of the book is to give a scientific account of how religion may have developed among creatures such as us. . . . Dennett here does an admirable job of dismantling the common (yet almost never argued for) prejudice that those who believe in the supernatural have some sort of moral superiority."
—*San Francisco Chronicle*

"In *Breaking the Spell*, the blasting rhetoric of Dawkins and Harris is absent, replaced by provocative, often humorous examples and thought experiments."
—*Wired*

"Mr. Dennett's main argument is that religious belief—especially in the United States—is often sheltered from the cut and thrust of intellectual argument and scientific scrutiny, and it should not be."
—*The Economist*

"Dennett describes the various stages of the long historical evolution of religion, beginning with primitive tribal myths and rituals, and ending with the market driven evangelical megachurches of modern America. . . . He emphasizes that his explanation of the evolution of religion is testable with the methods of science. It could be tested by quantitative measurements of the transmissibility and durability of various belief systems. . . . [His] account is on the whole fair and well balanced."
—*The New York Review of Books*

PENGUIN BOOKS

BREAKING THE SPELL

Daniel C. Dennett is University Professor, professor of philosophy, and co-director of the Center for Cognitive Studies at Tufts University. His books include *Brainstorms*, *Elbow Room*, *Consciousness Explained*, *Darwin's Dangerous Idea* (a finalist for the National Book Award and the Pulitzer Prize), and *Freedom Evolves*. He lives in North Andover, Massachusetts.

PENGUIN BOOKS

DANIEL C. DENNETT

BREAKING THE SPELL

RELIGION AS A NATURAL PHENOMENON

PENGUIN BOOKS

Published by the Penguin Group
Penguin Group (USA) Inc., 375 Hudson Street, New York, New York 10014, U.S.A.
Penguin Group (Canada), 90 Eglinton Avenue East, Suite 700, Toronto,
Ontario, Canada M4P 2Y3 (a division of Pearson Penguin Canada Inc.)
Penguin Books Ltd, 80 Strand, London WC2R 0RL, England
Penguin Ireland, 25 St Stephen's Green, Dublin 2, Ireland (a division of Penguin Books Ltd)
Penguin Group (Australia), 250 Camberwell Road, Camberwell,
Victoria 3124, Australia (a division of Pearson Australia Group Pty Ltd)
Penguin Books India Pvt Ltd, 11 Community Centre, Panchsheel Park, New Delhi – 110 017, India
Penguin Group (NZ), 67 Apollo Drive, Mairangi Bay, Auckland 1311, New Zealand
(a division of Pearson New Zealand Ltd)
Penguin Books (South Africa) (Pty) Ltd, 24 Sturdee Avenue,
Rosebank, Johannesburg 2196, South Africa

Penguin Books Ltd, Registered Offices:
80 Strand, London WC2R 0RL, England

First published in the United States of America by Viking Penguin,
a member of Penguin Group (USA) Inc. 2006
Published in Penguin Books 2007

10   9   8   7   6   5   4   3   2   1

THE LIBRARY OF CONGRESS HAS CATALOGED THE HARDCOVER EDITION AS FOLLOWS:
Dennett, Daniel Clement.
Breaking the spell : religion as a natural phenomenon / Daniel C. Dennett.
p.   cm.
Includes bibliographical references and index.
ISBN 0-670-03472-X (hc.)
ISBN 978-0-14-303833-7 (pbk.)
1. Religion—Controversial literature.   I. Title.
BL2775.3.D46 2006
200—dc22   2005042415

Printed in the United States of America
Set in Scala with Berkeley • Designed by Carla Bolte

FOR SUSAN

Contents

Preface

Let me begin with an obvious fact: I am an American author, and this book is addressed in the first place to American readers. I shared drafts of this book with many readers, and most of my non-American readers found this fact not just obvious but distracting—even objectionable in some cases. Couldn't I make the book less provincial in outlook? Shouldn't I strive, as a philosopher, for the most universal target audience I could muster? No. Not in this case, and my non-American readers should consider what they can learn about the situation in America from what they find in this book. More compelling to me than the reaction of my non-American readers was the fact that so few of my American readers had any inkling of this bias—or, if they did, they didn't object. That is a pattern to ponder. It is commonly observed—both in America and abroad—that America is strikingly different from other First World nations in its attitudes to religion, and this book is, among other things, a sounding device intended to measure the depths of those differences. I decided I had to express the emphases found here if I was to have any hope of reaching my intended audience: the curious and conscientious citizens of my native land—as many as possible, not just the academics. (I saw no point in preaching to the choir.) This is an experiment, a departure from my aims in earlier books, and those who are disoriented or disappointed by the departure now know that I had my reasons, good or bad. Of course I may have missed my target. We shall see.

My focus on America is deliberate; when it comes to contemporary religion, on the other hand, my focus on Christianity first, and Islam and Judaism next, is unintended but unavoidable: I simply do not know enough about other religions to write with any confidence about them. Perhaps I should have devoted several more years to study before writing this book, but since the urgency of the message was borne in on me again and again by current events, I had to settle for the perspectives I had managed to achieve so far.

One of the departures from my previous stylistic practices is that for once I am using endnotes, not footnotes. Usually I deplore this practice, since it obliges the scholarly reader to keep an extra bookmark running while flipping back and forth, but in this instance I decided that a reader-friendly flow for a wider audience was more important than the convenience of scholars. This then let me pack rather more material than usual into rather lengthy endnotes, so the inconvenience has some recompense for those who are up for the extra arguments. In the same spirit, I have pulled four chunks of material meant mainly for academic readers out of the main text and deposited them at the end as appendixes. They are referred to at the point in the text where otherwise they would be chapters or chapter sections.

\\\

Once again, thanks to Tufts University, I have been able to play Tom Sawyer and the whitewashed fence with a remarkably brave and conscientious group of students, mostly undergraduates, who put their own often deeply held religious convictions on the line, reading an early draft in a seminar in the fall of 2004, correcting many errors, and guiding me into their religious worlds with good humor and tolerance for my gaffes and other offenses. If I do manage to find my target audience, their feedback deserves much of the credit. Thank you, Priscilla Alvarez, Jacquelyn Ardam, Mauricio Artinano, Gajanthan Balakaneshan, Alexandra Barker, Lawrence Bluestone, Sara Brauner, Benjamin Brooks, Sean Chisholm, Erika Clampitt, Sarah Dalglish, Kathleen Daniel, Noah Dock, Hannah Ehrlich, Jed Forman, Aaron Goldberg, Gena Gorlin, Joseph Gulezian, Christopher Healey, Eitan Hersh, Joe Keating, Matthew Kibbee, Tucker Lentz, Chris Lintz, Stephen Martin, Juliana McCanney, Akiko Noro,

David Polk, Sameer Puri, Marc Raifman, Lucas Recchione, Edward Rossel, Zack Rubin, Ariel Rudolph, Mami Sakamaki, Bryan Salvatore, Kyle Thompson-Westra, and Graedon Zorzi.

Thanks also to my happy team in the Center for Cognitive Studies, the teaching assistants, research assistants, research associate, and program assistant. They commented on student essays, advised students who were upset by the project, advised me; helped me devise, refine, copy, and translate questionnaires; entered and analyzed data; retrieved hundreds of books and articles from libraries and Web sites; helped one another, and helped keep me on track: Avery Archer, Felipe de Brigard, Adam Degen Brown, Richard Griffin, and Teresa Salvato. Thanks as well to Chris Westbury, Diana Raffman, John Roberts, John Symons, and Bill Ramsey for their participation at their universities in our questionnaire project, which is still under way, and to John Kihlstrom, Karel de Pauw, and Marcel Kinsbourne for steering me to valuable reading.

Special thanks to Meera Nanda, whose own brave campaign to bring scientific understanding of religion to her native India was one of the inspirations for this book, and also for its title. See her book *Breaking the Spell of Dharma* (2002) as well as the more recent *Prophets Facing Backwards* (2003).

The readers mentioned in the first paragraph include a few who have chosen to remain anonymous. I thank them, and also Ron Barnette, Akeel Bilgrami, Pascal Boyer, Joanna Bryson, Tom Clark, Bo Dahlbom, Richard Denton, Robert Goldstein, Nick Humphrey, Justin Junge, Matt Konig, Will Lowe, Ian Lustick, Suzanne Massey, Rob McCall, Paul Oppenheim, Seymour Papert, Amber Ross, Don Ross, Paul Seabright, Paul Slovak, Dan Sperber, and Sue Stafford. Once again, Terry Zaroff did an outstanding copyediting stint for me, picking up not just stylistic slips but substantive weaknesses as well. Richard Dawkins and Peter Suber are two who provided particularly valuable suggestions in the course of conversations, as did my agent, John Brockman, and his wife, Katinka Matson, but let me also thank, without naming them, the many other people who have taken an interest in this project over the last two years and provided much-appreciated suggestions, advice, and moral support.

Finally, I must once again thank my wife, Susan, who makes every book of mine a duet, not a solo, in ways I could never calculate.

Daniel Dennett

PART I

‖‖\‖‖‖

OPENING PANDORA'S BOX

CHAPTER ONE

||||\\\||||

Breaking Which Spell?

1 What's going on?

And he spake many things unto them in parables, saying, Behold, a sower went forth to sow; And when he sowed, some seeds fell by the way side, and the fowls came and devoured them up. —Matthew 13:3–4

If "survival of the fittest" has any validity as a slogan, then the Bible seems a fair candidate for the accolade of the fittest of texts.
  —Hugh Pyper, "The Selfish Text: The Bible and Memetics"

You watch an ant in a meadow, laboriously climbing up a blade of grass, higher and higher until it falls, then climbs again, and again, like Sisyphus rolling his rock, always striving to reach the top. Why is the ant doing this? What benefit is it seeking for itself in this strenuous and unlikely activity? Wrong question, as it turns out. No biological benefit accrues to the ant. It is not trying to get a better view of the territory or seeking food or showing off to a potential mate, for instance. Its brain has been commandeered by a tiny parasite, a lancet fluke *(Dicrocelium dendriticum)*, that needs to get itself into the stomach of a sheep or a cow in order to complete its reproductive cycle. This little brain worm is driving the ant into position

3

to benefit *its* progeny, not the ant's. This is not an isolated phenomenon. Similarly manipulative parasites infect fish, and mice,
among other species. These hitchhikers cause their hosts to behave
in unlikely—even suicidal—ways, all for the benefit of the guest,
not the host.[1]

Does anything like this ever happen with human beings? Yes indeed. We often find human beings setting aside their personal interests, their health, their chances to have children, and devoting
their entire lives to furthering the interests of an *idea* that has
lodged in their brains. The Arabic word *islam* means "submission,"
and every good Muslim bears witness, prays five times a day, gives
alms, fasts during Ramadan, and tries to make the pilgrimage, or
hajj, to Mecca, all on behalf of the idea of Allah, and Muhammad, the
messenger of Allah. Christians and Jews do likewise, of course, devoting their lives to spreading the Word, making huge sacrifices,
suffering bravely, risking their lives for an idea. So do Sikhs and
Hindus and Buddhists. And don't forget the many thousands of
secular humanists who have given their lives for Democracy, or Justice, or just plain Truth. There are many ideas to die for.

Our ability to devote our lives to something we deem more important than our own personal welfare—or our own biological
imperative to have offspring—is one of the things that set us aside
from the rest of the animal world. A mother bear will bravely defend a food patch, and ferociously protect her cub, or even her
empty den, but probably more people have died in the valiant attempt to protect sacred places and texts than in the attempt to protect food stores or their own children and homes. Like other
animals, we have built-in desires to reproduce and to do pretty
much whatever it takes to achieve this goal, but we also have
creeds, and the ability to transcend our genetic imperatives. This
fact does make us different, but it is itself a biological fact, visible
to natural science, and something that requires an explanation
from natural science. How did just one species, *Homo sapiens,*
come to have these extraordinary perspectives on their own lives?

Hardly anybody would say that the most important thing in life is having more grandchildren than one's rivals do, but this is the default *summum bonum* of every wild animal. They don't know any better. They can't. They're just animals. There is one interesting exception, it seems: the dog. Can't "man's best friend" exhibit devotion that rivals that of a human friend? Won't a dog even die if need be to protect its master? Yes, and it is no coincidence that this admirable trait is found in a domesticated species. The dogs of today are the offspring of the dogs our ancestors most loved and admired in the past; without even trying to breed for loyalty, they managed to do so, bringing out the best (by their lights, by our lights) in our companion animals.[2] Did we unconsciously model this devotion to a master on our own devotion to God? Were we shaping dogs in our own image? Perhaps, but then where did we get our devotion to God?

The comparison with which I began, between a parasitic worm invading an ant's brain and an idea invading a human brain, probably seems both far-fetched and outrageous. Unlike worms, ideas aren't alive, and don't *invade brains;* they are *created by minds.* True on both counts, but these are not as telling objections as they first appear. Ideas aren't alive; they can't see where they're going and have no limbs with which to steer a host brain even if they could see. True, but a lancet fluke isn't exactly a rocket scientist either; it's no more intelligent than a carrot, really; it doesn't even have a brain. What it has is just the good fortune of being endowed with features that affect ant brains in this useful way whenever it comes in contact with them. (These features are like the eye spots on butterfly wings that sometimes fool predatory birds into thinking some big animal is looking at them. The birds are scared away and the butterflies are the beneficiaries, but are none the wiser for it.) An inert idea, if it were designed just right, *might* have a beneficial effect on a brain without having to know it was doing so! And if it did, it might prosper because it had that design.

The comparison of the Word of God to a lancet fluke is unsettling, but the idea of comparing an idea to a living thing is not new. I have a page of music, written on parchment in the mid-sixteenth century, which I found half a century ago in a Paris bookstall. The text (in Latin) recounts the moral of the parable of the Sower (Matthew 13): *Semen est verbum Dei; sator autem Christus.* The Word of God is a seed, and the sower of the seed is Christ. These seeds take root in individual human beings, it seems, and get those human beings to spread them, far and wide (and in return, the human hosts get eternal life—*eum qui audit manebit in eternum*).

How are ideas created by minds? It might be by miraculous inspiration, or it might be by more natural means, as ideas are spread from mind to mind, surviving translation between different languages, hitchhiking on songs and icons and statues and rituals, coming together in unlikely combinations in particular people's heads, where they give rise to yet further new "creations," bearing family resemblances to the ideas that inspired them but adding new features, new powers as they go. And perhaps some of the "wild" ideas that first invaded our minds have yielded offspring that have been domesticated and tamed, as we have attempted to become their masters or at least their stewards, their shepherds. What are the ancestors of the domesticated ideas that spread today? Where did they originate and why? And once our ancestors took on the goal of spreading these ideas, not just harboring them but cherishing them, how did this *belief in belief* transform the ideas being spread?

The great ideas of religion have been holding us human beings enthralled for thousands of years, longer than recorded history but still just a brief moment in biological time. If we want to understand the nature of religion today, as a natural phenomenon, we have to look not just at what it is today, but at what it used to be. An account of the origins of religion, in the next seven chapters, will provide us with a new perspective from which to look, in the last three chapters, at what religion is today, why it means so much to so many people, and what they might be right and wrong about in their self-understanding as religious people. Then we can see better where religion might be heading in the near future, our future on this planet. I can think of no more important topic to investigate.

2 A working definition of religion

Philosophers stretch the meaning of words until they retain scarcely anything of their original sense; by calling "God" some vague abstraction which they have created for themselves, they pose as deists, as believers, before the world; they may even pride themselves on having attained a higher and purer idea of God, although their God is nothing but an insubstantial shadow and no longer the mighty personality of religious doctrine. —Sigmund Freud, *The Future of an Illusion*

How do I define religion? It doesn't matter *just* how I define it, since I plan to examine and discuss the neighboring phenomena that (probably) aren't religions—spirituality, commitment to secular organizations, fanatical devotion to ethnic groups (or sports teams), superstition. . . . So, wherever I "draw the line," I'll be going over the line in any case. As you will see, what we usually call religions are composed of a variety of quite different phenomena, arising from different circumstances and having different implications, forming a loose family of phenomena, not a "natural kind" like a chemical element or a species.

What is the essence of religion? This question should be considered askance. Even if there is a deep and important affinity between many or even most of the world's religions, there are sure to be variants that share some typical features while lacking one or another "essential" feature. As evolutionary biology advanced during the last century, we gradually came to appreciate the deep reasons for grouping living things the way we do—sponges are animals, and birds are more closely related to dinosaurs than frogs are—and new surprises are still being discovered every year. So we should expect—and tolerate—some difficulty in arriving at a counterexample-proof definition of something as diverse and complex as religion. Sharks and dolphins look very much alike and behave in many similar ways, but they are not the same sort of thing at all. *Perhaps,* once we understand the whole field better, we will see that Buddhism and Islam, for all their similarities, deserve to be considered two entirely different species of cultural phenomenon. We can start with common sense and tradition and consider them both to be religions, but we shouldn't blind ourselves to the prospect that our initial sorting may have to be adjusted as we learn more. Why is *suckling one's young* more fundamental than *living in the ocean?* Why is *having a backbone* more fundamental than *having wings?* It may be obvious now, but it wasn't obvious at the dawn of biology.

In the United Kingdom, the law regarding cruelty to animals draws an important moral line at whether the animal is a vertebrate: as far as the law is concerned, you may do what you like to a live worm or fly or shrimp, but not to a live bird or frog or mouse. It's a pretty good place to draw the line, but laws can be amended, and this one was. Cephalopods—octopus, squid, cuttlefish—were recently made *honorary vertebrates,* in effect, because they, unlike their close mollusc cousins the clams and oysters, have such strikingly sophisticated nervous systems. This seems to me a wise political adjustment, since the similarities that mattered to the law and morality didn't line up perfectly with the deep principles of biology.

We may find that drawing a boundary between *religion* and its nearest neighbors among cultural phenomena is beset with similar, but more vexing, problems. For instance, since the law (in the United States, at least) singles out religions for special status, declaring something that has been regarded as a religion to be really something else is bound to be of more than academic interest to those involved. Wicca (witchcraft) and other New Age phenomena have been championed as religions by their adherents precisely in order to elevate them to the legal and social status that religions have traditionally enjoyed. And, coming from the other direction, there are those who have claimed that evolutionary biology is really "just another religion," and hence its doctrines have no place in the public-school curriculum. Legal protection, honor, prestige, and a *traditional exemption from certain sorts of analysis and criticism*—a great deal hinges on how we define religion. How should I handle this delicate issue?

Tentatively, I propose to define religions as *social systems whose participants avow belief in a supernatural agent or agents whose approval is to be sought.* This is, of course, a circuitous way of articulating the idea that a religion without *God* or *gods* is like a vertebrate without a backbone.3 Some of the reasons for this roundabout language are fairly obvious; others will emerge over time—and the definition is subject to revision, a place to start, not something carved in stone to be defended to the death. According to this definition, a devout Elvis Presley fan club is not a religion, because, although the members may, in a fairly obvious sense, *worship* Elvis, he is not deemed by them to be literally supernatural, but just to have been a particularly superb human being. (And if some fan clubs decide that Elvis is truly immortal and divine, then they are indeed on the way to starting a new religion.) A supernatural agent need not be very *anthropomorphic.* The Old Testament Jehovah is definitely a sort of divine man (not a woman), who sees with eyes and hears with ears—and talks and acts in real time. (God *waited* to *see* what Job would do, and *then* he *spoke* to him.) Many contemporary Christians, Jews, and Muslims insist that God, or Allah, being omniscient, has no need for anything like sense organs, and, being eternal, does not act in real time. This is puzzling, since many of them continue to pray to God, to hope that God *will* answer their prayers tomorrow, to express gratitude to God for *creating* the universe, and to use such locutions as "what God intends us to do" and "God have mercy," acts that *seem* to be in flat contradiction to their insistence that their God is not at all anthropomorphic. According to a long-standing tradition, this tension between God as agent and God as eternal and immutable Being is one of those things that are simply beyond human comprehension, and it would be foolish and arrogant to try to understand it. That is as it may be, and this topic will be carefully treated later in the book, but we cannot proceed with my definition of religion (or any other definition, really) until we (tentatively, pending further illumination) get a *little* clearer about the spectrum of views that are discernible through this pious fog of modest incomprehension. We need to seek further interpretation before we can decide how to classify the doctrines these people espouse.

For some people, prayer is not literally *talking to God* but, rather, a "symbolic" activity, a way of talking *to oneself* about one's deepest concerns, expressed metaphorically. It is rather like beginning a diary entry with "Dear Diary." If what they call God is really *not* an agent in their eyes, a being that can *answer* prayers, *approve* and *disapprove*, *receive* sacrifices, and *mete out* punishment or forgiveness, then, although they may call this Being God, and stand in awe of *it* (not *Him*), their creed, whatever it is, is not really a religion according to my definition. It is, perhaps, a wonderful (or terrible) surrogate for religion, or a *former* religion, an offspring of a genuine religion that bears many family resemblances to religion, but it is another species altogether.4 In order to get clear about what religions *are*, we will have to allow that some religions may have turned into things that aren't religions any more. This has certainly happened to particular practices and traditions that used to be parts of genuine religions. The rituals of Halloween are no longer religious rituals, at least in America. The people who go to great effort and expense to participate in them are not, thereby, practicing religion, even though their activities can be placed in a clear line of descent from religious practices. Belief in Santa Claus has also lost its status as a religious belief.

For others, prayer really is talking to God, who (not *which*) really does listen, and forgive. Their creed *is* a religion, according to my definition, provided that they are part of a larger social system or community, not a congregation of one. In this regard, my definition is profoundly at odds with that of William James, who defined religion as "the feelings, acts, and experiences of individual men in their solitude, so far as they apprehend themselves to stand in relation to whatever they may consider the divine" (1902, p. 31). He would have no difficulty identifying a lone believer as a person with a religion; he himself was apparently such a one. This concentration on individual, private religious *experience* was a tactical choice for James; he thought that the creeds, rituals, trappings, and political hierarchies of "organized" religion were a distraction from the root phenomenon, and his tactical path bore wonderful fruit, but he could hardly deny that those social and cultural factors hugely affect the content and structure of the individual's experience. Today, there are reasons for trading in James's psychological microscope for a wide-angle biological and social telescope, looking at the factors, over large expanses of both space and time, that shape the experiences and actions of individual religious people.

But just as James could hardly deny the social and cultural factors, I could hardly deny the existence of individuals who very sincerely and devoutly take themselves to be the lone communicants of what we might call private religions. Typically these people have had considerable experience with one or more world religions and have chosen not to be joiners. Not wanting to ignore them, but needing to distinguish them from the much, much more typical religious people who identify themselves with a particular creed or church that has many other members, I shall call them *spiritual* people, but not *religious*. They are, if you like, honorary vertebrates.

There are many other variants to be considered in due course— for instance, people who pray, and believe in the efficacy of prayer, but don't believe that this efficacy is channeled through an agent God who literally hears the prayer. I want to postpone consideration of all these issues until we have a clearer sense of where these doctrines sprang from. The core phenomenon of religion, I am proposing, invokes gods who are effective agents in real time, and who play a central role in the way the participants think about what they ought to do. I use the evasive word "invokes" here because, as we shall see in a later chapter, the standard word "belief" tends to distort and camouflage some of the most interesting features of religion. To put it provocatively, religious belief isn't always *belief*. And why is the approval of the supernatural agent or agents to be sought? That clause is included to distinguish religion from "black magic" of various sorts. There are people—very few, actually, although juicy urban legends about "satanic cults" would have us think otherwise—who take themselves to be able to command demons with whom they form some sort of unholy alliance. These (barely existent) social systems are on the boundary with religion, but I think it is appropriate to leave them out, since our intuitions recoil at the idea that people who engage in this kind of tripe deserve the special status of the devout. What apparently grounds the widespread respect in which religions of all kinds are held is the sense that those who are religious are well intentioned, trying to lead morally good lives, earnest in their desire not to do evil, and to make amends for their transgressions. Somebody who is both so selfish and so gullible as to try to make a pact with evil supernatural agents in order to get his way in the world lives in a comic-book world of superstition and deserves no such respect.[5]

3 To break or not to break

Science is like a blabbermouth who ruins a movie by telling you how it ends. —Ned Flanders (fictional character on *The Simpsons*)

You're at a concert, awestruck and breathless, listening to your favorite musicians on their farewell tour, and the sweet music is lifting you, carrying you away to another place . . . and then somebody's cell phone starts ringing! Breaking the spell. Hateful, vile, inexcusable. This inconsiderate jerk has ruined the concert for you, stolen a precious moment that can never be recovered. How evil it is to break somebody's spell! I don't want to be that person with the cell phone, and I am well aware that I will seem to many people to be courting just that fate by embarking on this book.

The problem is that there are good spells and then there are bad spells. If only some timely phone call could have interrupted the proceedings at Jonestown in Guyana in 1978, when the lunatic Jim Jones was ordering his hundreds of spellbound followers to commit suicide! If only we could have broken the spell that enticed the Japanese cult Aum Shinrikyo to release sarin gas in a Tokyo subway, killing a dozen people and injuring thousands more! If only we could figure out some way today to break the spell that lures thousands of poor young Muslim boys into fanatical madrassahs where they are prepared for a life of murderous martyrdom instead of being taught about the modern world, about democracy and history and science! If only we could break the spell that convinces some of our fellow citizens that they are commanded by God to bomb abortion clinics!

Religious cults and political fanatics are not the only casters of evil spells today. Think of the people who are addicted to drugs, or gambling, or alcohol, or child pornography. They need all the help they can get, and I doubt if anybody is inclined to throw a protective mantle around these entranced ones and admonish, "Shhh! Don't break the spell!" And it may be that the best way to break these bad spells is to introduce the spellbound to a good spell, a god spell, a gospel. It may be, and it may not. We should try to find out. Perhaps, while we're at it, we should inquire whether the world would be a better place if we could snap our fingers and cure the workaholics, too—but now I'm entering controversial waters. Many workaholics would claim that theirs is a benign addiction, useful to society and to their loved ones, and, besides, they would insist, it is their right, in a free society, to follow their hearts wherever they lead, so long as no harm comes to anyone else. The principle is unassailable: we others have no right to intrude on their private practices *so long as we can be quite sure that they are not injuring others.* But it is getting harder and harder to be sure about when this is the case.

People make themselves dependent on many things. Some think they cannot live without daily newspapers and a free press, whereas others think they cannot live without cigarettes. Some think a life without music would not be worth living, and others think a life without religion would not be worth living. Are these addictions? Or are these genuine needs that we should strive to preserve, at almost any cost?

Eventually, we must arrive at questions about ultimate values, and no factual investigation could answer them. Instead, we can do no better than to sit down and reason together, a political process of mutual persuasion and education that we can try to conduct in good faith. But in order to do that we have to know what we are choosing between, and we need to have a clear account of the reasons that can be offered for and against the different visions of the participants. Those who refuse to participate (because they already *know* the answers in their hearts) are, from the point of view of the rest of us, part of the problem. Instead of being participants in our democratic effort to find agreement among our fellow human beings, they place themselves in the inventory of obstacles to be dealt with, one way or another. As with El Niño and global warming, there is no point in trying to argue with them, but every reason to study them assiduously, whether they like it or not. They may change their minds and rejoin our political congregation, and assist us in the exploration of the grounds for their attitudes and practices, but whether or not they do, it behooves the rest of us to learn everything we can about them, for they put at risk what *we* hold dear.

It is high time that we subject religion as a global phenomenon to the most intensive multidisciplinary research we can muster, calling on the best minds on the planet. Why? Because religion is too important for us to remain ignorant about. It affects not just our social, political, and economic conflicts, but the very meanings we find in our lives. For many people, probably a majority of the people on Earth, nothing matters more than religion. For this very reason, it is imperative that we learn as much as we can about it. That, in a nutshell, is the argument of this book.

\\\

Wouldn't such an exhaustive and invasive examination damage the phenomenon itself? Mightn't it *break the spell*? That is a good question, and I don't know the answer. *Nobody knows the answer.* That is why I raise the question, to explore it carefully now, so that we (1) don't rush headlong into inquiries we would all be much better off not undertaking, and yet (2) don't hide facts from ourselves that could guide us to better lives for all. The people on this planet confront a terrible array of problems—poverty, hunger, disease, oppression, the violence of war and crime, and many more—and in the twenty-first century we have unparalleled powers for doing something about all these problems. But what shall we do?

Good intentions are not enough. If we learned anything in the twentieth century, we learned this, for we made some colossal mistakes with the best of intentions. In the early decades of the century, communism seemed to many millions of thoughtful, well-intentioned people to be a beautiful and even obvious solution to the terrible unfairness that all can see, but they were wrong. An obscenely costly mistake. Prohibition also seemed like a good idea at the time, not just to power-hungry prudes intent on imposing their taste on their fellow citizens, but to many decent people who could see the terrible toll of alcoholism and figured that nothing short of a total ban would suffice. They were proven wrong, and we still haven't recovered from all the bad effects that well-intentioned policy set in motion. There was a time, not so long ago, when the idea of keeping blacks and whites in separate communities, with separate facilities, seemed to many sincere people to be a reasonable solution to pressing problems of interracial strife. It took the civil-rights movement in the United States, and the painful and humiliating experience of Apartheid and its eventual dismantling in South Africa, to show how wrong those well-intentioned people were to have ever believed this. Shame on them, you may say. They should have known better. That is my point. We *can* come to know better if we try our best to find out, and we have no excuse for not trying. Or do we? Are some topics off limits, no matter what the consequences?

Today, billions of people pray for peace, and I wouldn't be surprised if most of them believe with all their hearts that the best path to follow to peace throughout the world is a path that runs through their particular religious institution, whether it is Christianity, Judaism, Islam, Hinduism, Buddhism, or any of hundreds of other systems of religion. Indeed, many people think that the best hope for humankind is that we can bring together all of the religions of the world in a mutually respectful conversation and ultimate agreement on how to treat one another. They may be right, but *they don't know.* The fervor of their belief is no substitute for good hard evidence, and the evidence in favor of this beautiful hope is hardly overwhelming. In fact, it is not persuasive at all, since just as many people, apparently, sincerely believe that world peace is less important, in both the short run and the long, than the global triumph of their particular religion over its competition. Some see religion as the best hope for peace, a lifeboat we dare not rock lest we overturn it and all of us perish, and others see religious self-identification as the main source of conflict and violence in the world, and believe just as fervently that religious conviction is a terrible substitute for calm, informed reasoning. Good intentions pave both roads.

Who is right? I don't know. Neither do the billions of people with their passionate religious convictions. Neither do those atheists who are sure the world would be a much better place if all religion went extinct. There is an asymmetry: atheists in general welcome the most intensive and objective examination of their views, practices, and reasons. (In fact, their incessant demand for self-examination can become quite tedious.) The religious, in contrast, often bristle at the impertinence, the lack of respect, the *sacrilege*, implied by anybody who wants to investigate their views. I respectfully demur: there is indeed an ancient tradition to which they are appealing here, but it is mistaken and should not be permitted to continue. *This* spell must be broken, and broken now. Those who are religious and believe religion to be the best hope of humankind cannot reasonably expect those of us who are skeptical to refrain from expressing our doubts if they themselves are unwilling to put their convictions under the microscope. If they are right—especially if they are obviously right, on further reflection—we skeptics will not only concede this but enthusiastically join the cause. We want what they (mostly) say they want: a world at peace, with as little suffering as we can manage, with freedom and justice and well-being and meaning for all. If the case for their path cannot be made, *this is something that they themselves should want to know.* It is as simple as that. They claim the moral high ground; maybe they deserve it and maybe they don't. Let's find out.

4 Peering into the abyss

Philosophy is questions that may never be answered. Religion is answers that may never be questioned. —Anonymous

The spell that I say *must* be broken is the taboo against a forthright, scientific, no-holds-barred investigation of religion as one natural phenomenon among many. But certainly one of the most pressing and plausible reasons for resisting this claim is the fear that if that spell is broken—if religion is put under the bright lights and the microscope—there is a serious risk of breaking a different and much more important spell: the life-enriching enchantment of religion itself. If interference caused by scientific investigation somehow disabled people, rendering them incapable of states of mind that are the springboards for religious experience or religious conviction, this *could* be a terrible calamity. You can only lose your virginity once, and some are afraid that imposing too much knowledge on some topics could rob people of their innocence, crippling their hearts in the guise of expanding their minds. To see the problem, one has only to reflect on the recent global onslaught of secular Western technology and culture, sweeping hundreds of languages and cultures to extinction in a few generations. Couldn't the same thing happen to your religion? Shouldn't we leave well enough alone, just in case? What arrogant nonsense, others will scoff. The Word of God is invulnerable to the puny forays of meddling scientists. The presumption that curious infidels need tiptoe around to avoid disturbing the faithful is laughable, they say. But in that case, there would be no harm in looking, would there? And we might learn something important.

The first spell—the taboo—and the second spell—religion itself—are bound together in a curious embrace. *Part* of the strength of the second may be—may be—the protection it receives from the first. But who knows? If we are enjoined by the first spell not to investigate this possible causal link, then the second spell has a handy shield, whether it needs it or not. The relationship between these two spells is vividly illustrated in Hans Christian Andersen's charming fable "The Emperor's New Clothes." Sometimes falsehoods and myths that are "common wisdom" can survive indefinitely simply because the prospect of exposing them is itself rendered daunting or awkward by a taboo. An indefensible mutual presumption can be kept aloft for years or even centuries because each person assumes that *somebody else* has some very good reasons for maintaining it, and nobody dares to challenge it.

Up to now, there has been a largely unexamined mutual agreement that scientists and other researchers will leave religion alone, or restrict themselves to a few sidelong glances, since people get so upset at the mere thought of a more intensive inquiry. I propose to disrupt this presumption, and examine it. If we shouldn't study all the ins and outs of religion, I want to know why, and I want to see good, factually supported reasons, not just an appeal to the tradition I am rejecting. If the traditional cloak of privacy or "sanctuary" is to be left in place, we should know why we're doing this, since a compelling case can be made that we're paying a terrible price for our ignorance. This sets the order of business: First, we must look at the issue of *whether* the first spell—the taboo—should be broken. Of course, by writing and publishing this book I am jumping the gun, leaping in and *trying* to break the first spell, but one has to start somewhere. Before continuing further, then, and possibly making matters worse, I am going to pause to defend my decision to try to break that spell. Then, having mounted my defense for *starting* the project, I am going to start the project! Not by answering the big questions that motivate the whole enterprise but by asking them, as carefully as I can, and pointing out what we already know about how to answer them, and showing why we need to answer them.

I am a philosopher, not a biologist or an anthropologist or a sociologist or historian or theologian. We philosophers are better at asking questions than at answering them, and this may strike some people as a comical admission of futility—"He says his specialty is just *asking* questions, not answering them. What a puny job! And they pay him for this?" But anybody who has ever tackled a truly tough problem knows that one of the most difficult tasks is finding the right questions to ask and the right order to ask them in. You have to figure out not only what you don't know, but what you *need* to know and *don't* need to know, and what you need to know in order to *figure out* what you need to know, and so forth. The form our questions take opens up some avenues and closes off others, and we don't want to waste time and energy barking up the wrong trees. Philosophers can sometimes help in this endeavor, but of course they have often gotten in the way, too. Then some other philosopher has to come in and try to clean up the mess. I have always liked the way John Locke put it, in the "Epistle to the Reader" at the beginning of his *Essay Concerning Human Understanding* (1690):

> . . . it is ambition enough to be employed as an under-labourer in clearing the ground a little, and removing some of the rubbish that lies in the way to knowledge;—which certainly had been very much more advanced in the world, if the endeavours of ingenious and industrious men had not been much cumbered with the learned but frivolous use of uncouth, affected, or unintelligible terms, introduced into the sciences, and there made an art of, to that degree that Philosophy, which is nothing but the true knowledge of things, was thought unfit or incapable to be brought into well-bred company and polite conversation.

Another of my philosophical heroes, William James, recognized as well as any philosopher ever has the importance of enriching your philosophical diet of abstractions and logical arguments with large helpings of hard-won fact, and just about a hundred years ago, he published his classic investigation, *The Varieties of Religious Experience*. It will be cited often in this book, for it is a treasure trove of insights and arguments, too often overlooked in recent times, and I will begin by putting an old tale he recounts to a new use:

> A story which revivalist preachers often tell is that of a man who found himself at night slipping down the side of a precipice. At last he caught a branch which stopped his fall, and remained clinging to it in misery for hours. But finally his fingers had to loose their hold, and with a despairing farewell to life, he let himself drop. He fell just six inches. If he had given up the struggle earlier, his agony would have been spared. [James, 1902, p. 111]

Like the revivalist preacher, I say unto you, O religious folks who fear to break the taboo: Let go! Let go! You'll hardly notice the drop! The sooner we set about studying religion scientifically, the sooner your deepest fears will be allayed. But that is just a plea, not an argument, so I must persist with my case. I ask just that you try to keep an open mind and refrain from prejudging what I say because I am a godless philosopher, while I similarly do my best to understand you. (I am a *bright*. My essay "The Bright Stuff," in the *New York Times*, July 12, 2003, drew attention to the efforts of some agnostics, atheists, and other adherents of naturalism to coin a new term for us nonbelievers, and the large positive response to that essay helped persuade me to write this book. There was also a negative response, largely objecting to the term that had been chosen [not by me]: *bright*, which seemed to imply that others were dim or stupid. But the term, modeled on the highly successful hijacking of the ordinary word "gay" by homosexuals, does not have to have that implication. Those who are not gays are not necessarily glum; they're *straight*. Those who are not brights are not necessarily dim. They might like to choose a name for themselves. Since, unlike us brights, they believe in the supernatural, perhaps they would like to call themselves *supers*. It's a nice word with positive connotations, like *gay* and *bright* and *straight*. Some people would not willingly associate with somebody who was openly gay, and others would not willingly read a book by somebody who was openly bright. But there is a first time for everything. Try it. You can always back out later if it becomes too offensive.)

As you can already see, this is going to be something of a roller-coaster ride for both of us. I have interviewed many deeply religious people in the last few years, and most of these volunteers had never conversed with anybody like me about such topics (and I had certainly never before attempted to broach such delicate topics with people so unlike myself), so there were more than a few awkward surprises and embarrassing miscommunications. I learned a lot, but in spite of my best efforts I will no doubt outrage some readers, and display my ignorance of matters they consider of the greatest importance. This will give them a handy reason to discard my book without considering just which points in it they disagree with and why. I ask that they resist hiding behind this excuse and soldier on.

They will learn something, and then they may be able to teach us all something.

Some people think it is deeply immoral even to *consider* reading such a book as this! For them, *wondering whether* they should read it would be as shameful as wondering whether to watch a porno-graphic videotape. The psychologist Philip Tetlock (1999, 2003, 2004) identifies values as *sacred* when they are so important to those who hold them that the very act of considering them is offen-sive. The comedian Jack Benny was famously stingy—or so he pre-sented himself on radio and television—and one of his best bits was the skit in which a mugger puts a gun in his back and barks, "Your money or your life!" Benny just stands there silently. "Your money or your life!" repeats the mugger, with mounting impa-tience. "I'm thinking, I'm thinking," Benny replies. This is funny because most of us—religious or not—think that nobody should even think about such a trade-off. Nobody should *have to* think about such a trade-off. It should be unthinkable, a "no-brainer." Life is sacred, and no amount of money would be a fair exchange for a life, *and if you don't already know that, what's wrong with you?* "To transgress this boundary, to attach a monetary value to one's friendships, children, or loyalty to one's country, is to disqualify oneself from the accompanying social roles" (Tetlock et al., 2004, p. 5). That is what makes life a sacred value.

Tetlock and his colleagues have conducted ingenious (and some-times troubling) experiments in which subjects are obliged to con-sider "taboo trade-offs," such as whether or not to purchase live human body parts for some worthy end, or whether or not to pay somebody to have a baby that you then raise, or pay somebody to perform your military service. As their model predicts, many sub-jects exhibit a strong "mere contemplation effect": they feel guilty and sometimes get angry about being lured into even thinking about such dire choices, even when they make all the right choices. When given the opportunity by the experimenters to engage in "moral cleansing" (by volunteering for some relevant community service, for instance), subjects who have had to think about taboo trade-offs are significantly more likely than control subjects to volunteer—for real—for such good deeds. (Control subjects had been asked to think about purely secular trade-offs, such as whether to hire a housecleaner or buy food instead of something else.) So this book may do *some* good by just increasing the level of charity in those who feel guilty reading it! If you feel yourself contaminated by reading this book, you will perhaps feel resentful, but also more eager than you otherwise would be to work off that resentment by engaging in some moral cleansing. I hope so, and you needn't thank me for inspiring you.

\\\

In spite of the religious connotations of the term, even atheists and agnostics can have sacred values, values that are simply not up for re-evaluation at all. I have sacred values—in the sense that I feel vaguely guilty even thinking about whether they are defensible and would *never* consider abandoning them (I like to think!) in the course of solving a moral dilemma. My sacred values are obvious and quite ecumenical: democracy, justice, life, love, and truth (in alphabetical order). But since I'm a philosopher, I've learned how to set aside the vertigo and embarrassment and ask myself what in the end supports even them, what should give when they conflict, as they often tragically do, and whether there are better alternatives. It is this traditional philosophers' open-mindedness to *every* idea that some people find immoral in itself. They think that they *should* be closed-minded when it comes to certain topics. They know that they share the planet with others who disagree with them, but they *don't* want to enter into dialogue with those others. They want to discredit, suppress, or even kill those others. While I recognize that many religious people could never bring themselves to read a book like this—that is part of the problem the book is meant to illuminate—I intend to reach as wide an audience of believers as possible. Other authors have recently written excellent books and articles on the scientific analysis of religion that are directed primarily to their fellow academics. My goal here is to play the role of ambassador, introducing (and distinguishing, criticizing, and defending) the main ideas of that literature. This puts *my* sacred values to work: I want the resolution to the world's problems to be as *democratic* and *just* as possible, and both democracy and justice depend on getting on the table for all to see as much of the *truth* as possible, bearing in mind that sometimes the truth hurts, and hence should sometimes be left concealed, out of *love* for those who would suffer were it revealed. But I'm prepared to consider alternative values and reconsider the priorities I find among my own.

5 Religion as a natural phenomenon

As every enquiry which regards religion is of the utmost importance, there are two questions in particular which challenge our attention, to wit, that concerning its foundation in reason, and that concerning its origin in human nature.

—David Hume, *The Natural History of Religion*

What do I mean when I speak of religion as a natural phenomenon?

I might mean that it's like natural food—not just tasty but healthy, unadulterated, "organic." (That, at any rate, is the myth.) So do I mean: "Religion is *healthy;* it's good for you!"? This might be true, but it is not what I mean.

I might mean that religion is not an artifact, not a product of human intellectual activity. Sneezing and belching are natural, reciting sonnets is not; going naked—*au naturel*—is natural; wearing clothes is not. But it is obviously false that religion is natural in this sense. Religions are transmitted culturally, through language and symbolism, not through the genes. You may get your father's nose and your mother's musical ability through your genes, but if you get your religion from your parents, you get it the way you get your language, through upbringing. So of course that is not what I mean by *natural*.

With a slightly different emphasis, I might mean that religion is *doing what comes naturally,* not an acquired taste, or an artificially groomed or educated taste. In this sense, speaking is natural but writing is not; drinking milk is natural but drinking a dry martini is not; listening to tonal music is natural but listening to atonal music is not; gazing at sunsets is natural but gazing at late Picasso paintings is not. There is some truth to this: religion is not an unnatural act, and this will be a topic explored in this book. But it is not what I mean.

I might mean that religion is natural as opposed to *supernatural,* that it is a human phenomenon composed of events, organisms, objects, structures, patterns, and the like that all obey the laws of physics or biology, and hence do not involve miracles. And that *is* what I mean. Notice that it could be true that God exists, that God is indeed the intelligent, conscious, loving creator of us all, and yet *still* religion itself, as a complex set of phenomena, is a perfectly natural phenomenon. Nobody would think it was presupposing atheism to write a book subtitled *Sports as a Natural Phenomenon* or *Cancer as a Natural Phenomenon*. Both sports and cancer are widely recognized as natural phenomena, not supernatural, in spite of the well-known exaggerations of various promoters. (I'm thinking, for instance, of two famous touchdown passes known respectively as the Hail Mary and the Immaculate Reception, to say nothing of the weekly trumpetings by researchers and clinics around the world of one "miraculous" cancer cure or another.)

Sports and cancer are the subject of intense scientific scrutiny by researchers working in many disciplines and holding many different religious views. They all assume, tentatively and for the sake of science, that the phenomena they are studying are natural phenomena. This doesn't prejudge the verdict that they are. Perhaps there *are* sports miracles that actually defy the laws of nature; perhaps some cancer cures *are* miracles. If so, the only hope of ever demonstrating this to a doubting world would be by adopting the scientific method, with its assumption of no miracles, and showing that science was utterly unable to account for the phenomena. Miracle-hunters must be scrupulous scientists or else they are wasting their time—a point long recognized by the Roman Catholic Church, which at least goes through the motions of subjecting the claims of miracles made on behalf of candidates for sainthood to objective scientific investigation. So no deeply religious person should object to the scientific study of religion with the presumption that it is an entirely natural phenomenon. If it isn't entirely natural, if there really are miracles involved, the best way—indeed, the only way—to show that to doubters would be to demonstrate it scientifically. Refusing to play by these rules only creates the suspicion that one doesn't really believe that religion is supernatural after all.

In assuming that religion is a natural phenomenon, I am not prejudging its value to human life, one way or the other. Religion, like love and music, is natural. But so are smoking, war, and death. In this sense of *natural,* everything artificial is natural! The Aswan Dam is no less natural than a beaver's dam, and the beauty of a skyscraper is no less natural than the beauty of a sunset. The natural sciences take everything in Nature as their topic, and that includes both jungles and cities, both birds and airplanes, the good, the bad, the ugly, the insignificant, and the all-important as well.

Over two hundred years ago, David Hume wrote two books on religion. One was about religion as a natural phenomenon, and its opening sentence is the epigraph of this section. The other was about the "foundation in reason" of religion, his famous *Dialogues Concerning Natural Religion* (1779). Hume wanted to consider whether there was any good reason—any *scientific* reason, we might say—for believing in God. *Natural* religion, for Hume, would be a creed that was as well supported by evidence and argument as Newton's theory of gravitation, or plane geometry. He contrasted it with *revealed* religion, which depended on the revelations of mystical experience or other extra-scientific paths to conviction. I gave Hume's *Dialogues* a place of honor in my 1995 book, *Darwin's Dangerous Idea*—Hume is yet another of my heroes—so you might think that I intend to pursue this issue still further in this book, but that is not in fact my intention. This time I am pursuing Hume's other path. Philosophers have spent two millennia and more concocting and criticizing arguments for the existence of God, such as the Argument from Design and the Ontological Argument, and arguments against the existence of God, such as the Argument from Evil. Many of us brights have devoted considerable time and energy at some point in our lives to looking at the arguments for and against the existence of God, and many brights continue to pursue these issues, hacking away vigorously at the arguments of the believers as if they were trying to refute a rival scientific theory. But not I. I decided some time ago that diminishing returns had set in on the arguments about God's existence, and I doubt that any breakthroughs are in the offing, from either side. Besides, many deeply religious people insist that all those arguments—on both sides—simply miss the whole point of religion, and their demonstrated lack of interest in the arguments persuades me of their sincerity. Fine. So what, then, *is* the point of religion?

What is this phenomenon or set of phenomena that means so much to so many people, and why—and how—does it command allegiance and shape so many lives so strongly? That is the main question I will address here, and once we have sorted out and clarified (not settled) some of the conflicting answers to this question, it will give us a novel perspective from which to look, briefly, at the traditional philosophical issue that some people insist is the *only* issue: whether or not there are good reasons for believing in God. Those who insist that they *know* that God exists and can prove it will have their day in court.[6]

\\\

Chapter 1 Religions are among the most powerful natural phenomena on the planet, and we need to understand them better if
we are to make informed and just political decisions. Although
there are risks and discomforts involved, we should brace ourselves
and set aside our traditional reluctance to investigate religious phenomena scientifically, so that we can come to understand how and
why religions inspire such devotion, and figure out how we should
deal with them all in the twenty-first century.

\\\

Chapter 2 There are obstacles confronting the scientific study of
religion, and there are misgivings that need to be addressed. A preliminary exploration shows that it is both possible and advisable for
us to turn our strongest investigative lights on religion.

CHAPTER TWO

||||\\\||||

Some Questions About Science

I *Can* science study religion?

To be sure, man is, zoologically speaking, an animal. Yet, he is a unique animal, differing from all others in so many fundamental ways that a separate science for man is well-justified.

—Ernst Mayr, *The Growth of Biological Thought*

There has been some confusion about whether the earthly manifestations of religion should count as a part of Nature. Is religion out-of-bounds to science? It all depends on what you mean. If you mean the religious experiences, beliefs, practices, texts, artifacts, institutions, conflicts, and history of *H. sapiens*, then this is a voluminous catalogue of unquestionably natural phenomena. Considered as psychological states, drug-induced hallucination and religious ecstasy are both amenable to study by neuroscientists and psychologists. Considered as the exercise of cognitive competence, memorizing the periodic table of elements is the same sort of phenomenon as memorizing the Lord's Prayer. Considered as examples of engineering, suspension bridges and cathedrals both obey the law of gravity and are subject to the same sorts of forces and stresses. Considered as salable manufactured goods, both mystery novels and Bibles fall under the regularities of economics. The logistics of holy wars do not differ from the logistics of entirely secular conflicts. "Praise the Lord and pass the ammunition!" as the World War II song said. A *crusade* or a *jihad* can be investigated by researchers in many disciplines, from anthropology and military history to nutrition and metallurgy.

In his book *Rocks of Ages* (1999), the late Stephen Jay Gould defended the political hypothesis that science and religion are two "non-overlapping *magisteria*"—two domains of concern and inquiry that can coexist peacefully as long as neither poaches on the other's special province. The *magisterium* of science is factual truth on all matters, and the *magisterium* of religion, he claimed, is the realm of morality and the meaning of life. Although Gould's desire for peace between these often warring perspectives was laudable, his proposal found little favor on either side, since in the minds of the religious it proposed abandoning all religious claims to factual truth and understanding of the natural world (including the claims that God created the universe, or performs miracles, or listens to prayers), whereas in the minds of the secularists it granted too much authority to religion in matters of ethics and meaning. Gould exposed some clear instances of immodest folly on both sides, but the claim that *all* conflict between the two perspectives is due to overreaching by one side or the other is implausible, and few readers were persuaded. But whether or not the case can be made for Gould's proposal, my proposal is different. There may be some domain that is religion's alone to command, some realm of human activity that science can't properly address and religion can, but that does not mean that science cannot or should not study this very fact. Gould's own book was presumably a product of just such a scientific investigation, albeit a rather informal one. He looked at religion with the eyes of a scientist and thought he could see a boundary that revealed two domains of human activity. Was he right? That is presumably a scientific, factual question, not a religious question. I am not suggesting that science should *try to do* what religion does, but that it should *study*, scientifically, what religion does.

One of the surprising discoveries of modern psychology is how easy it is to be ignorant of your own ignorance. You are normally oblivious of your own blind spot, and people are typically amazed to discover that we don't see colors in our peripheral vision. It *seems* as if we do, but we don't, as you can prove to yourself by wiggling colored cards at the edge of your vision—you'll see motion just fine but not be able to identify the color of the moving thing. It takes special provoking like that to get the *absence* of information to reveal itself to us. And the absence of information about religion is what I want to draw to everyone's attention. We have neglected to gather a wealth of information about something of great import to us.

This may come as a surprise. Haven't we been looking carefully at religion for a long time? Yes, of course. There have been centuries of insightful and respectful scholarship about the history and variety of religious phenomena. This work, like the bounty gathered by dedicated bird-watchers and other nature lovers before Darwin's time, is proving to be a hugely valuable resource to those pioneers who are now beginning, for the first time really, to study the natural phenomena of religion through the eyes of contemporary science. Darwin's breakthrough in biology was enabled by his deep knowledge of the wealth of empirical details scrupulously garnered by hundreds of pre-Darwinian, non-Darwinian natural historians. Their theoretical innocence was itself an important check on his enthusiasm; they had not gathered their facts with an eye to proving Darwinian theory correct, and we can be equally grateful that almost all the "natural history of religion" that has been accumulated to date is, if not theoretically innocent, at least oblivious to the sorts of theories that now may be supported or undercut by it.

The research to date has hardly been neutral, however. We don't just walk up to religious phenomena and study them point-blank, as if they were fossils or soybeans in a field. Researchers tend to be either respectful, deferential, diplomatic, tentative—or hostile, invasive, and contemptuous. It is just about impossible to be neutral in your approach to religion, because many people view neutrality in itself as hostile. If you're not for us, you're against us. And so, since religion so clearly matters so much to so many people, researchers have almost never even attempted to be neutral; they have tended to err on the side of deference, putting on the kid gloves. It is either that or open hostility. For this reason, there has been an unfortunate pattern in the work that has been done. People who want to study religion usually have an ax to grind. They either want to defend their favorite religion from its critics or want to demonstrate the irrationality and futility of religion, and this tends to infect their methods with bias. Such distortion is not inevitable. Scientists in every field have pet theories they hope to confirm, or target hypotheses they yearn to demolish, but, knowing this, they take a variety of tried-and-true steps to prevent their bias from polluting their evidence-gathering: double-blind experiments, peer review, statistical tests, and many other standard constraints of good scientific method. But in the study of religion, the stakes have often been seen to be higher. If you think that the disconfirmation of a hypothesis about one religious phenomenon or another would not be just an undesirable crack in the foundation of some theory but a moral calamity, you tend not to run all the controls. Or so, at least, it has often seemed to observers.

That impression, true or false, has created a positive feedback loop: scientists don't want to deal with second-rate colleagues, so they tend to shun topics where they see what they take to be mediocre work being done. This self-selection is a frustrating pattern that begins when students think about "choosing a major" in college. The best students typically shop around, and if they are unimpressed by the work they are introduced to in the first course in a field, they cross that field off their list for good. When I was an undergraduate, physics was still the glamour field, and then the race to the moon drew more than its share of talent. (A fossil trace is the phrase "Hey, it's not rocket science.") This was followed by computer science for a while, and all along—for half a century and more—biology, especially molecular biology, has attracted many of the smartest. Today, cognitive science and the various strands of evolutionary biology—bio-informatics, genetics, developmental biology—are on the rise. But through all this period, sociology and anthropology, social psychology, and my own home field, philosophy, have struggled along, attracting those whose interests match the field well, including some brilliant people, but having to combat somewhat unenviable reputations. As my old friend and former colleague, Nelson Pike, a respected philosopher of religion, once ruefully put it:

> If you are in a company of people of mixed occupations, and somebody asks what you do, and you say you are a college professor, a glazed look comes into his eye. If you are in a company of professors from various departments, and somebody asks what is your field, and you say philosophy, a glazed look comes into his eye. If you are at a conference of philosophers, and somebody asks you what you are working on, and you say philosophy of religion . . . [Quoted in Bambrough, 1980]

This is not just a problem for philosophers of religion. It is equally a problem for sociologists of religion, psychologists of religion, and other social scientists—economists, political scientists—and for those few brave neuroscientists and other biologists who have decided to look at religious phenomena with the tools of their trade. One of the factors is that people think they already know everything they need to know about religion, and this received wisdom is pretty bland, not provocative enough to inspire either refutation or extension. In fact, if you set out to design an impermeable barrier between scientists and an underexplored phenomenon, you could hardly do better than to fabricate the dreary aura of low prestige, backbiting, and dubious results that currently envelops the topic of religion. And since we know from the outset that many people think such research violates a taboo, or at least meddles impertinently in matters best left private, it is not so surprising that few good researchers, in any discipline, want to touch the topic. I myself certainly felt that way until recently.

These obstacles can be overcome. In the twentieth century, a lot was learned about how to study human phenomena, social phenomena. Wave after wave of research and criticism has sharpened our appreciation of the particular pitfalls, such as biases in data-gathering, investigator-interference effects, and the interpretation of data. Statistical and analytical techniques have become much more sophisticated, and we have begun setting aside the old oversimplified models of human perception, emotion, motivation, and control of action and replacing them with more physiologically and psychologically realistic models. The yawning chasm that was seen to separate the sciences of the mind (*Geisteswissenschaften*) from the natural sciences (*Naturwissenschaften*) has not yet been bridged securely, but many lines have been flung across the divide. Mutual suspicion and professional jealousy as well as genuine theoretical controversy continue to shake almost all efforts to carry insights back and forth on these connecting routes, but every day the traffic grows. The question is not whether good science of religion as a natural phenomenon is possible: it is. The question is whether we should do it.

2 *Should* science study religion?

Look before you leap. —Aesop, "The Fox and the Goat"

Research is expensive and sometimes has harmful side effects. One of the lessons of the twentieth century is that scientists are not above confabulating justifications for the work they want to do, driven by insatiable curiosity. Are there in fact good reasons, aside from sheer curiosity, to try to develop the natural science of religion? Do we need this for anything? Would it help us choose policies, respond to problems, improve our world? What do we know about the future of religion? Consider five wildly different hypotheses:

1. *The Enlightenment is long gone; the creeping "secularization" of modern societies that has been anticipated for two centuries is evaporating before our eyes.* The tide is turning and religion is becoming more important than ever. In this scenario, religion soon resumes something like the dominant social and moral role it had before the rise of modern science in the seventeenth century. As people recover from their infatuation with technology and material comforts, spiritual identity becomes a person's most valued attribute, and populations come to be ever more sharply divided among Christianity, Islam, Judaism, Hinduism, and a few other major multinational religious organizations. Eventually—it might take another millennium, or it might be hastened by catastrophe—one major faith sweeps the planet.

2. *Religion is in its death throes; today's outbursts of fervor and fanaticism are but a brief and awkward transition to a truly modern society in which religion plays at most a ceremonial role.* In this scenario, although there may be some local and temporary revivals and even some violent catastrophes, the major religions of the world soon go just as extinct as the hundreds of minor religions that are vanishing faster than anthropologists can record them. Within the lifetimes of our grandchildren, Vatican City becomes the European Museum of Roman Catholicism, and Mecca is turned into Disney's Magic Kingdom of Allah.

3. *Religions transform themselves into institutions unlike anything seen before on the planet: basically creedless associations selling self-help and enabling moral teamwork, using ceremony and tradition to cement relationships and build "long-term fan loyalty."* In this scenario, being a member of a religion becomes more and more like being a Boston

Red Sox fan, or a Dallas Cowboys fan. Different colors, different songs and cheers, different symbols, and vigorous competition—would you want your daughter to marry a Yankees fan?—but aside from a rabid few, everybody appreciates the importance of peaceful coexistence in a Global League of Religions. Religious art and music flourish, and friendly rivalry leads to a degree of specialization, with one religion priding itself on its environmental stewardship, providing clean water for the world's billions, while another becomes duly famous for its concerted defense of social justice and economic equality.

4. *Religion diminishes in prestige and visibility, rather like smoking; it is tolerated, since there are those who say they can't live without it, but it is discouraged, and teaching religion to impressionable young children is frowned upon in most societies and actually outlawed in others.* In this scenario, politicians who still practice religion can be elected if they prove themselves worthy in other regards, but few would advertise their religious affiliation—or affliction, as the politically incorrect insist on calling it. It is considered as rude to draw attention to the religion of somebody as it is to comment in public about his sexuality or whether she has been divorced.

5. *Judgment Day arrives. The blessed ascend bodily into heaven, and the rest are left behind to suffer the agonies of the damned, as the Antichrist is vanquished.* As the Bible prophecies foretold, the rebirth of the nation of Israel in 1948 and the ongoing conflict over Palestine are clear signs of the End Times, when the Second Coming of Christ sweeps all the other hypotheses into oblivion.

Other possibilities are describable, of course, but these five hypotheses highlight the extremes that are taken seriously. What is remarkable about the set is that just about anybody would find at least one of them preposterous, or troubling, or even deeply offensive, but every one of them is not just anticipated but yearned for. People act on what they yearn for. We are at cross-purposes about religion, to say the least, so we can anticipate problems, ranging from wasted effort and counterproductive campaigns if we are lucky to all-out war and genocidal catastrophe if we are not.

Only one of these hypotheses (at most) will turn out to be true; the rest are not just wrong but wildly wrong. Many people think they know which is true, but nobody does. Isn't that fact, all by itself, enough reason to study religion scientifically? Whether you want religion to flourish or perish, whether you think it should transform itself or stay just as it is, you can hardly deny that whatever happens will be of tremendous significance to the planet. It would be useful to your hopes, whatever they are, to know more about what is likely to happen and why. In this regard, it is worth noting how assiduously those who firmly believe in number 5 scan the world news for evidence of prophecies fulfilled. They sort and evaluate their sources, debating the pros and cons of various interpretations of those prophecies. They think there is a reason to investigate the future of religion, and they don't even think the course of future events lies within human power to determine. The rest of us have all the more reason to investigate the phenomena, since it is quite obvious that complacency and ignorance could lead us to squander our opportunities to steer the phenomena in what we take to be the benign directions.

Looking ahead, anticipating the future, is the crowning achievement of our species. We have managed in a few short millennia of human culture to multiply the planet's supply of look-ahead by many orders of magnitude. We know when eclipses will occur centuries in advance; we can predict the effects on the atmosphere of adjustments in how we generate electricity; we can anticipate in broad outline what will happen as our petroleum reserves dwindle in the next decades. We do this not with miraculous prophecy but with basic perception. We gather information from the environment, using our senses, and then we use science to cobble together anticipations based on that information. We mine the ore, and then refine it, again and again, and it lets us see into the future—dimly, with lots of uncertainty, but much better than a coin toss. In every area of human concern, we have learned how to anticipate and then avoid catastrophes that used to blindside us.[1] We have recently forestalled a global disaster due to a growing hole in the ozone layer because some far-seeing chemists were able to prove that some of our manufactured compounds were causing the problem. We have avoided economic collapses in recent years because our economic models have shown us impending problems.

A catastrophe averted is an anticlimax, obviously, so we tend not to appreciate how valuable our powers of look-ahead are. "See?" we complain. "It wasn't going to happen after all." The flu season in the winter of 2003–2004 was predicted to be severe, since it arrived earlier than usual, but the broadcast recommendations for inoculation were so widely heeded that the epidemic collapsed as rapidly as it began. Ho-hum. It has become something of a tradition in recent years for the meteorologists on television to hype an oncoming hurricane or other storm, and then for the public to be underwhelmed by the actual storm. But sober evaluations show that many lives are saved, destruction is minimized. We accept the value of intensely studying El Niño and the other cycles in ocean currents so that we can do better meteorological forecasting. We keep exhaustive records of many economic events so that we can do better economic forecasting. We should extend the same intense scrutiny, for the same reasons, to religious phenomena. Few forces in the world are as potent, as influential, as religion. As we struggle to resolve the terrible economic and social inequities that currently disfigure our planet, and minimize the violence and degradation we see, we have to recognize that if we have a blind spot about religion our efforts will almost certainly fail, and may make matters much worse. We wouldn't permit the world's food-producing interests to deflect us from studying human agriculture and nutrition, and we have learned not to exempt the banking-and-insurance world from intense and continuous scrutiny. Their effects are too important to take on faith. So what I am calling for is a concerted effort to achieve a mutual agreement under which religion—all religion—becomes a proper object of scientific study.

Here I find that opinion is divided among those who are already convinced that this would be a good idea, those who are dubious and inclined to doubt that it would be of much value, and those who find the proposal evil—offensive, dangerous, and stupid. Not wanting to preach to the converted, I am particularly concerned to address those who hate this idea, in hopes of persuading them that their repugnance is misplaced. This is a daunting task, like trying to persuade your friend with the cancer symptoms that she really ought to see a doctor *now*, since her anxiety may be misplaced and the sooner she learns that the sooner she can get on with her life, and if she does have cancer, timely intervention may make all the difference. Friends can get quite annoyed when you interfere with their denial at times like that, but perseverance is called for. Yes, I want to put religion on the examination table. If it is fundamentally benign, as many of its devotees insist, it should emerge just fine; suspicions will be put to rest and we can then concentrate on the few peripheral pathologies that religion, like every other natural phenomenon, falls prey to. If it is not, the sooner we identify the problems clearly the better. Will the inquiry itself generate some discomfort and embarrassment? Almost certainly, but that is a small price to pay. Is there a risk that such an invasive examination will make a healthy religion ill, or even disable it? Of course. There are always risks. Are they worth taking? Perhaps not, but I haven't yet seen an argument that persuades me of this, and we will soon consider the best of them. The only arguments worth attending to will have to demonstrate that (1) religion provides net benefits to humankind, and (2) these benefits would be unlikely to survive such an investigation. I, for one, fear that if we *don't* subject religion to such scrutiny now, and work out together whatever revisions and reforms are called for, we will pass on a legacy of ever more toxic forms of religion to our descendants. I can't prove that, and those who are dead sure that this will not happen are encouraged to say what supports their conviction, aside from loyalty to their tradition, which goes without saying and doesn't count for anything here.

In general, knowing more improves your chances of getting what you value. That's not quite a truth of logic, since uncertainty is not the only factor that can lower the probability of achieving one's goals. The costs of knowing (such as the cost of *coming* to know) must be factored in, and these costs may be high, which is why "Wing it!" is sometimes good advice. Suppose there is a limit on how much knowledge about some topic is good for us. If so, then, whenever that limit is reached (if that is possible—the limit may be unreachable for one reason or another), we should prohibit or at least strongly discourage any further seeking of knowledge on that topic, as antisocial activity. This may be a principle that never comes into play, but we don't know that, and we should certainly accept the principle. It may be, then, that some of our major disagreements in the world today are about whether we've reached such a limit. This reflection puts the Islamist[2] conviction that Western science is a bad thing in a different light: it may not be an ignorant mistake so much as a profoundly different view of where the threshold is. Sometimes ignorance *is* bliss. We need to consider such possibilities carefully.

3 Might music be bad for you?

Music, the greatest good that mortals know,
*   And all of heaven we have below.*
   —Joseph Addison

Is it not strange that sheep's guts should hale souls out of men's bodies?
   —William Shakespeare

It is not that I don't sympathize with the distaste of those who resist my proposal. Trying to imagine what their emotional response to my proposal would be, I have come up with an unsettling thought experiment that seems to me to do the trick. (I am speaking now to those who, like me, are *not* appalled by the idea of this examination.) Imagine how you would feel if you were to read in the science section of the *New York Times* that new research conducted at Cambridge University and Caltech showed that music, long viewed as one of the unalloyed treasures of human culture, is actually bad for your health, a major risk factor for Alzheimer's and heart disease, a mood-distorter that impairs judgment in subtle but clearly deleterious ways, a significant contributor to aggressive tendencies, xenophobia, and weakness of will. Early and habitual exposure to music, both performing and listening, makes you 40 percent more likely to suffer serious depression, knocks an average of ten points off your IQ, and nearly doubles the probability that you will commit an act of violence at some time in your life. A panel of researchers recommends that people restrict their music intake to no more than an hour a day (including everything from elevator music and background music on television to symphony concerts) and that the widespread practice of music lessons for children be curtailed immediately.

Aside from the utter disbelief with which I would greet a report of such "findings," I can detect in my imagined reactions a visceral defensive surge, along the lines of "So much the worse for Cambridge and Caltech! What do *they* know about music?" and "I don't care if it *is* true! Anybody who tries to take away my music had better be prepared for a fight, because a life without music isn't worth living. I don't care if it 'hurts' me, and I don't even care if it 'hurts' others—we're going to have music, and that's all there is to it." That is how I would be tempted to respond. I would rather not live in a world without music. "But why?" someone might ask. "It's just some silly sawing away and making noise together. It doesn't feed the hungry or cure cancer or . . ." I answer: "But it brings great comfort and joy to hundreds of millions of people. Sure, there are excesses and controversies, but, still, can anybody doubt that music is by and large a good thing?" "Well, yes," comes the reply. There are religious sects—the Taliban, for instance, but also Puritan sects of yore in Christianity and no doubt others—that have held that music is an evil pastime, a sort of drug to be forbidden. The idea is not clearly insane, so we should accept the intellectual burden of showing that it is an error.

I recognize that many people feel about religion the way I feel about music. They may be right. Let's find out. That is, let's subject religion to the same sort of scientific inquiry that we have done with tobacco and alcohol and, for that matter, music. Let's find out why people love their religion, and what it's good for. And we should no more take the existing research to settle the issue than we took the tobacco companies' campaigns about the safety of cigarette smoking at face value. Sure, religion saves lives. So does tobacco— ask those GIs for whom tobacco was an even greater comfort than religion during World War II, the Korean War, and Vietnam.

I'm prepared to look hard at the pros and cons of music, and if it turns out that music causes cancer, ethnic hatred, and war, then I'll have to think seriously about how to live without music. It is only because I am so supremely confident that music *doesn't* do much harm that I can enjoy it with such a clear conscience. If I were told by credible people that music might be harmful to the world, all things considered, I would feel morally bound to examine the evidence as dispassionately as I could. In fact, I would feel guilty about my allegiance to music if I didn't check it out.

But isn't the hypothesis that the costs of religion outweigh the benefits even more ludicrous than the fantastic claim about music? I don't think so. Music may be what Marx said religion is: the opiate of the masses, keeping working people in tranquilized subjugation, but it may also be the rallying song of revolution, closing up the ranks and giving heart to all. On this point, music and religion have quite similar profiles. In other regards, music looks far less problematic than religion. Over the millennia, music has started a few riots, and charismatic musicians may have sexually abused a shocking number of susceptible young fans, and seduced many others to leave their families (and their wits) behind, but no crusades or jihads have been waged over differences in musical tradition, no pogroms have been instituted against the lovers of waltzes or ragas or tangos. Whole populations haven't been subjected to obligatory scale-playing or kept in penury in order to furnish concert halls with the finest acoustics and instruments. No musicians have had fatwas pronounced against them by musical organizations, not even accordionists.

The comparison of religion to music is particularly useful here, since music is another natural phenomenon that has been ably studied by scholars for hundreds of years but is only just beginning to be an object of the sort of scientific study I am recommending. There has been no dearth of professional research on music theory—harmony, counterpoint, rhythm—or the techniques of musicianship, or the history of every genre and instrument. Ethnomusicologists have studied the evolution of musical styles and practices in relation to social, economic, and other cultural factors, and neuroscientists and psychologists have rather recently begun studying the perception and creation of music, using all the latest technology to uncover the patterns of brain activity associated with musical experience, musical memory, and related topics. But most of this research still takes music for granted. It seldom asks: Why does music exist? There is a short answer, and it is true, so far as it goes: it exists because we love it, and hence we keep bringing more of it into existence. But why do we love it? Because we find that it is beautiful. But why is it beautiful to us? This is a perfectly good biological question, but it does not yet have a good answer. Compare it, for instance, with the question: Why do we love sweets? Here we know the evolutionary answer, in some detail, and it has some curious twists. It is no accident that we find sweet things to our liking, and if we want to adjust our policies regarding sweet things in the future, we had better understand the evolutionary basis of their appeal. We mustn't make the mistake of the man in the old joke who complained that, just when he'd finally succeeded in training his donkey not to eat, the stupid animal up and died on him.

Some things are necessary to life, and some things are at least so life-enhancing or life-enabling that we tamper with them at our peril, and we need to figure out these roles and needs. Ever since the Enlightenment in the eighteenth century, many quite well-informed and brilliant people have confidently thought that religion would soon vanish, the object of a human taste that could be satisfied by other means. Many are still waiting, somewhat less confidently. Whatever religion provides for us, it is something that many *think* they cannot live without. Let's take them seriously this time, for they might be right. But there is only one way to take them seriously: we need to study them scientifically.

4 Would neglect be more benign?

Sweet is the lore which Nature brings;
   Our meddling intellect
Misshapes the beauteous forms of things:—
   We murder to dissect.
   —William Wordsworth, "The Tables Turned"

Why then must science and scientists continue to be governed by fear—fear of public opinion, fear of social consequence, fear of religious intolerance, fear of political pressure, and, above all, fear of bigotry and prejudice—as much within as without the professional world?
   —William Masters and Virginia Johnson, *Human Sexual Response*

And ye shall know the truth, and the truth shall make you free.
   —Jesus of Nazareth, in John 8:32

It is time to confront the worry that such an investigation might actually kill all the specimens, destroying something precious in the name of discovering its inner nature. Wouldn't it be more prudent to leave well enough alone? As I have already noted, the case for curbing our curiosity here has two parts: it must show both (1) that religion provides net benefits to humankind, and (2) that these benefits would be unlikely to survive such an investigation. The tactical problem that confronts us is that there is really no way of showing the first point without actually engaging in the investigation. Religion *seems* to many people to be the source of many wonderful things, but others doubt this, for compelling reasons, and we shouldn't just concede the point out of a misplaced respect for tradition. Perhaps this very respect is like the protective outer shell that often conceals deadly viruses from our immune system, a sort of camouflage that disengages much-needed criticism. So the most we can say is that point 1 is not yet proven. We can, however, proceed tentatively, and consider how likely point 2 would be if we were to *assume for the sake of argument* that religion is indeed a thing of great value. We can assume it innocent until proven guilty— in other words, just the way our legal system operates.

Now, what about point 2? How much damage do we suppose an investigation might do, in the worst case? Might it not break the spell and disenchant us forever? This concern has been a favorite ground for resisting scientific curiosity for centuries, but although it is undeniable that taking apart particular instances of wonderful things—plants, animals, musical instruments—may sometimes destroy them beyond reconstruction, other wonderful things—poems, symphonies, theories, legal systems—thrive on analysis, however painstaking, and one can hardly deny the benefit to *other* plants, animals, and musical instruments derived from dissecting a few specimens. In spite of all the warnings over the centuries, I have been unable to come up with a case of some valuable phenomenon that has actually been destroyed, or even seriously damaged, by scientific scrutiny.

Field biologists often confront a terrible quandary when studying an endangered species: does their well-meant attempt at a census, involving live capture and release, actually hasten the extinction of the species? When anthropologists descend on a heretofore isolated and pristine people, their inquiries, however discreet and diplomatic, swiftly change the culture they are so eager to know. With regard to the former cases, *thou shalt not study* is a policy that may indeed be wisely invoked on occasion, but with regard to the latter, prolonging the isolation of people by putting them, in effect, in a cultural zoo, though it is sometimes advocated, does not bear scrutiny. These are people, and we have no right to keep them ignorant of the larger world they share with us. (Whether they have the right to keep themselves ignorant is one of the vexing questions to be considered later in this book.)

It is worth recalling that it took brave pioneers many years to overcome the powerful taboo against the dissection of human cadavers during the early years of modern medicine. And we should note that, notwithstanding the outrage and revulsion with which the idea of dissection was then received, overcoming that tradition has not led to the feared collapse of morality and decency. We live in an era in which human corpses are still treated with due respect—indeed, with rather more respect and decorum than they were treated with at the time dissection was still disreputable. And which of us would choose to forgo the benefits of medicine made possible by the invasive, meddling science Wordsworth deplores?

More recently, another taboo was broken, with even greater outcry. Alfred C. Kinsey, in the 1940s and 1950s, began the scientific investigation of human sexual practices in America that led to the notorious Kinsey Reports, *Sexual Behavior in the Human Male* (1948) and *Sexual Behavior in the Human Female* (1953). There were substantial flaws in Kinsey's studies, but the weight of the evidence he amassed led to surprising conclusions that have needed only minor adjustments in the wake of the better-controlled investigations that followed. For the first time, boys and men could learn that over 90 percent of American males masturbate, and that around 10 percent are homosexual; girls and women could learn that orgasms were normal and achievable for them as well, both in coitus and in masturbation, and—not surprisingly, in retrospect—that lesbians were better at inducing orgasms in women than men were.

Kinsey's research tools were interviews and questionnaires, but soon William H. Masters and Virginia Johnson got up the nerve to subject human sexual arousal to scientific investigation in the laboratory, recording the physiological responses of volunteers engaged in sexual acts, using all the tools of science, including color cinematography (this was before the ready availability of videotape). Their pioneering work, *Human Sexual Response* (1966), was met with a wild mixture of hostility and outrage, amusement and prurient fascination—and cautious applause from the medical and scientific community. By shining the bright light of science on what had heretofore been conducted in the dark (with a huge measure of secrecy and shame), they dispelled a host of myths, revised the medical understanding of some kinds of sexual dysfunction, liberated untold numbers of anxious people whose tastes and practices had been under a cloud of socially inculcated disapproval, and—wonder of wonders—improved the sex lives of millions. It turns out that in this case, at least, you can break the spell and yet not break the spell at the same time. You can violate the taboo against dispassionate study of a phenomenon—there's one spell broken—and not destroy it in the process—there's a spell one can still blissfully fall under.

But at what cost? I deliberately draw attention to Masters and Johnson's still-controversial work, since it illustrates so clearly the difficult issues with which this book will be concerned. Many will agree with me when I say that, thanks to the pioneering work of Kinsey, and Masters and Johnson, the knowledge we have acquired has not only not destroyed sex, it has made sex better. But there are also many who will pounce on the comparison and declare that this is exactly why they oppose any scientific exploration of religion: there is a chance it might do for religion what Kinsey et al. did for sex—teach us more than is good for us. Let me put words in their mouths:

If masturbating without shame, tolerance for homosexuality, and greater knowledge of how to achieve female orgasm are examples of the benefits science can bring us, then so much the worse for science. By treating sex as something natural (in the sense of nothing to be ashamed about), it has contributed to an explosion of pornography and degradation, defiling the sacred act of procreative union between husband and wife. We were better off not knowing all these facts, and we should take whatever steps we can to shelter our children from this contaminating information!

This is a very serious objection. There is no denying that the matter-of-fact candor about sex that was fostered by this research has had some terrible side effects, opening up new and fertile fields for exploitation by those who are always looking for ways to prey upon their fellow citizens. The sexual revolution of the sixties was not the glorious and all-benign liberation that it is often portrayed as being. The explorations of "free love" and "open marriage" broke many hearts, and robbed many young people of a deep sense of the moral importance of sexual relations by encouraging a shallow vision of sex as mere entertainment of the senses. Although it is widely believed that the sexual revolution contributed to the negligence and casual promiscuity that have heightened the scourge of sexually transmitted diseases, this may not be the case. Most evidence suggests that when information about sex is widespread, sexual behavior becomes more responsible (Posner, 1992), but anyone raising a child today has to worry about the surfeit of information about sex that now engulfs us.

Knowledge really is power, for good and for ill. Knowledge can have the power to disrupt ancient patterns of belief and action, the power to subvert authority, the power to change minds. It can interfere with trends that may or may not be desirable. In a notorious memorandum to President Richard Nixon, Daniel Patrick Moynihan wrote:

The time may have come when the issue of race could benefit from a period of "benign neglect." The subject has been too much talked about. The forum has been too much taken over by hysterics, paranoids, and boodlers on all sides. We may need a period in which Negro progress continues and racial rhetoric fades. The administration can help bring this about by paying close attention to such progress—as we are doing—while seeking to avoid situations in which extremists of either race are given opportunities for martyrdom, heroics, histrionics or whatever. [Moynihan, 1970]

We will probably never know if Moynihan was right, but he may have been. Those who suspect that he was right may hope that we follow his advice this time, postponing vigorous attention to religion as long as possible, deflecting inquiry, and hoping for the best. But it is hard to see how this policy could be achieved in any case. Since the Enlightenment, we have already had more than two hundred years of deferential, muted curiosity, and it doesn't seem to have led to the fading of religious rhetoric, does it? Recent history strongly suggests that religion is going to garner more and more attention, not less, in the immediate future. If it is going to receive attention, it had better be high-quality attention, not the sort that hysterics, paranoids, and boodlers on all sides engage in.

\\\

The problem is that it is just too hard nowadays to keep secrets. Whereas in earlier centuries ignorance was the default condition of most of the human race, and it took a considerable exercise of inquiry to learn about the wide world, today we are all swimming in a sea of information and misinformation, on every topic, from masturbation to how to build a nuclear weapon to Al Qaeda. As we deplore the attempt by some religious leaders in the Muslim world to keep their girls and women uneducated and uninformed about the world, we can hardly approve of similar embargoes on knowledge in our own sphere.

Or can we? Perhaps this point of disagreement is the continental divide in Opinion Space, between those who think our best hope is to try to nail the lid on Pandora's box and keep ourselves forever ignorant, and those who think that this is politically impossible and immoral in the first place. The former already pay a heavy price for their self-imposed factual poverty: they can't imagine in detail the consequences of their own chosen policy. Can they not see that nothing short of a police state, bristling with laws prohibiting inquiry and the dissemination of knowledge, or the sequestration of the population in a windowless world, could accomplish the feat? Is that really what they want? Do they think that they have methods undreamt of by the conservative mullahs for halting the inexorable flow of liberating information to their flock? *Think ahead.*

There is a trap here lying in wait of those without foresight. Perhaps no parents are immune to a twinge of regret when they see the first evidence of loss of innocence in their child, and the urge to shelter a child from the tawdry world is strong, but reflection should show anybody that it just won't work. We need to let our children grow up to face the world armed with knowledge, with much more knowledge than we ourselves had at their age. It is scary, but the alternative is worse.

There are some people—millions, apparently—who proudly declare that they do not have to foresee the consequences: they know in their hearts that this is the right path, whatever the details. Since Judgment Day is just around the corner, there is no reason to plan for the future. If you are one of these, here is what I hope will be a sobering reflection: have you *considered* that you are perhaps being irresponsible? You would willingly risk not only the lives and future well-being of your loved ones, but also the lives and future well-being of all the rest of us, without hesitation, without due diligence, guided by one revelation or another, a conviction that you have no good way of checking for soundness. "Every prudent man dealeth with knowledge: but a fool layeth open his folly" (Proverbs 13:16). Yes, I know, the Bible has a contrary text as well: "For it is written, I will destroy the wisdom of the wise, and will bring to nothing the understanding of the prudent" (1 Corinthians 1:19). Anybody can quote the Bible to prove anything, which is why you ought to worry about being overconfident.

Do you ever ask yourself: *What if I'm wrong?* Of course there is a large crowd of others around you who share your conviction, and this distributes—and, alas, dilutes—the responsibility, so, if you ever get a chance to breathe a word of regret, you will have a handy excuse: you got swept up by a crowd of enthusiasts. But surely you have noticed a troubling fact. History gives us many examples of large crowds of deluded people egging one another on down the primrose path to perdition. How can you be so sure you're not part of such a group? I for one am not in awe of your faith. I am appalled by your arrogance, by your unreasonable certainty that you have all the answers. I wonder if any believers in the End Times will have the intellectual honesty and courage to read this book through.

What we dimly imagine in dreaded anticipation often turns out to be much worse than reality. Before we lament our inability to hold back the rising tide of information, we should consider its likely consequences calmly. They may not be so bad. Imagine, if you can, that we had never had the Santa Claus myth at all, that Christmas was just another Christian feast, like Palm Sunday or Pentecost, celebrated but barely anticipated in the wide world. And imagine that the fans of J. K. Rowling's Harry Potter stories were to attempt to start a new tradition: every year, on the anniversary of the publication date of the first Harry Potter book, children shall receive gifts from Harry Potter, who flies in through the window on his magic broomstick, accompanied by his owl. Let's make Harry Potter Day a worldwide day for children! Toy manufacturers (and Rowling's publishers) would all be in favor, presumably, but imagine the doomsayers who would oppose it:

What a terrible idea! Think of the traumatic effects on young children when they learn, as they eventually would, that their innocence and trust had been exploited by a gigantic public conspiracy of grown-ups. The psychic and social toll of such a massive deception would be cynicism, despair, paranoia, and grief that might cripple children for life. Could there be anything more evil than deliberately concocting a seductive set of lies to spread to our children? They will hate us bitterly, and we will deserve their fury.

Had this quite compelling concern been effectively raised in the early days of the evolving Santa Claus mythology, it might well have prevented the Great Santa Claus Catastrophe of 1985! But we know better. There was no such catastrophe and never will be. Some children do suffer relatively brief bouts of embarrassment and bitterness on learning that there is no Santa Claus, but others take delicious pride in their Sherlock Holmes triumph of detection, and relish their new status among Those in the Know, eagerly contributing to the ruse next year, and soberly answering all the innocent questions put to them by their younger siblings.

So far as we know,3 the Santa Claus disillusionment does no harm. More to the point, it is likely (but not yet investigated, to the best of my knowledge) that part of the enduring appeal of the Santa Claus myth is that adults, who can no longer directly experience the innocent joys of Santa-anticipation, settle for the vicarious thrill of enjoying their children's excitement. People do go to a great deal of effort and expense to perpetuate the Santa Claus mythology. Why? Are they trying to recapture the lost innocence of childhood? Are they more directly motivated by their own gratification than by generosity? Or are the pleasures of conspiracy with community absolution (untarnished by the guilt that accompanies the conspiracies of adultery, embezzlement, or tax evasion, for example) enough on their own to pay for the substantial costs? Such impertinent ways of thinking will loom large in subsequent chapters, when we turn to the more upsetting questions about why religion is so popular. They are not rhetorical questions. They can be answered, if we try.

I appreciate that many readers will be profoundly distrustful of the tack I am taking here. They will see me as just another liberal professor trying to cajole them out of some of their convictions, and they are dead right about that—that's what I am, and that's exactly what I'm trying to do. Why, then, should they pay any attention? They are appalled by the moral decay they see on all sides, and are sincerely convinced that the protection of their religion from all inquiry and criticism is the best way to turn the tide. I wholeheartedly agree with them that there is a moral crisis, and that nothing is more important than working together on finding paths out of our current dilemmas, but I think I have a better way. Prove it, they will say. Let me try, I respond. That's what this book is about, and I ask them to try to read it with an open mind.

\\\

Chapter 2 Religion is not out-of-bounds to science, in spite of propaganda to the contrary from a variety of sources. Moreover, scientific inquiry is needed to inform our most momentous political decisions. There is risk and even pain involved, but it would be irresponsible to use that as an excuse for ignorance.

\\\

Chapter 3 If we want to know why we value the things we love, we need to delve into the evolutionary history of the planet, uncovering the forces and constraints that have generated the glorious array of things we treasure. Religion is not exempt from this survey, and we can sketch out a variety of promising avenues for further research, while coming to understand how we can achieve a perspective on our own inquiries that all can share, regardless of their different creeds.

CHAPTER THREE

||||\\\||||

Why Good Things Happen

1 Bringing out the best

Religious allegory has become a part of the fabric of reality. And living in that reality helps millions of people cope and be better people.

—Langdon, hero of *The Da Vinci Code*, by Dan Brown

When I began working on this book, I conducted interviews with quite a few people to try to get a sense of the different roles that religion plays in their lives. This was not scientific data-gathering (though I have also done some of that) but, rather, an attempt to set theories and experiments aside and go directly to real people and let them tell me in their own words why religion was so important to them. These were strictly confidential interviews, almost all one-on-one,[1] and although I was persistently inquisitive, I didn't challenge or argue with my informants. These occasions were often moving, to say the least, and I learned a lot. Some people had endured hardships that I could not readily imagine myself surviving, and some had found in their religion the strength to make, and hold fast to, decisions that were nothing short of heroic. Less dramatic, but even more impressive in retrospect, were the people of modest talent and accomplishment who were, in one way or another, simply *much better people* than one might expect them to be; it wasn't just that their lives had meaning to them—though this was certainly true—but that they were actually making the world better by their efforts, inspired by their conviction that their lives were not their own to dispose of as they chose.

Religion can certainly bring out the best in a person, but it is not the only phenomenon with that property. Having a child often has a wonderfully maturing effect on a person. Wartime, famously, gives people an abundance of occasions to rise to, as do natural disasters like floods and hurricanes. But for day-in, day-out lifelong bracing, there is probably nothing so effective as religion: it makes powerful and talented people more humble and patient, it makes average people rise above themselves, it provides sturdy support for many people who desperately need help staying away from drink or drugs or crime. People who would otherwise be self-absorbed or shallow or crude or simply quitters are often ennobled by their religion, given a perspective on life that helps them make the hard decisions that we all would be proud to make.

No all-in value judgment can be based on such a limited and informal survey, of course. Religion does all this good and more, no doubt, but something else we could devise might do it as well or better. There are many wise, engaged, morally committed atheists and agnostics, after all. Perhaps a survey would show that as a group atheists and agnostics are more respectful of the law, more sensitive to the needs of others, or more ethical than religious people. Certainly no reliable survey has yet been done that shows otherwise. It *might* be that the best that can be said for religion is that it helps some people achieve the level of citizenship and morality typically found in brights. If you find that conjecture offensive, you need to adjust your perspective.

Among the questions that we need to consider, objectively, are whether Islam is more or less effective than Christianity at keeping people off drugs and alcohol (and whether the side effects in either case are worse than the benefit), whether sexual abuse is more or less of a problem among Sikhs than among Mormons, and so forth. You don't get to advertise all the good that your religion does without first scrupulously subtracting all the harm it does and considering seriously the question of whether some other religion, or no religion at all, does better. World War II certainly brought out the best in many people, and those who lived through it often say that it was the most important thing in their lives, without which their lives would have no meaning, but it certainly doesn't follow from this that we should try to have another world war. The price you must pay for *any* claim about the virtue of your religion or any other religion is the willingness to see your claim put squarely to the test. My point here at the outset is just to acknowledge that we already know enough about religion to know that, however terrible its negative effects are—bigotry, murderous fanaticism, oppression, cruelty, and enforced ignorance, to cite the obvious—the people who view religion as the most important thing in life have many good reasons for thinking so.

2 *Cui bono?*

Blessed be the Lord, who daily loadeth us with benefits, even the God of our salvation. Selah. —Psalm 68:19

The more we learn about the details of natural processes, the more evident it becomes that these processes are themselves creative. Nothing transcends Nature like Nature itself. —Loyal Rue

Good things don't just happen by chance. There are "strokes of luck," but sustaining a good thing isn't just luck. It might be Providence, of course. It might be that God makes sure that the good thing happens and that it sustains itself when it wouldn't otherwise, without God's intervening. But any such account will have to wait its turn, for the same reason that cancer researchers are unwilling to treat unexpected remissions as just "miracles" that needn't be explored any further. What *natural,* nonmiraculous set of processes could produce and sustain this phenomenon that is so highly valued? The only way to take the hypothesis of miracles seriously is to eliminate the nonmiraculous alternatives.

The stinginess of Nature can be seen wherever we look, if we know what to look for. For instance, coyotes are emerging as a welcome addition to the wildlife of New England, howling eerily in the winter nights, but these beautiful, wily predators are wary of humans, and seldom seen. How can you tell their footprints in the snow from those of their cousins, domestic dogs? Even up close, it can be hard to tell the paw print of a coyote from the paw print of a similarly sized dog—a dog's claws tend to be longer, since they spend scant time digging—but even from afar, a coyote's track can be readily distinguished from a dog's—the coyote's prints fall in an uncannily straight and single-file line, with hind paws in almost perfect registration with forepaws, whereas a dog's track is typically a mess, as the dog galumphs exuberantly hither and yon, indulging every curious whim (David Brown, 2004). The dog is well fed and knows it will get its supper no matter what, whereas the coyote is on a very tight budget and needs to conserve every calorie for the job at hand: self-preservation. Its methods of locomotion have been ruthlessly optimized for efficiency. But, then, what explains the pack's characteristic howling? What good accrues to the coyote from that conspicuous expenditure of energy? Hardly a low profile. Doesn't it serve to scare away their supper and draw their presence to the attention of their own predators? Such costs would not be lightly recouped, one would think. These are good questions. Biologists are working on them, and even though they don't yet have definitive answers, they are surely right to seek them.[2] Any such pattern of conspicuous outlay demands an accounting.

Consider, for instance, the huge outlay of human effort devoted worldwide to sugar: not just the planting and harvesting of sugarcane and sugar beets, and the refining and transporting of the basic product, but the larger surrounding world of manufacturing candy, publishing cookbooks full of dessert recipes, advertising soft drinks and chocolates, commercializing Halloween, as well as the counterbalancing parts of the system: obesity clinics, government-sponsored research on the epidemic of early-onset diabetes, dentists and the inclusion of fluoride in toothpaste and drinking water. Over a hundred million metric tons of sugar are produced and consumed each year. To explain the thousands of features of this huge system, which provides the lifework of millions of people and can be discerned at every level of society, we need many different scientific and historical investigations, only a small fraction of which are *biological*. We need to study the chemistry of sugar, the physics of crystallization and caramelization, human physiology, and the history of agriculture, but also the history of engineering, manufacturing, transportation, banking, geopolitics, advertising, and much more.

None of these sugar-related expenditures of time and energy would exist if it weren't for the bargain that was struck about fifty million years ago between plants blindly "seeking" a way of dispersing their pollinated seeds, and animals similarly seeking efficient sources of energy to fuel their own reproductive projects. There are other ways to get your seeds dispersed, such as windborne gliders and whirligigs, and each method has its associated costs and benefits. Heavy, fleshy fruits full of sugar are a high-investment strategy, but they can have a bonanza payoff: the animal not only carries away the seed, but deposits it on a suitable bit of ground wrapped in a large helping of fertilizer. The strategy almost never works—not even once in a thousand tries—but it only has to work once or twice in the lifetime of a plant for it to replace itself on the planet and keep its lineage going. This is a good example of Mother Nature's stinginess in the final accounting combined with absurd profligacy in the methods. Not one sperm in a billion accomplishes its life mission—thank goodness—but each is designed and equipped as if everything depended on its success. (Sperm are like e-mail spam, so cheap to make and deliver that a vanishingly small return rate is sufficient to underwrite the project.)

Coevolution endorsed the bargain between plant and animal, sharpening our ancestors' capacity to discriminate sugar by its "sweetness." That is, evolution provided animals with specific receptor molecules that respond to the concentration of high-energy sugars in anything they taste, and hard-wired these receptor molecules to the seeking machinery, to put it crudely. People generally say that we like some things because they are sweet, but this really puts it backward: it is more accurate to say that some things are sweet (to us) because we like them! (And *we* like them because our ancestors who were wired up to like them had more energy for reproduction than their less fortunately wired-up peers.) There is nothing "intrinsically sweet" (whatever that would mean) about sugar molecules, but they are intrinsically *valuable* to energy-needing organisms, so evolution has arranged for organisms to have a built-in and powerful preference for anything that tickles their special-purpose high-energy detectors. That is why we are born with an instinctual liking for sweets—and, in general, the sweeter the better.

Both parties—plants and animals—benefited, and the system improved itself over the eons. What paid for all the design and manufacture (of the initial plant and animal equipment) was the differential reproduction of frugivorous and omnivorous animals and edible-fruit-bearing plants. Not all plants "chose" the edible-fruit-making bargain, but those that did had to make their fruits attractive in order to compete. It all made perfectly good sense, economically; it was a *rational* transaction, conducted at a slower-than-glacial pace over the eons, and of course no plant or animal had to understand any of this in order for the system to flourish. This is an example of what I call a *free-floating rationale* (Dennett, 1983, 1995b). Blind, directionless evolutionary processes "discover" designs that work. They work because they have various features, and these features can be described and evaluated in retrospect *as if* they were the intended brainchildren of intelligent designers who had worked out the rationale for the design in advance. This is not controversial in the general run of cases. The lens of an eye, for instance, is exquisitely well-designed to do its job, and the engineering rationale for the details is unmistakable, but no designer ever articulated it until the eye was reverse-engineered by scientists. The economic rationality of the *quid pro quo* bargains of coevolution is unmistakable, but until very recently, with the advent of human trade a few millennia ago, the rationales of such good deals were never represented in any minds.

Digression: This is a sticking point for those who don't yet appreciate just how well established the theory of evolution by natural selection is. According to a recent survey, only about a quarter of the population of the United States understands that evolution is about as well established as the fact that water is H_2O. This embarrassing statistic requires some explanation, since other scientifically advanced nations don't show the same pattern. Could so many people be wrong? Well, there was a time not so long ago when only a small minority of Earth's inhabitants believed that it was round and that it traveled around the sun, so we know that majorities *can* be flat wrong. But how, in the face of so much striking confirmation and massive scientific evidence, could so many Americans disbelieve in evolution? It is simple: they have been solemnly *told* that the theory of evolution is false (or at least unproven) by people they trust more than they trust scientists. Here is an interesting question: who is to blame for this widespread misinforming of the population? Suppose the ministers of your faith, who are wise and good people, assure you that evolution is a false and dangerous theory. If you are a layperson, you may be innocent in taking them at their authoritative word and then passing it on, authoritatively, to your children. We all trust the experts about many things, and these are your experts. But where, then, did your ministers get this misinformation?

If they claim to have gotten it from scientists, they have been duped, since there are no reputable scientists who claim this. Not a one. There are plenty of frauds and charlatans, though. As you see, I will not mince words. What about the Scientific Creationists and Intelligent Design proponents who are so vocal and visible in well-publicized campaigns? They have all been carefully and patiently rebutted by conscientious scientists who have taken the trouble to penetrate their smoke screens of propaganda and expose both their shoddy arguments and their apparently deliberate misrepresentations and evasions.[3] If you disagree heartily with this flat dismissal, you have two good choices to consider at this point:

1. Educate yourself in evolutionary theory and its critics and see for yourself whether what I say is true before proceeding. (The endnotes to this chapter provide all the references you will need to get going, and it should take only a few months of hard work.)
2. Suspend disbelief temporarily in order to learn what an evolutionist makes of religion as a natural phenomenon. (Perhaps your time and energy as a skeptic would be better spent trying to get to the heart of this evolutionist's perspective in search of a fatal flaw.)

Alternatively, you may believe that you don't need to consider the scientific evidence at all, since "the Bible says" that evolution is false, and that's all there is to it. This is a more extreme position than is sometimes recognized. Even if you believe that the Bible is the last and perfect word on every topic, you must recognize that there are people in the world who do not share your interpretation of the Bible. For instance, many take the Bible to be the Word of God but *don't* read it to rule out evolution, so it is just a plain everyday fact that the Bible does not speak clearly and unmistakably to all. Since that is so, the Bible is not a plausible candidate as common ground to be shared *without further discussion* in a reasonable conversation. If you insist it is, you are thumbing your nose at the whole inquiry. (Good-bye, and I hope to see you back again someday.)

But isn't there an unjustified asymmetry here, with me refusing to defend my anticreationism here and now, while sending the biblical inerrantist off for not playing by the rules of rational discussion? No, because I have directed everyone to the literature that *defends* the dismissal of creationism against all objections, whereas the inerrantist is refusing to take on even that obligation. To be symmetrical, the inerrantist should encourage me to consult the literature, if it exists, that purports to *demonstrate,* against all objections, that the Bible is indeed the Word of God and that it rules out evolution. I haven't yet been directed to any such literature, and haven't found it on any Web site, but if it exists, it would indeed warrant consideration as a topic for another day and another project— just like creationism and its critics. Those readers who remain will not demand any further consideration *by me* of creationism and its variants, since I have told them where to find the answers I endorse, for better or for worse. *End of Digression.*

Lawyers have a stock Latin phrase, *cui bono?*, which means "Who benefits from this?," a question that is even more central in evolutionary biology than in the law (Dennett, 1995b). Any phenomenon in the living world that *apparently* exceeds the functional cries out for explanation. The suspicion is always that we must be missing something, since a gratuitous outlay is, in a word, uneconomical, and as the economists are forever reminding us, there is no such thing as a free lunch. We don't marvel at an animal doggedly grubbing in the earth with its nose, for we figure it is seeking its food, but if it regularly interrupts its rooting with somersaults, we want to know why. Since accidents do happen, it is always *possible* that some feature of a living thing that appears to be a pointless excess is just as pointless as it appears (rather than a deep and baffling ploy in some game we don't understand). But evolution is remarkably efficient at sweeping pointless accidents off the scene, so if we find a *persistent pattern* of expensive equipment or activity, we can be quite sure that something benefits from it in the only stocktaking that evolution honors: differential reproduction. We should cast our nets widely when hunting for the beneficiaries, since they are often elusive. Suppose you find rats that extravagantly risk their lives in the presence of cats, and ask the *cui bono?* question. What good accrues to these rats from this foolhardy behavior? Are they showing off to impress potential mates, or does their extravagant behavior somehow improve their access to good food sources? Conceivably, but probably you are looking in the wrong place for the beneficiary. Like the lancet fluke that has taken up residence in the strenuous ant with which I began this book, there is a parasite, *Toxoplasma gondii,* that can live in many mammals but needs to get into a cat's stomach to reproduce, and when it infects rats, it has the useful property of interfering with their nervous systems and making them hyperactive and relatively fearless—and hence much more likely to be eaten by any cat in the vicinity! *Cui bono?* The benefit is to the fitness—the reproductive success—of *Toxoplasma gondii,* not the rats it infects (Zimmer, 2000).

Every bargain in nature has its rationale, free-floating unless it happens to be a bargain devised by human bargainers, the only rationale-representers yet to have evolved on the planet. But a rationale can become obsolete. As the opportunities and perils in the environment change, a good bargain can lapse. It takes time for evolution to "recognize" this. Our sweet tooth is a good example. Like the coyotes, our hunter-gatherer ancestors lived on very tight energy budgets, and had to avail themselves of every practical opportunity to store away calories for emergency use. A practically insatiable appetite for sweets made good sense then. Now that we have developed methods for creating a superabundance of sugar, that insatiability has become a serious design flaw. Recognizing the evolutionary source of this glitch helps us figure out how to deal with it. Our sweet tooth is not just an accident or a pointless bug in an otherwise excellent system; it was *designed* to do the work it does, and if we underestimate its resourcefulness, its resistance to perturbation and suppression, our efforts to cope with it are apt to be counterproductive. There is a reason why we love sugar, and it is—or used to be—a very good reason. We may find other superannuated loves that need our attention.

I mentioned music in the previous chapter, and we will eventually turn to a more detailed examination of its possible evolutionary sources, but I want to warm up first on some easier things we love. What about alcohol? What about money? What about sex? Sex presents some of the most interesting and challenging problems in evolutionary theory, because, on the face of it, sexual reproduction is a bad bargain indeed. Forget—for the moment—about our human kind of sex (sexy sex), and consider the most basic varieties of sexual reproduction in the living world: the sexual reproduction of almost all multicellular life-forms, from insects and clams to apple trees, and even many single-celled organisms. The great evolutionary biologist François Jacob once quipped that the dream of every cell is to become two cells. Every time this fission happens, a complete copy of the cell's genome is copied into the offspring. The parent clones itself, in other words; the resulting organism shares 100 percent of its genes. If you can make perfect genetic copies of yourself, why would you go to the expense of reproducing sexually, which involves not just finding a mate but, much more important, passing on only *half* of your genes to your offspring?4 This 50 percent reduction (from the gene's point of view) is known as the *cost of meiosis* (the kind of fission that occurs in sex cells, to distinguish it from the cloning fission of *mitosis*). Something must pay for this cost, and it must pay on delivery, not at some future date, since evolution lacks foresight and cannot approve bargains on the speculative basis of eventual return at some distant time.

Sexual reproduction is thus a costly investment that has to pay for itself in the short run. The details of theory and experiment on this topic are fascinating (see, e.g., Maynard Smith, 1978; Ridley, 1993), but for our purposes a few highlights from the currently front-running theory are most instructive: sex (in vertebrates like us, at least) pays for itself by making our offspring relatively inscrutable to the parasites we endow them with from birth. Parasites have short lifespans compared with their hosts, and typically reproduce many times during their host's lifetime. Mammals, for instance, are hosts to trillions of parasites. (Yes, right now, no matter how healthy and clean you are, there are trillions of parasites of thousands of different species inhabiting your gut, your blood, your skin, your hair, your mouth, and every other part of your body. They have been rapidly evolving to survive against the onslaught of your defenses since the day you were born.) Before a female can mature to reproductive age, her parasites evolve to fit her better than any glove. (Meanwhile, her immune system evolves to combat them, a standoff—if she is healthy—in an ongoing arms race.) If she gave birth to a clone, her parasites would leap to it and find themselves at home from the outset. They would be already optimized to their new surroundings. If instead she uses sexual reproduction to endow her offspring with a mixed set of genes (half from her mate), many of these genes—or, more directly, their products, in the offspring's internal defenses—will be alien or cryptic to the ship-jumping parasites. Instead of *home sweet home,* the parasites will find themselves in *terra incognita.* This gives the offspring a big head start in the arms race.

Could such a bargain pay for itself? That is the question at the heart of current research in evolutionary biology, and if the positive answer holds up to further scrutiny, then we will have found the ancient but ongoing source, in evolution, of the huge system of activities and products that we normally think of when we think of sex: marriage rituals and taboos against adultery, clothing and hairstyles, breath fresheners and pornography and condoms and HIV and all the rest. To explain why each and every facet of this huge complex exists, we will have to resort to many different kinds and levels of theory, not all of it biological. But none of this would exist if we weren't sexually reproducing creatures, and we need to understand the biological underpinnings first if we are to have a clear view of what is optional or mere historical accident, what is highly resistant to perturbation, what is exploitable. There are reasons why we love sex, and they are more complicated than you might think.

With alcohol, a somewhat different perspective emerges. What pays for the breweries, the vineyards and distilleries, and the massive delivery systems that bring alcoholic beverages within easy reach of almost every human being on the planet? We know that alcohol, like nicotine, caffeine, and the active ingredients in chocolate, has quite specific effects on receptor molecules in our brains. Let us suppose that these effects are just coincidences at the outset. That *some* large molecules in *some* plants happen to be biochemically similar to large molecules that play important modulating roles in animal brains is, let us suppose, as likely as not. Evolution must always begin with an element of brute chance. But, then, it is not surprising that, over millions of years of exploratory ingestion, our species and others should discover the plants with psychoactive ingredients and develop preferential or aversive dispositions regarding them. Elephants—and baboons and other African animals— have been known to get falling-down drunk eating fermenting fruit from marula trees, and there is evidence that elephants will travel great distances to arrive at the marula trees just when their fruits ripen. It seems that the fruit ferments in their stomachs when yeast cells resident on the fruit undergo a population explosion, consuming the sugar and excreting carbon dioxide and alcohol. The alcohol happens to create the same sort of pleasurable effects in the elephants' brains that it does in ours.

It may be that the basic bargain struck between fruit trees and frugivores—the seed-spreading-for-sugar deal—is enhanced by an additional partnership of yeast and fruit tree. This would create an added attraction that pays off by enhancing the reproductive prospects of both the yeast and the trees, or it may be just an accident in the wild. In any case, another species, *Homo sapiens*, has closed the loop and initiated just such a coevolutionary bargain: we domesticated both the yeast and the fruit, and for thousands of years we have been artificially selecting for the varieties that best engender the effects we love. The yeast cells provide a service for which they are paid off in protection and nutrients. That means that the yeast cultures carefully husbanded by brewers, vintners, and bakers are human symbionts just as much as the *E. coli* bacteria that inhabit our intestines. Unlike *endosymbiont* bacteria, such as *Toxoplasma gondii,* which have to get into the bodies of both rat and cat, the yeast cells are a sort of *ectosymbiont,* like the "cleaner" fish that groom larger fish, depending on another species, us, but not having to enter our bodies. They may—like a wayward cleaner fish—get ingested by us more or less by accident, but it is really only their excretions that need to get inside us for them to prosper!

Now consider a strikingly different sort of good thing: money. Unlike the other goods we have considered, it is restricted (so far) to a single species, us, and its design is transmitted through culture, not genes. I will have more to say about cultural evolution in later chapters. In this introductory overview I want to point out just a few striking similarities between money and the "more biological" treasures we have just surveyed. Like eyesight and flight, money has evolved more than once,[5] and hence is a compelling candidate for what I call a Good Trick—a move in design space that will be "discovered" again and again by blind evolutionary processes simply because so many different adaptive paths lead to it and thereby endorse it (Dennett, 1995b). Economists have worked out the rationale for money in some detail.

Money is clearly one of the most effective "inventions" of our clever species, but that rationale was free-floating until very recently. We used, and relied on, and valued money, and occasionally killed and died for money, long before the rationale of its value was made explicit in any minds. Money is not the only cultural invention to lack a specific inventor or author. Nobody invented language or music either.[6] An entertaining coincidence is that an old term for money in the form of coin and paper issue is *specie* (from the same Latin root as *species*), and, as many have noted, the free-floating rationale of specie could lapse in the foreseeable future, and it could go extinct in the wake of credit cards and other forms of electronic funds transfer. Specie, like a virus, travels light, and doesn't carry its own reproductive machinery with it, but, rather, depends for the persistence of its kind on provoking a host (us) to make copies of it using our expensive reproduction machinery (printing presses, stamps and dies).7 Individual coins and pieces of paper money wear out over time, and unless more are made and adopted, the whole system may go extinct. (You may confirm this by trying to buy a boat with a pile of cowrie shells.) But since money is a Good Trick, expect some other *species* of money to take over the niche left vacant by the departing *specie*.

I have another, ulterior motive for bringing up money. The goods being surveyed—sugar, sex, alcohol, music, money—are all problematic because in each case we can develop an obsession, and crave too much of a good thing, but money has perhaps the worst reputation as a good thing. Alcohol is condemned by many—by the Muslims in particular—but among those who appreciate it—such as the Roman Catholics—a person who loves it in moderation is not considered ignoble or a fool. But we are all *supposed* to despise money *as a thing in itself*, and value it only instrumentally. Money is "filthy lucre," something to be enjoyed *only* for what it can provide in the way of more worthy things of value, things with "intrinsic" value.8 As the old song says, not entirely convincingly, the best things in life are free. Is this because money is "artificial" and the others are all "natural"? Not likely. Is a string quartet or a single-malt whisky or a chocolate truffle any less artificial than a gold coin?

What we should make of this theme in human culture is an interesting question, about which I will say more later, but in the meantime we should note that the only anchor we have seen so far for "intrinsic" value is the capacity of something to provoke a preference response in the brain quite directly. Pain is "intrinsically bad," but this negative valence is just as dependent on an evolutionary rationale as the "intrinsic goodness" of satisfied hunger. A rose by any other name would smell as sweet, no doubt, but it is also true that if poking around in rotting elephant carcases was as good for our reproductive prospects as it is for those of vultures, such a dead elephant would smell as sweet as a rose to us.9 Biology insists on delving beneath the surface of "intrinsic" values and asking why they exist, and any answer that is supported by the facts has the effect of showing that the value in question is—or once was—really instrumental, not intrinsic, even if we don't see it that way. A *truly* intrinsic value couldn't have such an explanation of course. It would be good just because it was good, not because it was good *for* something. A hypothesis to consider seriously, then, is that *all* our "intrinsic" values started out as instrumental values, and now that their original purpose has lapsed, at least in our eyes, they remain as things we like just because we like them. (That would *not* mean that we are wrong to like them! It would mean—by definition— that we like them *without needing any ulterior reason* to like them.)

3 Asking what pays for religion

But what are the benefits; why do people want religion at all? They want it because religion is the only plausible source of certain rewards for which there is a general and inexhaustible demand.
   —Rodney Stark and Roger Finke, *Acts of Faith*

Whatever else religion is as a human phenomenon, it is a hugely costly endeavor, and evolutionary biology shows that nothing so costly just happens. Any such regular expenditure of time and energy has to be balanced by something of "value" obtained, and the ultimate measure of evolutionary "value" is *fitness*: the capacity to replicate more successfully than the competition does. (This does *not* mean that we ought to value replication above all! It means only that nothing can evolve and persist for long in this demanding world unless it somehow provokes its own replication better than the replication of its rivals.) Since money is such a recent innovation from the perspective of evolutionary history, it is weirdly anachronistic to ask *what pays for* one evolved biological feature or another as if there were actual transactions and ledgers in Darwin's countinghouse. But this metaphor nevertheless nicely captures the underlying balance of forces observed everywhere in nature, and *we know of no exceptions to the rule.* So, risking offense but shrugging off that risk as just one more aspect of the taboo that must be broken, I ask: what pays for religion? Abhor the language if you must, but that gives you no good reason to ignore the question. Any claim to the effect that religion—your religion or all religion—stands above the biosphere and does not have to answer to this demand is simply bluster. It might be that God implants each human being with an immortal soul that thirsts for opportunities to worship God. That would indeed explain the bargain struck, the exchange of human time and energy for religion. The only honest way to defend that proposition, or anything like it, is to give fair consideration to alternative theories of the persistence and popularity of religion and rule them out by showing that they are unable to account for the phenomena observed. Besides, you might want to defend the hypothesis that God set up the universe so that we would evolve to have a love of God. If so, we would want to understand how that evolution occurred.

The same sort of investigation that has unlocked the mysteries of sweetness and alcohol and sex and money can be undertaken for the many facets of religion. There was a time, not so very long ago by evolutionary standards, when there was no religion on this planet, and now there is lots of it. Why? It may have one primary evolutionary source or many, or it may defy evolutionary analysis altogether, but we won't know until we look. Do we really need to inquire about this? Can't we just accept the obvious fact that religion is a human phenomenon and that humans are mammals, and hence products of evolution, and then leave the biological underpinnings of religion at that? People make religions but they also make automobiles and literature and sports, and surely we don't need to look deep into biological prehistory to understand the differences between a sedan, a poem, and a tennis tournament. Aren't most of the religious phenomena that need investigation *cultural* and *social*—ideological, philosophical, psychological, political, economic, historical—and hence somehow "above" the biological level?

This is a familiar presumption among researchers in the social sciences and humanities, who often deem it "reductionistic" (and very bad form) even to *pose* questions about the biological bases of these delightful and important phenomena. I can see some cultural anthropologists and sociologists rolling their eyes in disdain—"Oh, no! Here comes Darwin again, butting in where he isn't needed!"—while some historians and philosophers of religion and theologians snicker at the philistinism of anybody who could ask with a straight face about the evolutionary underpinnings of religion. "What next, a search for the Catholicism gene?" This negative response is typically unthinking, but it isn't foolish. It is supported in part by unpleasant memories of past campaigns that failed: naïve and ill-informed forays by biologists into the thickets of cultural complexity. There is a good case to be made that the social sciences and humanities—the *Geisteswissenschaften,* or mind sciences—have their own "autonomous" methodologies and subject matters, independent of the natural sciences. But in spite of all that can be said in favor of this idea (and I will spend some time looking at the best case for it in due course), the disciplinary isolation it motivates has become a major obstacle to good scientific practice, a poor excuse for ignorance, an ideological crutch that should be thrown away.[10]

We have particularly compelling reasons for investigating the biological bases of religion *now*. Sometimes—rarely—religions go bad, veering into something like group insanity or hysteria, and causing great harm. Now that we have created the technologies to cause global catastrophe, our jeopardy is multiplied to the maximum: a toxic religious mania could end human civilization overnight. We need to understand what makes religions work, so we can protect ourselves in an informed manner from the circumstances in which religions go haywire. What is religion composed of? How do the parts fit together? How do they mesh? Which effects depend on which causes? Which features, if any, invariably occur together? Which exclude each other? What constitutes the health and pathology of religious phenomena? These questions can be addressed by anthropology, sociology, psychology, history, and any other variety of cultural studies that you like, but it is simply inexcusable for researchers in these fields to let disciplinary jealousy and fear of "scientific imperialism" create an ideological iron curtain that could conceal important underlying constraints and opportunities from them.

Consider our current controversies regarding nutrition and diet. Understanding the design rationale of the machinery in our bodies that drives us to overindulge in sweets and fats is the key to finding the corrective measures that will actually work. For many years, nutritionists thought that the key to preventing obesity was simply cutting fat out of the diet. Now it is emerging that this simplistic approach to dieting is counterproductive: when you strenuously keep your fat-craving system unsatisfied, this intensifies your body's compensatory efforts, leading to overindulgence in carbohydrates. The evolutionarily naïve thinking of the recent past helped build and put in motion the low-fat bandwagon, which then became self-sustaining under the solicitous care of the low-fat-food manufacturers and advertisers. Taubes (2001) is an eye-opening account of the political processes that created and sustained this "low-fat gospel," and it provides a timely warning for the enterprise I am proposing here: "It's a story of what can happen when the demands of public health policy—and *the demands of the public for simple advice* [emphasis added]—run up against the confusing ambiguity of real science" (p. 2537). Even if we do the science of religion right (for the first time), we must strenuously guard the integrity of the next process, the boiling down of the complex results of the research into political decisions. This will not be easy at all. Basil Rifkind, one of the nutritionists who were pressured into a premature verdict on low dietary fat, puts it succinctly: "There comes a point when, if you don't make a decision, the consequences can be great as well. If you just allow Americans to keep on consuming 40% of calories from fat, there's an outcome to that as well" (Taubes, 2001, p. 2541). Good intentions are not enough. This is the sort of misguided campaign that we want to avoid when we try to correct what we take to be the toxic excesses of religion. One recoils in horror at the possible effects of trying to impose one misguided "crash diet" or another on those hungry for religion.

It may be tempting to argue that we'd all have been better off if there hadn't been any know-it-all nutritionists meddling with our diets in the first place. We'd have eaten what was good for us by just relying on our evolution-shaped instincts, the way other animals do. But this is simply mistaken, in the case of both diet and religion. Civilization—agriculture in particular and technology in general—has hugely and swiftly altered our ecological circumstances compared with the circumstances of our quite recent ancestors, and this renders many of our instincts out of date. Some of them may still be valuable in spite of their obsolescence, but it is likely that some are positively harmful. We can't return to the blissful ignorance of our animal past with any confidence. We're stuck being the *knowing* species, and that means we'll have to use our knowledge as best we can to adapt our policies and practices to cope with our biological imperatives.

4 A Martian's list of theories

If you were God, would you have invented laughter?
—Christopher Fry, *The Lady's Not for Burning*

We may be too close to religion to be able to see it clearly at first. This has been a familiar theme among artists and philosophers for years. One of their self-appointed tasks is to "make the familiar strange,"[11] and some of the great strokes of creative genius get us to break through the crust of excessive familiarity and look at ordinary, obvious things with fresh eyes. Scientists couldn't agree more. Sir Isaac Newton's mythic moment was asking himself the weird question about why the apple fell *down* from the tree. ("Well, why *wouldn't* it?" asks the everyday nongenius; "It's *heavy!*"—as if this were a satisfactory explanation.) Albert Einstein asked a similarly weird question: everyone knows what "now" means, but Einstein asked whether you and I mean the same thing by "now" when we are leaving each other's company at near the speed of light. Biology has some strange questions as well. "Why don't male animals lactate?" asks the late great evolutionary biologist John Maynard Smith (1977), vividly awakening us from our dogmatic slumbers to confront a curious prospect. "Why do we blink with both eyes simultaneously?" asks another great evolutionary biologist, George Williams (1992). Good questions, not yet answered by biology. Here are some more. Why do we laugh when something funny happens? We may think it is just obvious that laughter (as opposed to, say, scratching one's ear or belching) is the appropriate response to humor, but why is it? Why are some female shapes sexy and others not? Isn't it obvious? Just look at them! But that is not the end of it. The regularities and trends in our responses to the world do indeed guarantee, trivially, that they are part of "human nature," but that still leaves the question of why. Curiously, it is this very feature of evolutionary questioning that is often viewed with deep aversion

by . . . artists and philosophers. The philosopher Ludwig Wittgenstein famously said that explanation has to stop somewhere, but
this undeniable truth misleads us if it discourages us from asking
such questions, prematurely terminating our curiosity. Why does
music exist, for instance? "Because it's *natural!*" comes the complacent everyday reply, but science takes nothing natural for granted.
People around the world devote many hours—often their professional lives—to making, and listening and dancing to, music. Why?
Cui bono? Why does music exist? Why does religion exist? To say
that it is natural is only the beginning of the answer, not the end.

The remarkable autistic author and animal expert Temple
Grandin gave neurologist Oliver Sacks a great title for one of his
collections of case studies of unusual human beings: *An Anthropologist on Mars* (1995). That's what she felt like, she told Sacks,
when dealing with other people right here on Earth. Usually such
alienation is a hindrance, but getting some distance from the ordinary world helps focus our attention on what is otherwise too obvious to notice, and it will help if we temporarily put ourselves into
the (three bright green) shoes of a "Martian," one of a team of alien
investigators who can be imagined to be unfamiliar with the phenomena they are observing here on Planet Earth.

What they see today is a population of over six billion people,
almost all of whom devote a significant fraction of their time and
energy to some sort of religious activity: rituals such as daily prayer
(both public and private) or frequent attendance at ceremonies, but
also costly sacrifices—not working on certain days no matter what
looming crisis needs prompt attention, deliberately destroying valuable property in lavish ceremonies, contributing to the support of
specialist practitioners within the community and the maintenance
of elaborate buildings, and abiding by a host of strenuously observed prohibitions and requirements, including not eating certain
foods, wearing veils, taking offense at apparently innocuous behaviors in others, and so forth. The Martians would have no doubt that

all of this was "natural" in one sense: they observe it almost every-where in nature, in one species of vocal bipeds. Like the other phe-nomena of nature, it exhibits both breathtaking diversity and striking commonalities, ravishingly ingenious design (rhythmic, poetic, architectural, social . . .) and yet baffling inscrutability. Where did all this design come from, and what sustains it? In addition to all the contemporary expenditures of time and effort, there is all the implied design work that preceded it. Design work—R & D, re-search and development—is costly, too.

Some of the R & D can be observed by the Martians directly: de-bates among religious leaders about whether to abandon awkward elements of their orthodoxy, decisions by building committees to accept a winning architectural proposal for a new temple, com-posers executing commissions for new anthems, theologians writ-ing tracts, televangelists meeting with advertising agencies and other consultants to plan their new season of broadcasts. In the de-veloped world, in addition to the time and energy spent in religious observance, there is a huge enterprise of public and private criti-cism and defense, interpretation and comparison, of every aspect of religion. If the Martians just focus on this, they will form the im-pression that religion, like science or music or professional sport, consists of systems of social activity that are designed and re-designed by conscious, deliberate agents who are aware of the points or purposes of the enterprises, the problems that need solv-ing, the risks and costs and benefits. The National Football League was created and designed by identifiable individuals to fulfill a set of human purposes, and so was the World Bank. These institutions show clear evidence of design, but they are not "perfect." People make mistakes, errors get identified and corrected over time, and when there are substantial disagreements among those who have the power and responsibility for maintaining such a system, com-promises are sought and often achieved. Some of the R & D that has shaped and is still shaping religion falls clearly into this cate-gory. An extreme case would be Scientology, a whole religion that is unquestionably the deliberately designed brainchild of a single author, L. Ron Hubbard, though of course he borrowed elements that had proved themselves in existing religions.

At the other extreme, there is no doubt that the equally intricate, equally designed folk religions or tribal religions found all over the world have never been subjected by their practitioners to anything like the "design review board" processes exemplified by the Council of Trent or Vatican II. Like folk music and folk art, these religions have acquired their aesthetic properties and other design features by a less self-conscious system of influences. And, whatever these influences are or were, they exhibit deep commonalities and patterns. How deep? As deep as the genes? Are there "genes for" the similarities among religions around the world? Or are the patterns that matter more geographical or ecological than genetic?

The Martians don't need to invoke genes to explain why people in equatorial climates don't wear fur coats, or why watercraft all over the world are both elongated and symmetrical around the long axis (aside from Venetian gondolas and a few other specialized craft). The Martians, having mastered the world's languages, will soon notice that there is huge variation in sophistication among boatbuilders around the world. Some of them can give articulate and accurate explanations of just why they insist that their vessels be symmetrical, explanations that any naval architect with a Ph.D. in engineering would applaud, but others have a simpler answer: we build boats this way because this is the way we have always built them. They copy the designs they learned from their fathers and grandfathers, who did the same in their day. This more or less mindless copying, the Martians will notice, is a tempting parallel with the other transmission medium they have identified, the genes. If boatbuilders or potters or singers are in the habit of copying old designs "religiously," they may preserve design features over hundreds or even thousands of years. Human copying is variable, so slight variations in the copies will often appear, and although most of these promptly disappear, since they are deemed defective or "seconds" or in any event not popular with the customers, every now and then a variation will engender a new lineage, in some sense an improvement or innovation for which there is a "market niche." And, lo and behold, *without anybody's realizing it, or intending it,* this relatively mindless process over long periods of time can shape designs to an exquisite degree, optimizing them for local conditions.

A culturally transmitted design can, in this way, have a free-floating rationale in exactly the same way a genetically transmitted design does. The boatbuilders and boat owners no more need to understand the reasons why their boats are symmetrical than the fruit-eating bear needs to understand his role in propagating wild apple trees when he defecates in the woods. Here we have the design of a human artifact—culturally, not genetically transmitted—without a human designer, without an author or inventor or even a knowing editor or critic.[12] And the reason the process can work is exactly the same in human culture as it is in genetics: *differential replication.* When copies are made with variation, and some variations are in some tiny way "better" (just better enough so that more copies of *them* get made in the next batch), this will lead inexorably to the ratcheting process of design improvement Darwin called evolution by natural selection. What gets copied doesn't have to be genes. It can be anything at all that meets the basic requirements of the Darwinian algorithm.[13]

This concept of cultural replicators—items that are copied over and over—has been given a name by Richard Dawkins (1976), who proposed to call them *memes,* a term that has recently been the focus of controversy. For the moment, I want to make a point that should be uncontroversial: cultural transmission can *sometimes* mimic genetic transmission, permitting competing variants to be copied at different rates, resulting in gradual revisions in features of those cultural items, and *these revisions have no deliberate, foresighted authors.* The most obvious, and well-researched, examples are natural languages. The Romance languages—French, Italian,

Spanish, Portuguese, and a few other variants—all descend from Latin, preserving many of the basic features while revising others. Are these revisions *adaptations*? That is, are they in any sense *improvements* over their Latin ancestors in their environments? There is much to be said on this topic, and the "obvious" points tend to be simplistic and wrong, but at least this much is clear: once a shift starts to emerge in one locality, it generally behooves local people to go along with it, if they want to be understood. When in Rome, speak as the Romans do, or be ignored or misunderstood. Thus do idiosyncrasies in pronunciation, slang idioms, and other novelties "go to fixation," as a geneticist would say, in a local language. And none of this is genetic. What is copied is a *way of saying something*, a behavior or routine.

The gradual transformations that turned Latin into French and Portuguese and other offspring languages were not intended, planned, foreseen, desired, commanded by anyone. On rare occasions, a particular local celebrity's peculiar pronunciation of a word or sound may have caught on, a fad that eventually turned into a cliché and then into an established part of the local language, and in these instances we can plausibly identify the "Adam" or "Eve" at the root of that feature's family tree. On even rarer occasions, individuals may *set out* to invent a word or a pronunciation and actually succeed in coining something that eventually enters the language, but in general, the changes that accumulate have no salient human authors, deliberate or inadvertent.

Folk art and folk music, folk medicine, and other products of such folk processes are often brilliantly adapted to quite advanced and specific purposes, but, however wonderful these fruits of cultural evolution are, we should resist the strong temptation to postulate some sort of mythic folk genius or mystical shared consciousness to explain them. These excellent designs often *do* owe some of their features to deliberate improvements by individuals along the way, but they *can* arise by exactly the same sort of blind, mechanical, foresightless sifting-and-duplicating process that has produced the exquisite design of organisms by natural selection, and in both cases the "judging" is harsh, austere, and unimaginative. Mother Nature is a philistine accountant who cares only about the immediate payoff in terms of differential replication, cutting no slack for promising candidates who can't measure up to the contemporary competition. Indeed, the tin-eared and forgetful singer who can hardly carry a tune and forgets almost every song he hears *but can remember this one memorable song* contributes as much quality control to the folk process (by replicating this classic-in-the-making at the expense of all the competing songs) as the most gifted tunesmith.

Words exist. What are they made of? Air under pressure? Ink? Some instances of the word "cat" are made of ink, and some are made of bursts of acoustic energy in the atmosphere, and some are made of patterns of glowing dots on computer screens, and some occur silently in thoughts, and what they have in common is just that they count as "the same" (*tokens* of the same *type,* as we philosophers say) in a system of symbols known as a language. Words are such familiar items in our language-drenched world that we tend to think of them as if they were unproblematically tangible things—as real as teacups and raindrops—but they are in fact quite abstract, even more abstract than voices or songs or haircuts or opportunities (and what are *they* made of?). What *are* words? Words are basically information packets of some sort, recipes for using one's vocal apparatus and ears (or hands and eyes)—and brains—in quite specific ways. A word is more than a sound or a spelling. For instance, *fast* sounds the same and is spelled the same in English and German, but has completely different meanings and roles in the two languages. Two different words, sharing only some of their surface properties. Words exist. Do memes exist? Yes, because words exist, and words are memes that can be pronounced. Other memes are *the same sort of thing*—information packets or recipes for doing something other than pronouncing—behaviors such as shaking hands or making a particular rude gesture, or taking off your shoes when you enter a house, or driving on the right, or making your boats symmetrical. These behaviors can be described and taught explicitly, but they don't have to be; people can just imitate the behaviors they see others perform. Variations in pronunciation can spread, and so can variations in cooking methods, doing the laundry, planting crops.

There are vexatious problems about just what the boundaries of a meme are—is wearing a baseball cap backward one meme or two (wearing a cap, and putting it on backward)?—but similar problems arise for word boundaries—should we count "copping out" as one word or two?—and, indeed, for genes. The boundary conditions are crisp for single molecules of DNA, or their constituent parts such as nucleotides or codons (triplets of nucleotides, such as AGC or AGA), but genes don't line up cleanly with these boundaries. They sometimes come apart into several separated pieces, and the reasons that biologists call the separated strings of codons parts of a single gene instead of two genes are very much the same reasons that linguists would identify "tickle [my, his, her] fancy" or "read [me, him, her] the riot act" as salient idioms, not just verb phrases composed of several words. Such yoked-together parts raise problems for anybody trying to count genes—not insurmountable, but not obvious, either. And what is copied and transmitted, in the case of both memes and genes, is *information*.

I will have more to say about memes in later chapters, and since overeager meme-enthusiasts and equally overeager meme-debunkers have made the topic a hot-button issue for many people, I need to protect a (relatively!) sober version of the concept from some of its friends and enemies. Not everybody need participate in this exercise of conceptual hygiene, however, so I have reprinted my basic introduction to memes—"The New Replicators," from the recent two-volume *Encyclopedia of Evolution* published by Oxford University Press in 2002—as appendix A at the back of this book.[14]

For our purposes now, the main reason for taking the memes perspective seriously is that it permits us to look at the *cui bono?* question for every designed feature of religion without prejudging the issue of whether we're talking about genetic or cultural evolution, and whether the rationale for a design feature is free-floating or explicitly *somebody's* rationale. This expands the space of possible evolutionary theories, opening up room for us to consider multilevel, mixed processes, getting us away from the simplistic ideas of "genes for religion" at one extreme and "a conspiracy of priests" at the other extreme and permitting us to consider much more interesting (and more probable) accounts of how and why religions evolve. Evolutionary theory is not a one-trick pony, and when the Martians set out to theorize about Earthly religion, they have lots of options to explore, which I will swiftly sketch, in extreme versions, just to give a sense of the terrain to be explored more carefully in later chapters.

Sweet-tooth theories: First, consider the variety of things we like to ingest or otherwise insert into our bodies: sugar, fat, alcohol, caffeine, chocolate, nicotine, marijuana, and opium for a start. In each case, there is an evolved receptor system in the body designed to detect substances (either ingested or constructed within the body, such as the endorphins or endogenously created morphine analogues) that these favorites have in high concentration. Over the ages, our clever species has gone prospecting, sampling just about everything in the environment, and after millennia of trial and error has managed to discover ways of gathering and concentrating these special substances so that we can use them to (over)stimulate our innate systems. The Martians may wonder if there are also genetically evolved systems in our bodies that are designed to respond to *something* that religions provide in intensified form. Many have thought so. Karl Marx may have been more right than he knew when he called religion the opiate of the masses. Might we have a god center in our brains along with our sweet tooth? What would it be for? What would pay for it? As Richard Dawkins puts it, "If neu-roscientists find a 'god center' in the brain, Darwinian scientists like me want to know why the god center evolved. Why did those of our ancestors who had a genetic tendency to grow a god center sur-vive better than rivals who did not?" (2004b, p. 14).

If any such evolutionary account is correct, then those with a god center not only survived better than those without one; they tended to have more offspring. But we should carefully set aside the anachronism involved in thinking of this hypothesized innate sys-tem as a "god center," since its original target may have been quite unlike the intense stuff that turns it on today—we don't have an in-nate chocolate-ice-cream center in the brain, after all, or a nicotine center. God may just be the latest and most intense confection that triggers the *whatsis* center in so many people. What benefit accrued to those who satisfied their *whatsis* craving? It could even be that there isn't and never has been any actual target in the world to ob-tain, but just an imaginary or virtual target, in effect: it's been the *seeking,* not the *getting,* that has had a fitness advantage. In any case, if the need, or at least the taste, for this still-unidentified treasure has become a genetically transmitted part of human nature, we tamper with it at our peril.

Theories in this family raise some interesting possibilities. Both sugar and saccharine trigger our sweet-tooth system. Are there reli-gion substitutes to be found or concocted by clever psychoengi-neers? Or—even more interesting—are religions themselves a kind of saccharine for the brain, less filling or debilitating or intoxi-cating than the original and potentially harmful target? Is religion itself a subspecies of folk medicine, in which we self-medicate for relief, using therapies honed by thousands of years of trial-and-error development? Is there genetic variation in religious sensi-tivity, like the huge genetic variation recently discovered among human beings in taste and olfaction? Those of us who can't stand cilantro have a gene for an olfactory receptor that cilantro lovers don't share. Cilantro "tastes" rather like soap to us. William James speculated a hundred years ago that he—but not everybody—had a brute need for religion: "Call this, if you like, my mystical *germ*. It is a very common germ. It creates the rank and file of believers. As it withstands in my case, so it will withstand in most cases, all purely atheistic criticism" (letter to Leuba, quoted in introduction to James, 1902, p. xxiv). James's mystical germ might actually be a mystical *gene*. *Or* it might be, just as he said, a mystical *germ*, something that spread from person to person not "vertically" (by descent from parents) but "horizontally," by infection.

Symbiont theories: Religions might turn out to be species of cultural symbionts that manage to thrive by leaping from human host to human host. They may be *mutualists*—enhancing human fitness and even making human life possible just as the bacteria in our gut do. Or *commensals*—neutral, neither good for us nor bad for us, but along for the ride. Or they might be *parasites:* deleterious replicators that we would be better off without—at least so far as our genetic interests are concerned—but that are hard to eliminate, since they have evolved so well to counter our defenses and enhance their own propagation. We can expect that cultural parasites, like microbial parasites, exploit whatever preexisting systems come in handy. The sneezing reflex, for instance, is in the first place an adaptation for ridding the nasal passages of foreign irritants, but when a germ provokes sneezing, it is typically not the sneezer but the germ that is the principal beneficiary, getting a high-energy launching into a neighborhood where other potential hosts can take it in. Spreading germs and spreading memes may exploit similar mechanisms, such as irresistible urges to impart stories or other items of information to others, enhanced by traditions that heighten the length, intensity, and frequency of encounters with others who might be likely hosts.

When we look at religion from this perspective, the *cui bono?* question changes dramatically. Now it is not *our* fitness (as reproducing members of the species *Homo sapiens*) that is presumed to be enhanced by religion, but *its* fitness (as a reproducing—self-replicating—member of the symbiont genus *Cultus religiosus*). It may thrive as a mutualist because it benefits its hosts quite directly, or it may thrive as a parasite even though it oppresses its hosts with a virulent affliction that leaves them worse off but too weak to combat its spread. And the main point to get clear about at the outset is that we *can't tell which of these is more likely to be true without doing careful, objective research.* Your religion probably seems obviously benign to you, and other religions may well seem to you to be just as obviously toxic to those infected by them, but appearances can deceive. Perhaps *their* religion is providing them with benefits that you just don't understand yet, and perhaps *your* religion is poisoning you in ways that you have never suspected. You really can't tell from the inside. That's how parasites work: quietly, unobtrusively, without disturbing their hosts any more than is absolutely necessary. *If* (some) religions are culturally evolved parasites, we can expect them to be insidiously well designed to conceal their true nature from their hosts, since this is an adaptation that would further their own spread.

\|\|\|

These two families of theories, sweet tooth and symbiont, are not exclusive. As we have already seen with the example of the alcohol-excreting yeast, there are symbiotic possibilities that may combine several of these phenomena together. It may be that an initial craving is exploited by cultural symbionts that include both mutualist and parasitical forms. A relatively benign or harmless symbiont may mutate under some conditions into something virulent and even deadly. For millennia, people have imagined that *other* religions might be a form of disease or sickness, and apostates often look back on their earlier days as a period of affliction which they have somehow survived, but the evolutionary perspective allows us to see that there are just as many positive as negative scenarios once we start looking at religion as possibly a cultural symbiont.

Friendly symbionts are everywhere. Your body is composed of perhaps a hundred trillion cells, and nine out of ten of them are not human cells (Hooper et al., 1998)! Most of these trillions of microscopic guests are either harmless or helpful; only a minority are worth worrying about. Many of them, indeed, are valuable helpers that we inherit from our mothers and would be quite defenseless without. *These inheritances are not genetic.* Some of them may be passed on via the shared bloodstream of mother and fetus, but others are picked up by bodily contact or proximity. (A surrogate mother who makes no *genetic* contribution to the fetus implanted in her womb nevertheless makes a major contribution to the microflora that the infant will carry with it for the rest of its life.)

Cultural symbionts—memes—are similarly passed on to one's offspring by nongenetic pathways. Speaking one's "mother tongue," singing, being polite, and many other "socializing" skills are transmitted culturally from parents to offspring, and infant human beings deprived of these sources of inheritance are often profoundly disabled. It is well known that the parent-offspring link is the major pathway of transmission of religion. Children grow up speaking their parents' language and, in almost all cases, identifying with their parents' religion. Religion, not being genetic, *can* be spread "horizontally" to nondescendants, but such conversions play a negligible role under most circumstances. A dim appreciation of this has led in the past to some crude and cruel programs of "hygiene." If you think that religion is, all things considered, a malignant feature of human culture, a childhood disease of sorts with lingering aftereffects, the public-health policy to deal with it would be politically drastic but quite simple: inoculation and isolation. Don't let parents give their own children a religious upbringing! This policy has been tried, on a major scale, in the former Soviet Union, with dire consequences. The rebound of religion in post-USSR Russia suggests that religion has roles to play and resources undreamt of by this simple vision.

A completely different sort of evolutionary possibility is represented by *sexual-selection* theories. Perhaps religion is like a bower-bird's bower. Male bowerbirds devote extraordinary time and effort to building and decorating elaborate structures designed to impress females of the species, who choose a mate only after assessing rival bowers carefully. This is an example of *runaway sexual selection,* the subvariety of natural selection in which the pivotal selective role is played by the choosy female, whose preferences may, over many generations, snowball into highly specific and onerous demands, such as the whims of peahens that oblige peacocks to grow spectacular—and spectacularly expensive and awkward—tails. (See Cronin, 1991, for a fine overview.) The bright coloration of male birds is the best-studied example of sexual selection. In these cases, an initial bias in the innate whims of females, such as a preference for blue over yellow, gets amplified by positive feedback into intensely blue males, the bluer the better. Had a majority of the females in an isolated population of the species just happened to prefer yellow over blue, the runaway selection would have ended up with bright-yellow males. There is nothing in the environment that makes yellow better than blue or vice versa *except for the reigning taste of the species' females,* which exerts a powerful, if arbitrary, selection pressure.

How might something like the runaway sexual selection process shape the extravagances of religion? In several ways. First, there might have been straightforward sexual selection by human females for religion-enhancing psychological traits. Perhaps they preferred males who demonstrated a sensitivity to music and ceremony, which could then have snowballed into a proclivity for elaborate rapture. The females who had this preference wouldn't have had to understand why they had it; it could just have been a whim, a blind personal taste that prompted them to choose, but if the mates they chose just happened to be better providers, more faithful family men, these mothers and fathers would tend to raise more children

and grandchildren than others, and both the sensitivity to ceremony and the taste for those who loved ceremony would spread. Or
the same whim could have had a selective advantage *only* because
more females shared that whim, so that sons who lacked the fash-
ionable sensitivity to ceremony were passed over by the choosy females. (And if an influential sample of our female ancestors had
happened, for no good reason, to have a taste for males who
jumped up and down in the rain, we guys would now find ourselves unable to sit still whenever it rained. Girls might or might
not share our tendency to jump under these conditions, but they
would definitely go for guys who did—that is the implication of the
classic sexual-selection hypothesis.) The idea that *musical talent* is
the royal road to the embrace of a woman is certainly familiar; it
probably sells a million guitars a year. And there may well be something to it. This could be a genetically transmitted proclivity, with
significant variation in the population, but we should also consider
cultural analogues of sexual selection. The potlatch ceremonies
found among the Native Americans of the Northwest are striking:
ceremonial demonstrations of conspicuous generosity, in which individuals compete with one another to see who can give away the
most, sometimes to the point of ruin. These customs bear the
marks of having been created by a positive-feedback escalator like
those that establish peacock tails and giant Irish-elk antlers. Other
social phenomena also exhibit inflationary spirals of expensive and
essentially arbitrary competition: tail fins on cars of the 1950s, teen-
agers' fashions, and outdoor lighting displays at Christmas are
among those most often discussed, but there are others as well.

For more than a million years, our ancestors made beautiful
"Acheulean handaxes," pear-shaped stone implements of varying
size, lovingly finished and seldom showing any sign of wear and
tear. Clearly our ancestors spent a lot of time and energy making
these, and the design hardly changed over the eons. Large caches of
hundreds and even thousands of these have been found (Mithen,
1996). The archeologist Thomas Wynne (1995) has opined that "it

would be difficult to over-emphasize just how strange the handaxe is when compared to the products of modern culture." "They're *bio-facts*," said one archeologist, coining a new term, and inspiring the science writer Marek Kohn (1999) to come up with a striking hypothesis. *Geofacts* are what archeologists call stones that look like artifacts but aren't—they are just the unintended product of some geological process. Kohn proposes that these handaxes may not be *artifacts* so much as biofacts, more like a bowerbird's bower than a hunter's bow and arrow, conspicuously expensive advertisements of male superiority, a ploy that was transmitted culturally, not genetically, in a tradition that dominated the battle of the sexes for a million years. The hominoids who worked so hard to participate in this competition no more needed to understand the rationale of the enterprise than do the male spiders who catch an insect and wrap it neatly in silk to present as a "nuptial gift" to females during courtship. This is a highly speculative and controversial claim, but it is not yet disproven, and it usefully alerts us to the possibilities that might otherwise elude us. Whatever the reasons for it, our ancestors lavished time and effort on apparently unused artifacts whenever they could, a precedent worth remembering when we marvel at the expense of tombs, temples, and sacrifices.

The interplay of cultural and genetic transmission should also be explored. Consider the well-studied case of lactose tolerance in adults, for instance. Many of us adults can drink and digest raw milk without difficulty, but many others, who of course had no difficulty consuming milk when they were babies, can no longer digest milk after infancy, since their bodies switch off the gene for making lactase, the necessary enzyme, after they are weaned, which is the normal pattern in mammals. Who is lactose-tolerant and who isn't? There is a clear pattern discernible to geneticists: lactose tolerance is concentrated in human populations that have descended from dairying cultures, whereas lactose intolerance is common in those whose ancestors were never herders of dairy animals, such as the Chinese and Japanese.[15] Lactose tolerance is genetically transmitted, but pastoralism, the disposition to tend herds of animals, on which the genetic trait depends, is culturally transmitted. Presumably it *could* have been genetically transmitted, but, so far as we know, it hasn't been. (Border collies, *unlike* the children of Basque shepherds, have had herding instincts bred into them, after all [Dennett, 2003c,d].)

Then there are *money* theories, according to which religions are cultural artifacts rather like monetary systems: communally developed systems that have evolved, culturally, several times. Their presence in every culture is readily explained and even justified: it's a Good Trick that one would expect to be rediscovered again and again, a case of convergent social evolution. *Cui bono?* Who benefits? Here we can consider several answers:

A. Everybody in the society benefits, because religion makes life in society more secure, harmonious, efficient. Some benefit more than others, but nobody would be wise to wish the whole away.
B. The elite who control the system benefit, at the expense of the others. Religion is more like a pyramid scheme than a monetary system; it thrives by preying on the ill-informed and powerless, while its beneficiaries pass it along gladly to their heirs, genetic or cultural.
C. Societies *as wholes* benefit. Whether or not the individuals benefit, the perpetuation of *their social or political groups* is enhanced, at the expense of rival groups.

This last hypothesis, *group selection,* is tricky, since the conditions under which genuine group selection can exist are hard to specify.[16] The schooling of fish and flocking of birds, for instance, are certainly phenomena involving grouping, but they are *not* explained as group-*selection* phenomena. In order to see how individuals (or their individual genes) are benefited by the dispositions to school or flock, you have to understand the ecology of groups, but the groups aren't the primary beneficiaries; the individuals that compose them are. Some biological phenomena masquerade as group selection but are better dealt with as instances of individual-level selection that depend on certain environmental phenomena (such as grouping) or even as instances of symbiont-selection phenomena. As we have already noted, a symbiont meme needs to be spread to new hosts, and if it can drive people into groups (the way *Toxoplasma gondii* drives rats into the jaws of cats) where it can readily find alternate hosts, the explanation is not group selection after all.

If the Martians can't make any of these theories fit the facts, they should consider a default theory of sorts that we may call the *pearl* theory: religion is simply a beautiful by-product. It is created by a genetically controlled mechanism or family of mechanisms that are meant (by Mother Nature, by evolution) to respond to irritations or intrusions of one sort or another. These mechanisms were designed by evolution for certain purposes, but then, one day, along comes something novel, or a novel convergence of different factors, something never before encountered and of course never foreseen by evolution, that happens to trigger the activities that generate this amazing artifact. According to pearl theories, religion isn't *for* anything, from the point of view of biology; it doesn't benefit any gene, or individual, or group, or cultural symbiont. But once it exists, it can be an *objet trouvé,* something that just happens to captivate us human agents, who have an indefinitely expandable capacity for delighting in novelties and curiosities. A pearl begins with a meaningless speck of foreign matter (or, more likely, a parasite), and once the oyster has added layer after beautiful layer, it can become something of coincidental value to members of a species who just happen to prize such things, whether or not this coveting is wise *from the point of view of biological fitness.* There are other standards of value that may emerge, for reasons good or bad, free-floating or highly articulated. In much the way the oyster responds to the initial irritant and then incessantly responds to the results of its first response and then to the results of that response and so on, human beings may be unable to leave off reacting to their own reactions, incorporating ever more elaborate layers into a production that then takes on shapes and features unimaginable from its modest beginnings.

What explains religion? Sweet tooth, symbiont, bower, money, pearl, or none of the above? Religion may include phenomena of human culture that have no remote analogue in genetic evolution, but if so, we will still have to answer the *cui bono?* question, because it is undeniable that the phenomena of religion are *designed* to a very significant degree. There are few signs of randomness or arbitrariness, so some differential replication has to pay for the R & D responsible for the design. These hypotheses do not all pull in the same direction, but the truth about religion might well be an amalgam of several of them (plus others). If this is so, we will not get a clear vision of why religion exists until we have clearly distinguished these possibilities and put each of them to the test.

If you think you already know which theory is right, you are either a major scientist who has been concealing a vast mountain of unpublished research from the rest of the world, or else you are confusing wishful thinking with knowledge. Perhaps it seems to you that I am somewhat willfully ignoring the obvious explanation of why *your* religion exists and has the features it does: it exists because it is the inevitable response of enlightened human beings to the obvious fact that God exists! Some would add: we engage in these religious practices *because God commands us to do so,* or *because it pleases us to please God.* End of story. But that could not be the end of the story. Whichever religion is yours, there are more people in the world who don't share it than who do, and it falls to you—to all of us, really—to explain why so many have gotten it wrong, and to explain how those who know (if there are any) have managed to get it right. Even if it is obvious to you, it isn't obvious to everyone, or even to most people.

If you have come this far in the book, you are willing to inquire into the sources and causes of *other* religions. Wouldn't it be hypocritical to claim that your own religion was somehow out of bounds? Just to satisfy your own intellectual curiosity, you might wish to see how your own religion measures up to the sort of scrutiny we will be directing at others. But, you may well wonder, can science be truly nonpartisan? Isn't science, in fact, "just another religion"? Or, to put it the other way around, aren't religious perspectives just as valid as the scientific perspective? How can we find any common, objective ground from which to conduct our inquiries? These questions concern many readers, especially academics who have invested heavily in the answers to them, but others, I find, are impatient with them, and not all that concerned. The questions *are* important—indeed, crucial to my whole project— since they put into doubt the very possibility of conducting the inquiry I am embarking on, but they *can* be postponed until after the theory sketch is completed. If you disagree, then before continuing with chapter 4 you should turn directly to appendix B, "Some More Questions About Science," which deals with these questions, spelling out in more detail, and defending, the path by which we can work together to find mutual agreement about how to proceed and what matters.

\\\

Chapter 3 Everything we value—from sugar and sex and money to music and love and religion—we value for reasons. Lying behind, and distinct from, *our* reasons are evolutionary reasons, free-floating rationales that have been endorsed by natural selection.

\\\

Chapter 4 Like all animal brains, human brains have evolved to deal with the specific problems of the environments in which they must operate. The social and linguistic environment that coevolved with human brains gives human beings powers that no other species enjoys, but also created problems that folk religions apparently evolved to handle. The apparent extravagance of religious practices can be accounted for in the austere terms of evolutionary biology.

PART II

||| \\\\|||

THE EVOLUTION
OF RELIGION

||| \\\ |||

CHAPTER FOUR

The Roots of Religion

1 The births of religions

Everything is what it is because it got that way. —D'Arcy Thompson

Among Hindus, there is disagreement over whether Shiva or Vishnu is the higher Lord, and many have been killed for their belief in this matter. "The *Liṅgapurāṇa* promises Śiva's heaven to one who kills or tears out the tongue of someone who reviles Śiva" (Klostermaier, 1994).

Among the Zulus, when a pregnant woman is about to give birth, sometimes the "spirit-snake of an old woman" makes an angry appearance (according to the shamans), indicating that a goat or some other animal should be sacrificed to the tribe's ancestors so that the child may be born healthy (Lawson and McCauley, 1990, p. 116).

The Jivaro of Ecuador believe that you have three souls, the true soul you have from birth (it returns to your birthplace when you die, then turns into a demon, which dies in turn, becoming a giant moth, which becomes mist when it dies); the *arutam,* a soul you obtain by fasting, bathing in a waterfall, and partaking of hallucinogenic juice (it makes you invincible but has the unfortunate habit of leaving you when you're in a jam); and the *musiak,* the avenging soul which tries to escape a victim's head and kill the victim's murderer. This is why you must shrink the head of your victim (Harris, 1993).

These curious beliefs and practices have not existed "forever"—no matter what their devotees may say. Marcel Gauchet begins his book on the political history of religion by noting, "As far as we know, religion has without exception existed at all times and in all places" (1997, p. 22), but this is a historian's pinched perspective, and simply isn't true. There was a time before religious beliefs and practices had occurred to anyone. There was a time, after all, before there were any *believers* on the planet, before there were any beliefs about anything. Some religious beliefs are truly ancient (by historical standards), and the advent of others can be read about in newspaper archives. How did they all arise?

Sometimes the answer seems obvious enough, especially when we have reliable historical records from the recent past. When Europeans in their magnificent sailing ships first visited the islands of the South Pacific in the eighteenth century, the Melanesians living on these islands were awestruck by these vessels, and by the remarkable gifts they were given by the white men who lived in them: steel tools and bolts of cloth and glass you could see through, and other cargo beyond their ken. They reacted much as we would probably react today if visitors from outer space showed up capable of overwhelming us at will, and bearing technologies we hadn't even dreamt of: "We must get ourselves some of this cargo, and learn how to harness the magical powers of these visitors." And our puny efforts to use what we did know to take control of the situation and restore our security and sense of power would probably amuse these technologically superior aliens as much as we are amused by the Melanesians' conclusion that the Europeans must be their ancestors in disguise, coming back from the realm of the dead with untold wealth, demigods to be worshiped. When Lutheran missionaries arrived in Papua New Guinea in the late nineteenth century to try to convert the Melanesians to Christianity, they met stubborn suspicion: why were these stingy ancestors in disguise withholding the cargo and trying to make them sing hymns?

Cargo cults have sprung up again and again in the Pacific. During World War II, American forces arrived at the island of Tana to recruit a thousand men to help build an airfield and army base on neighboring Efate Island. When the workers returned with tales of white and black men who had possessions beyond the dreams of the people of Tana, the whole society was thrown into turmoil. The islanders, many of whom had earlier been converted to Christianity by British missionaries,

stopped going to church and began to build landing strips, warehouses and radio masts out of bamboo, in the belief that if it worked on Efate for the Americans, it would work for them on Tana. Carved figurines of American warplanes, helmets and rifles were made from bamboo and used as religious icons. Islanders began to march in parades with USA painted, carved or tattooed on their chests and backs. John Frum emerged as the name of their Messiah, although there are no records of an American soldier with that name.

When the last American GI left at the end of the war, the islanders predicted John Frum's return. The movement continued to flourish and on 15 February, 1957, an American flag was raised in Sulphur Bay to declare the religion of John Frum. It is on this date every year that John Frum Day is celebrated. They believe that John Frum is waiting in the volcano Yasur with his warriors to deliver his cargo to the people of Tana. During the festivities the elders march in an imitation army, a kind of military drill mixed with traditional dancing. Some carry imitation rifles made of bamboo and wear American army memorabilia such as caps,

T-shirts and coats. They believe that their annual rituals will draw the god John Frum down from the volcano and deliver the cargo of prosperity to all of the islanders. [MotDoc, 2004]

Still more recently, around 1960, on New Britain Island in Papua New Guinea, the Pomio Kivung cult was founded. It still flourishes.

Pomio Kivung doctrine holds that adherence to the Ten Laws (a modified version of the Decalogue [Ten Commandments]) and the faithful performance of an extensive set of rituals, including the payment of fines for the purpose of gaining absolution, are essential to the moral and spiritual improvement that is necessary to hasten the return of the ancestors. The most important of these rituals aims at placating the ancestors, who make up the so-called "Village Government." Headed by God, the Village Government includes those ancestors whom God has forgiven and perfected.

The spiritual leaders of the Pomio Kivung have been its founder, Koriam, his principal assistant, Bernard, and Koriam's successor, Kolman. Followers have regarded all three as already members of the Village Government and, hence, as divinities. All three have resided on earth physically (specifically in the Pomio region of the province), but their souls have dwelt with the ancestors all along.

Achieving sufficient collective purification is the decisive condition for inducing the return of the ancestors and inaugurating the "Period of the Companies." The Period of the Companies will be an era of unprecedented prosperity, which will result from the transfer of knowledge and an industrial infrastructure for the production of technological wonders and material wealth like that of the Western world. [Lawson and McCauley, 2002, p. 90]

These cases may be exceptional. Your religion, you may believe, came into existence when its fundamental truth was revealed by

God to somebody, who then passed it along to others. It flourishes today because you and the others of your faith know that it is the truth, and God has blessed you and encouraged you to keep the faith. It is as simple as that, for you. And why do all the other religions exist? If those people are just wrong, why don't their creeds crumble as readily as false ideas about farming or obsolete building practices? They will crumble in due course, you may think, leaving only true religion, your religion, standing. Certainly there is some reason to believe this. In addition to the few dozen major religions in the world today—those whose adherents number in the hundreds of thousands or millions—there are thousands of less populous religions recognized. Two or three religions come into existence every day, and their typical lifespan is less than a decade.[1] There is no way of knowing how many distinct religions have flourished for a while during the last ten or fifty or a hundred thousand years, but it might even be millions, of which all traces are now lost forever.

Some religions have confirmed histories dating back for several millennia—but only if we are generous with our boundaries. The Mormon Church is less than two hundred years old, as its official name reminds us: the Church of Jesus Christ of Latter-day Saints. Protestantism is less than five hundred years old, Islam is less than fifteen hundred years old, Christianity is less than two thousand years old. Judaism is not even twice as old as that, and the Judaisms of today have evolved significantly from the earliest identifiable Judaism, though the varieties of Judaism are as nothing compared with the riotous blossoming of variations that Christianity has spawned in the last two millennia.

These are short periods of time, biologically speaking. They are not even long compared with the ages of other features of human culture. Writing is more than five thousand years old, agriculture is more than ten thousand years old, and language is—who knows?—maybe "only" forty thousand years old and maybe ten or twenty times older than that. It's a contentious research topic, and since it's widely agreed that fully articulated natural languages must have developed out of some kind of proto-languages (which may have evolved over hundreds of thousands of years), there is no consensus about what would even count as the birthdate of language. Is language older than religion? However we date its beginnings, language is much, much older than any existing religion, or even any religion of which we have any historical or archeological knowledge. The earliest impressive archeological evidence of religion is the elaborate Cro-Magnon burial sites in the Czech Republic, and they are about twenty-five thousand years old.[2] It is hard to tell, but *something like* religion may well have existed from the early days of language, however, or even before that. What were our ancestors like before there was anything like religion? Were they like bands of chimpanzees? What, if anything, did they talk about, aside from food and predators and the mating game? The weather? Gossip? What was the psychological and cultural soil in which religion first took root?

We can tentatively work backward, extrapolating under the guidance of our fundamental biological constraint: each innovative step had to "pay for itself" somehow, in the *existing* environment in which it first occurred, independently of whatever its role might become in later environments. What, then, could explain both the diversity and the similarities in the religious ideas we observe around the world? Are the similarities due to the fact that all religious ideas spring from a common ancestor idea, passed on over the generations as people spread around the globe, or are such ideas independently rediscovered by just about every culture because they are simply the truth, and obvious enough to occur to people in due course? These are obviously naïve oversimplifications, but at least they are attempts to ask and answer explicit questions often left unexamined by people who lose interest once they have found a *purpose* or *function* for religion that strikes them as plausible: responding to a suitably grand "human need" to account for the manifest outlay of time and energy that religion requires. The three favorite purposes or *raisons d'être* for religion are to *comfort* us in our suffering and allay our fear of death

to *explain* things we can't otherwise explain

to encourage group *cooperation* in the face of trials and enemies

Thousands of books and articles have been written defending these claims, and such compelling and familiar ideas are probably at least partly right, but if you settle for one of them, or even all three taken together, you succumb to a disorder often encountered in the humanities and social sciences: premature curiosity satisfaction. There is so much more to ask about, and so much more to understand. Why would *these* ideas comfort people? (And why are they comforting, exactly? Might there be better, more comforting ideas to be found?) Why would *these* ideas appeal to people as explanations of baffling events? (And how could they have arisen? Did some would-be proto-scientist hit upon a supernatural theory and enthusiastically proselytize her neighbors?) How do *these* ideas actually manage to enhance cooperation in the face of suspicion and defection? (And once more, how could they have arisen? Did some wise tribal leader invent religion to give her tribe a teamwork edge over the rival tribes?)

Some people suppose that we can never do better than such simple speculations about these processes and outcomes from the remote past. Some insist on it, in fact, and their vehemence betrays the fact that they are afraid they are wrong. They are. Today, thanks to advances in a variety of sciences, we can sharpen the questions and *begin* to answer them. In this and the next four chapters, I will try to tell *the best current version* of the story science can tell about how religions have become what they are. I am not at all claiming that this is what science has already established about religion. The main point of this book is to insist that we *don't* yet know—but we can discover—the answers to these important questions if we make a concerted effort. Probably some of the features of the story I tell will prove in due course to be mistaken. Maybe many of them are wrong. The purpose of trying to sketch a *whole* story now is to get something on the table that is both testable and worth testing. It is usually easier to fix something that has flaws than to build something from scratch. Trying to bridge the gaps in our knowledge forces us to frame questions we haven't framed before, and puts the issues in a perspective that enables further questions to be posed and answered. And that in itself can undercut the defeatist proclamation that these are mysteries beyond all human comprehension. Many people may wish that these were unanswerable questions. Let's see what happens when we defy their defensive pessimism and give it a try.

2 The raw materials of religion

We may conclude, therefore, that, in all nations, which have embraced polytheism, the first ideas of religion arose not from a contemplation of the works of nature, but from a concern with regard to the events of life, and from the incessant hopes and fears, which actuate the human mind.
—David Hume, *The Natural History of Religion*

My guides are the pioneering scientists who have begun to tackle these questions with both imagination and discipline. An evolutionary biologist or a psychologist who knows only one religion at all well and has a smattering of (mis)information about the others (like most of us) is almost certain to overgeneralize from idiosyncratic familiarity when it comes to framing questions. A social historian or an anthropologist who knows a great deal about the beliefs and practices of people all around the world but is naïve about evolution is equally unlikely to frame the issues well. Fortunately, a few well-informed researchers have recently begun to pull these distant perspectives together, with tantalizing results. Their books and articles are well worth reading in their entirety, as I hope I will convince you by introducing a few highlights.

Jared Diamond's *Guns, Germs, and Steel* (1997) is an eye-opening exploration of very specific effects of geography and biology on the early development of agriculture in different parts of the world at different times. When the first agriculturalists domesticated animals, they naturally began living in close proximity to them, and this enhanced the likelihood of species-jumping by the animals' parasites. The most serious infectious diseases known to humanity, such as smallpox and influenza, all derive from domesticated animals, and our farming ancestors lived through a horrific pruning in which untold millions succumbed to early versions of these diseases, leaving only those fortunate enough to have some natural immunity to reproduce. Many generations of this evolutionary bottleneck guaranteed that their later descendants would be relatively immune to, or have a high tolerance for, the descendants of those virulent strains of parasite. When these grand-offspring, living mainly in Europe, developed the technology to cross the oceans, they brought their germs with them, and it was the germs, more than the guns and steel, that wiped out large fractions of the indigenous populations they encountered. The role of agriculture in spawning infectious diseases, and the relative immunity to them that had evolved among the peoples who had lived through the ravages of the early days of agriculture, can be studied with some precision now that we can extrapolate backward from the genomes of existing species of plants, animals, and germs. Accidents of geography gave European nations a head start that goes a long way to explain why they were the colonizers rather than the colonized in later centuries.

Diamond's Pulitzer Prize–winning book is deservedly well known, but not alone. There is a new generation of interdisciplinary researchers working to put together the biology with the evidence gleaned by centuries of work by historians, anthropologists, and archeologists. Pascal Boyer and Scott Atran are anthropologists who have done extensive fieldwork in Africa and Asia but who are also trained in evolutionary theory and cognitive psychology. Their recent books, *Religion Explained: The Evolutionary Origins of Religious Thought* (Boyer, 2001) and *In Gods We Trust* (Atran, 2002), develop largely harmonious accounts of the major steps into the swamp that they and others have been taking. Then there is David Sloan Wilson, an evolutionary biologist who has been devoting himself in recent years to analyses that systematically exploit the Human Relations Area File, a database of all the world's cultures compiled by anthropologists. His recent book *Darwin's Cathedral: Evolution, Religion, and the Nature of Society* (2002) makes the best case to date for the hypothesis that religion is a social phenomenon designed (by evolution) to improve cooperation within (not *among*!) human groups. According to Wilson, religion emerged by a process of *group selection*, a controversial wrinkle in evolutionary theory that is dismissed by many evolutionary theorists as at best a marginal process whose conditions for success are unlikely to arise and persist for long. There are deep reasons to be skeptical about group selection, especially in our species, and precisely because Wilson's thesis—religion as a cooperation-enhancer—is deeply attractive to many people, we need to brace ourselves to avoid wishful thinking. It is quite generally agreed among his critics that he has not (yet) succeeded in making the case for his radical thesis of group selection, but even a roundly refuted scientific theory can make a major contribution to the steady accumulation of scientific understanding if the evidence marshaled for and against it has been scrupulously gathered. (For more on this point, see appendix B.) Here I will introduce the main points of agreement, as well as acknowledging the continuing points of contention, packing off most of the controversial details into the endnotes and appendixes, where those with a taste for them can (begin to) pursue their own deeper consideration of them.

Both Boyer and Atran present the work of a small but growing community of researchers in relatively accessible terms.[3] Their central thesis is that in order to explain the hold that various religious ideas and practices have on people, we need to understand the evolution of the human mind. For many centuries, most philosophers and theologians contended that the human mind (or soul) was an immaterial, incorporeal thing, what René Descartes called a *res cogitans* (thinking thing). It was in some sense infinite, immortal, and utterly inexplicable by material means. We now understand that the mind is not, as Descartes confusedly supposed, *in communication with* the brain in some miraculous way; it *is* the brain, or, more specifically, a system or organization within the brain that has evolved in much the way our immune system or respiratory system or digestive system has evolved. Like many other natural wonders, the human mind is something of a bag of tricks, cobbled together over the eons by the foresightless process of evolution by natural selection. Driven by the demands of a dangerous world, it is deeply biased in favor of noticing the things that mattered most to the reproductive success of our ancestors.4

Some of the features of our minds are endowments we share with much simpler creatures, and others are specific to our lineage, and hence much more recently evolved. These features sometimes overshoot, sometimes have curious by-products, sometimes are ripe for exploitation by other replicators. Of all the quirky effects generated by the whole bag of tricks—our set of "gadgets," as Boyer calls them—a few happen to interact with one another in mutually reinforcing ways, creating patterns observable in all cultures, with interesting variations. Some of these patterns look rather like religions, or pseudo-religions, or proto-religions. The by-products of the various gadgets are what Boyer calls *concepts:*

> Some concepts happen to connect with inference systems in the brain in a way that makes recall and communication easy. Some concepts happen to trigger our emotional programs in particular ways. Some concepts happen to connect to our social mind. Some of them are represented in such a way that they soon become plausible and direct behavior. The ones that do *all* this are the religious ones we actually observe in human societies. [p. 50]

Boyer lists more than half a dozen distinct cognitive systems that feed effects into this recipe for religion—an agent-detector, a memory-manager, a cheater-detector, a moral-intuition-generator, a sweet tooth for stories and storytelling, various alarm systems, and what I call the intentional stance. Any mind with this particular set of thinking tools and biases is bound to harbor something like a religion sooner or later, he claims. Atran and others offer largely concurring accounts, and the details are well worth exploring, but I am just going to sketch some of the big picture so that we can see the overall *shape* of the theory, not (yet) assess it for truth. It will take decades of research to secure any of this theory, but right now we can get a sense of what the possibilities are, and hence what questions we ought to be trying to answer.

3   How Nature deals with the problem of other minds

We find human faces in the moon, armies in the clouds; and by a natural propensity, if not corrected by experience and reflection, ascribe malice and good-will to every thing, that hurts or pleases us.

—David Hume, *The Natural History of Religion*

"I saw you take his kiss!" " 'Tis true"
"Oh Modesty!" " 'Twas strictly kept:
He thought me asleep; at least I knew
He thought I thought he thought I slept"

—Coventry Patmore, "The Kiss"

The first thing we have to understand about human minds as suitable homes for religion is how our minds understand *other* minds! Everything that moves needs something like a mind, to keep it out of harm's way and help it find the good things; even a lowly clam, which tends to stay in one place, has one of the key features of a mind—a harm-avoiding retreat of its feeding "foot" into its shell when something alarming is detected. Any vibration or bump is apt to set it off, and probably most of these are harmless, but *better safe than sorry* is the clam's motto (the free-floating rationale of the clam's alarm system). More mobile animals have evolved more discriminating methods; in particular, they tend to have the ability to divide detected motion into the banal (the rustling of the leaves, the swaying of the seaweed) and the potentially vital: the "*animate* motion" (or "biological motion") of another *agent,* another animal with a mind, who might be a predator, or a prey, or a mate, or a rival conspecific. This makes economic sense, of course. If you startle at every motion you detect, you'll never find supper, and if you don't startle at the dangerous motions, you'll soon be somebody else's supper. This is another Good Trick, an evolutionary innovation— like eyesight itself, or flight—that is so useful to so many different ways of life that it evolves over and over again in many different species. Sometimes this Good Trick can be too much of a good thing; then we have what Justin Barrett (2000) calls a *hyperactive agent detection device,* or HADD. This overshooting is not restricted to human beings. When your dog leaps up and growls when some snow falls off the eaves with a thud that rouses him from his nap, he is manifesting a "false positive" orienting response triggered by his HADD.

Recent research on animal intelligence (Whiten and Byrne, 1988, 1997; Hauser, 2000; Sterelny, 2003; see also Dennett, 1996) has shown that some mammals and birds, and perhaps some other creatures as well, carry these agent-discriminations into more sophisticated territory. Evidence shows that they not only distinguish the animate movers from the rest, but draw distinctions between the likely *sorts* of motions to anticipate from the animate ones: will it attack me or flee, will it move left or right, will it back down if I threaten, does it see me yet, does it want to eat me or would it prefer to go after my neighbor? These cleverer animal minds have discovered the further Good Trick of *adopting the intentional stance* (Dennett, 1971, 1983, 1987): they treat some other things in the world as

- *agents* with
- limited *beliefs* about the world,
- specific *desires,* and
- enough common sense to do the *rational* thing given those beliefs and desires.

Once animals began adopting the intentional stance, something of an arms race ensued, with ploy and counterploy, deceptive move and intelligent detection of deceptive move, carrying animal minds to greater subtlety and power. If you have ever tried to catch or trap a wild animal, you have some appreciation of the wiliness that has evolved. (Clam-digging, in contrast, is child's play. Clams have not evolved the intentional stance, though they do have simple hair-trigger HADDs.)

The utility of the intentional stance in describing and predicting animal behavior is undeniable, but that doesn't mean that the animals themselves are clued in about what they are doing. When a low-nesting bird leads the predator away from her nestlings by doing a *distraction display,* she is making a convincing sham of a broken wing, creating the tempting illusion of an easy supper for the observing predator, but she need not understand this clever ruse. She *does* need to understand the conditions of likely success, so that she can adjust her behavior the better to fit the variations encountered, but she no more needs to be aware of the deeper rationale for her actions than does the fledgling cuckoo when it pushes the rival eggs out of the nest in order to maximize the food it will get from its foster parents.

Researchers have several other terms for the intentional stance. Some call it "theory of mind" (Premack and Woodruff, 1978; Leslie, 1987; Gopnik and Meltzoff, 1997), but there are problems with that formulation, so I am going to stick with my more neutral terminology.[5] Whenever an animal treats something as an *agent,* with beliefs and desires (with knowledge and goals), I say that it is *adopting the intentional stance* or treating that thing as an *intentional system.*

The intentional stance is a useful perspective for an animal to take in a hostile world (Sterelny, 2003), since there are things out there that may *want* it and may have *beliefs* about where it is and where it is heading. Among the species that have evolved the intentional stance, there is considerable variation in sophistication. Faced with a threatening rival, many animals can make an informationally sensitive decision either to retreat or to call the other's bluff, but there is scant evidence that they have any sense of what they are doing or why. There is some (controversial) evidence that a chimpanzee can *believe* that another agent—a chimpanzee or a human being, say—*knows* that the food is in the box rather than in the basket. This is *second-order* intentionality (Dennett, 1983), involving *beliefs about beliefs* (or *beliefs about desires,* or *desires about beliefs,* etc.), but there is no evidence (yet) that any nonhuman animal can *want* you to *believe* that it *thinks* you are hiding behind the tree on the left, not the right (*third-order* intentionality). But even preschool children delight in playing games in which one child *wants* another to *pretend* not to *know* what the first child *wants* the other to *believe* (*fifth-order* intentionality): "You be the sheriff, and ask me which way the robbers went!"

Whatever the situation is with nonhuman animals—and this is a topic of vigorous and hotly debated research[6]—there is no doubt at all that normal human beings do not have to be *taught* how to conceive of the world as containing lots of agents who, like themselves, have beliefs and desires, as well as beliefs and desires about the beliefs and desires of others, and beliefs and desires about the beliefs and desires that others have about them, and so forth. This virtuoso use of the intentional stance comes naturally, and it has the effect of saturating the human environment with *folk psychology* (Dennett, 1981). We experience the world as not just full of moving human bodies but of *rememberers* and *forgetters, thinkers* and *hopers* and *villains* and *dupes* and *promise-breakers* and *threateners* and *allies* and *enemies.* Indeed, those human beings who find perceiving the world from this perspective difficult—those suffering from autism are the best-studied category—have a more significant disability than those who are born blind or deaf (Baron-Cohen, 1995; Dunbar, 2004).

So powerful is our innate urge to adopt the intentional stance that we have real difficulty turning it off when it is no longer appropriate. When somebody we love or even just know well dies, we suddenly are confronted with a major task of cognitive updating: revising all our habits of thought to fit a world with one less familiar intentional system in it. "I wonder if she'd like . . . ," "Does she know I'm . . . ," "Oh, look, this is something she always wanted. . . ." A considerable portion of the pain and confusion we suffer when confronting a death is caused by the frequent, even obsessive, reminders that our intentional-stance habits throw up at us like annoying pop-up ads but much, much worse. We can't just *delete the file* in our memory banks, and, besides, we wouldn't want to be able to do so. What keeps many habits in place is the pleasure we take from indulging in them.7 And so we dwell on them, drawn to them like a moth to a candle. We preserve relics and other reminders of the deceased persons, and make images of them, and tell stories about them, to prolong these habits of mind even as they start to fade.

But there is a problem: a corpse is a potent source of disease, and we have evolved a strong compensatory innate disgust mechanism to make us keep our distance. Pulled by longing and pushed back by disgust, we are in turmoil when we confront the corpse of a loved one. Small wonder that this crisis should play so central a role in the birth of religions everywhere. As Boyer (2001, p. 203) stresses, *something must be done* with a corpse, and it has to be something that satisfies or allays competing innate urges of dictatorial power. What seems to have evolved everywhere, a Good Trick for dealing with a desperate situation, is an elaborate ceremony that removes the dangerous body from the daily environment either by burial or burning, combined with the interpretation of the persistent firing of the intentional-stance habits shared by all who knew the deceased as the unseen presence of the agent as a *spirit,* a sort of *virtual person* created by the survivors' troubled mind-sets, and almost as vivid and robust as a live person.

What role, if any, does language play in this? Are we the only species of mammal that buries its dead because we're the only species that can talk about what we share when we confront a fresh corpse? Do the burial practices of Neanderthals show that they must have had fully articulate language? These are among the questions we should try to answer. The world's languages are well stocked with verbs for the basic varieties of belief-desire manipulation: we *pretend* and *lie,* but also *bluff* and *suspect* and *flatter* and *brag* and *tempt* and *dissuade* and *command* and *prohibit* and *disobey,* for instance. Was our virtuosity as natural psychologists a prerequisite for our linguistic ability, or is it the other way around: did our use of language make our psychological talents possible? This is another controversial area of current research, and probably the truth is, as it so often is, that there was a coevolutionary process, with each talent feeding off the other. Plausibly, the very act of verbal communication requires *some* appreciation of third-order intentionality: I have to *want* you to *recognize* that I am trying to *inform* you, to get you to *believe* what I'm *saying* (Grice, 1957, 1969; Dennett, 1978; but see also Sperber and Wilson, 1986). But, like the fledgling cuckoo, a child can get under way quite cluelessly, achieving successful communication without having any reflective appreciation of the structure that underlies all intentional communication, without even recognizing, really, that she is communicating at all.

Once you've started talking (with other people), you will be bathed in new words, some of which you more or less understand; some of these objects of perception, such as the words "pretend" and "brag" and "tempt," will help draw and focus your attention on cases of *pretending* and *bragging* and *tempting,* giving you plenty of inexpensive practice in folk psychology. Whereas chimpanzees and some other mammals may also be "natural psychologists," as

Nicholas Humphrey (1978) has called them, since they lack language they never get to compare notes or discuss cases with other natural psychologists. The articulation of the intentional stance in verbal communication not only heightens the sensitivity, discrimination, and versatility of individual folk psychologists, but also magnifies and complicates the folk-psychological phenomena they are attending to. A fox may be cunning, but a person who can flatter you by declaring that you are cunning as a fox has more tricks up his sleeve than the fox does, by a wide margin.

Language gave us the power to remind ourselves of things not currently present to our senses, to *dwell on* topics that would otherwise be elusive, and this brought into focus a virtual world of imagination, populated by the agents that mattered the most to us, both the living but absent and the dead who were gone but not forgotten. Released from the corrective pressure of further actual encounters in the real world, these virtual agents were free to evolve in our minds to amplify our yearnings or our dreads. Absence makes the heart grow fonder, or—if the absent one was somewhat frightening in reality—more terrified. This still doesn't get our ancestors to religion, but it gets them to persistent—even obsessive—rehearsal and elaboration of some of their habits of thought.

\\\

Chapter 4 Extrapolating back to human prehistory with the aid of biological thinking, we can surmise how folk religions emerged without conscious and deliberate design, just as languages emerged, by interdependent processes of biological and cultural evolution. At the root of human belief in gods lies an instinct on a hair trigger: the disposition to attribute *agency*—beliefs and desires and other mental states—to anything complicated that moves.

\\\

Chapter 5 The false alarms generated by our overactive disposition to look for agents wherever the action is are the irritants around which the pearls of religion grow. Only the best, most mind-friendly variants propagate, by meeting—or seeming to meet—deep psychological and physical needs, and then these are further refined by the incessant pruning of selection processes.

CHAPTER FIVE

||||\\\||||

Religion, the Early Days

1 Too many agents: competition for rehearsal space

I might repeat to myself slowly and soothingly, a list of quotations beautiful from minds profound—if I can remember any of the damn things.
   —Dorothy Parker

What start as useful luxuries that give you an edge in a fast-moving world have a way of evolving into necessities. Today, we all wonder how we could live without our telephones, our driver's licenses, our credit cards, our computers. So it once was with language, and the intentional stance. What started as a Good Trick rapidly became a practical necessity of human life, as our ancestors became more and more social, more and more linguistic. And, as already noted for the simpler case of the HADD, there is the possibility of too much of a good thing. The continued experience of the presence of departed acquaintances as ghosts is not the only overshooting of the intentional stance in the lives of our ancestors. The practice of overattributing intentions to moving things in the environment is called *animism,* literally giving a soul (Latin, *anima*) to the mover. People who lovingly cajole their cranky automobiles or curse at their computers are exhibiting fossil traces of animism. They probably don't take their own speech acts entirely seriously, but are just indulging in something that makes them feel better. The fact that it *does* tend to make them feel better, and is apparently indulged in by people of every culture, suggests how deeply rooted in human biology is the urge to treat things—especially frustrating things—as agents with beliefs and desires. But if our bouts of animism today tend to be ironic and attenuated, there was a time when the desire of the river to flow to the sea, and the benign or evil intent of the rain clouds, were taken so literally and seriously that they could become a matter of life or death—for instance, to those poor souls who were sacrificed to appease the insatiable desires of the rain god.

Simple forms of what we might call *practical animism* are arguably not mistakes at all, but extremely useful ways of keeping track of the tendencies of designed things, living or artifactual. The gardener who tries to discover what her different flowers and vegetables *prefer*, or *tricks* a dogwood branch into *thinking it's spring* and opening its buds by bringing it indoors, where it is warm, doesn't have to go overboard and wonder what her petunias are daydreaming about. Even undesigned physical systems can sometimes be usefully described in intentional or animistic terms: the river doesn't *literally* want to return to the ocean, but water *seeks its own level*, as they say, and lightning *searches* for the best path to ground. It is not surprising that the attempt to explain patterns discerned in the world has often hit upon animism as a good—actually predictive—approximation of some unimaginably complex underlying phenomenon.

But sometimes the tactic of seeking an intentional-stance perspective comes up dry. Much as our ancestors would have loved to predict the weather by figuring out what it *wanted* and what *beliefs* it harbored about them, it simply didn't work. It no doubt often *seemed* to work, however. Every now and then the rain dances were rewarded by rain. What would the effect be? Many years ago, the behaviorist psychologist B. F. Skinner (1948) showed a striking "superstition" effect in pigeons that were put on a *random* schedule of reinforcement. Every so often, no matter what the pigeon was doing at the moment, a click and a food-pellet reward were delivered. Soon the pigeons put on this random schedule were doing elaborate "dances," bobbing and whirling and craning their necks. It's hard to resist putting a soliloquy into these birds' brains: "Now, let's see: the last time I got the reward, I'd just spun around once and craned my neck. Let's try it again. . . . Nope, no reward. Perhaps I didn't spin enough. . . . Nope. Perhaps I should bob once before spinning and craning. . . . YESSS! OK, now, what did I just do? . . ." You don't have to have language, of course, to be vulnerable to such enticing illusions. The soliloquy dramatizes the dynamics that produce the effect, which doesn't require conscious reflection, just reinforcement. But in a species that *does* represent both itself and other agents to itself, the effect can be multiplied. If such a strikingly extravagant behavioral effect can be produced in pigeons by making them wander into a random-reinforcement trap, it is not hard to believe that similar effects could be inculcated by happy accident in our ancestors, whose built-in love for the intentional stance would tend to encourage them to add invisible agents or other homunculi to be the secret puppeteers behind the perplexing phenomena. Clouds certainly don't *look* like agents with beliefs and desires, so it is no doubt natural to suppose that they are indeed inert and passive things being manipulated by hidden agents that *do* look like agents: rain gods and cloud gods and the like—if only we could see them.

This curiously paradoxical idea—something *invisible* that *looks like* a person (has a head, eyes, arms and legs, perhaps wears a special helmet)—is different from other self-contradictory combinations. Consider the idea of a box that has no interior space to put things in, or a liquid that isn't wet. To put it crudely, these ideas are not interesting enough to be puzzling for very long. Some nonsense is more attention-grabbing than other nonsense. Why? Just because our memories are not indifferent to the *content* of what they store. Some things we find more memorable than others, and some things are so interesting that they are well nigh unforgettable, and still others, such as the random string of words "volunteers trainer regardless court exercise" (pulled by me "at random" from the first newspaper story I could lay my hands on just now), could be remembered for more than a few seconds only if you either deliberately repeated it to yourself dozens of times or made up some interesting story that somehow made sense of these words in just this sequence.

We are all painfully aware today that our attention is a limited commodity with many competitors vying for more than their share. This information overload, with advertisements bombarding us on all sides, plus a host of other distractions, is nothing new; we've just become self-conscious about it, now that there are thousands of people who specialize in designing novel attention-grabbers and attention-holders. We—and, indeed, all animate species—have always had to have filters and biases built into our nervous systems to screen the passing show for things worth hanging on to, and these filters favor certain *sorts* of exceptions or anomalies. Pascal Boyer (2001) calls these exceptions *counterintuitive,* but he means this in a rather circumscribed technical sense: counterintuitive anomalies are especially attention-worthy and memorable if they violate *just one or two* of the basic default assumptions about a fundamental category like *person* or *plant* or *tool.* Concoctions that aren't readily classifiable at all because they are *too* nonsensical can't hold their own in the competition for attention, and concoctions that are too bland are just not interesting enough. An invisible ax with no handle and a spherical head is just irritating nonsense, an ax made of cheese is a bit titillating (there are conceptual artists who make a good living coming up with such japes), but a *talking* ax—ah, now we've got something to hold the attention!

Put these two ideas together—a hyperactive agent-seeking bias and a weakness for certain sorts of memorable combos—and you get a kind of fiction-generating contraption. Every time something puzzling happens, it triggers a sort of curiosity startle, a *"Who's there?"* response that starts churning out "hypotheses" of sorts: "Maybe it's Sam, maybe it's a wolf, maybe it's a falling branch, maybe it's . . . a tree that can walk—*hey, maybe it's a tree that can walk!"* We can suppose that this process *almost never* generates anything with any staying power—millions or billions of little stretches of fantasy that almost instantly evaporate beyond recall until, one day, one happens to occur at just the right moment, with just the right sort of zing, to get rehearsed not just once and not just twice, but many times. A lineage of ideas—the walking-tree lineage—is born. Every time the initiator's mind is led to review the curious idea, not deliberately but just idly, the idea gets a little stronger—in the sense of a little more likely to occur in the initiator's mind again. And again. It has a little self-replicative power, a little *more* self-replicative power than the other fantasies it competes with for time in the brain. It is not yet a meme, an item that escapes an individual mind and spreads through human culture, but it is a good proto-meme: a slightly obsessional—that is, oft-recurring, oft-rehearsed—little hobbyhorse of an idea.

(Evolution is all about processes that *almost never* happen. Every birth in every lineage is a potential speciation event, but speciation almost never happens, not once in a million births. Mutation in DNA almost never happens—not once in a trillion copyings—but evolution depends on it. Take the set of infrequent accidents—things that almost never happen—and sort them into the happy accidents, the neutral accidents, and the fatal accidents; amplify the effects of the happy accidents—which happens automatically when you have replication and competition—and you get evolution.)

Would-be memeticists often ignore the fact that part of the "life" cycle of a meme is its moment-to-moment competition with other ideas—not just other memes, but every other idea anybody can think about—within a host brain. Rehearsal, deliberate or involuntary, is replication. We can try to *make* something a meme—or just a memory—by rehearsing it deliberately (a phone number, a rule to follow), or, if we just let "nature take its course," our innate brain biases will automatically churn out rehearsals of the things that tickle them. This is probably the source, in fact, of *episodic memory*, our ability to recollect events in our lives. What did you have for breakfast on your last birthday? You probably can't remember. What did you wear at your wedding? You probably can remember, because you've gone over it many times, before, during, and since the wedding. Unlike computer memory, which is an equal-opportunity storehouse that can readily record whatever is thrown into it, human brain memory is both competitive and biased. It has been designed by eons of evolution to remember some sorts of things more readily than others. It does this in part by differential rehearsal, dwelling on what is vital and tending to discard the trivia after a single pass. It does a pretty good job, keying on features that happen to have lined up with what tended to be vital in the past. Good advice to a potential meme is: if you want lots of rehearsals (replications), try to *look important*!

Human memory is biased in favor of vital combinations, but so, presumably, is the memory found in the brains of all other animals. Animal memory has probably been relatively impervious to fantasy, however, for a simple reason: lacking language, animal brains have not had a way of inundating themselves with an explosion of combinations not found in the natural environment. How is an anxious ape going to *concoct* the counterintuitive combination of a walking tree or an invisible banana—ideas that might indeed captivate an ape mind if only they could be presented to it?

Do we know that something like this fantasy-generation process has been taking place in our species (and our species alone) for thousands of years? No, but it is a serious possibility to investigate further. Using only materials that would have been put in place by evolution for other purposes, this hypothesis could explain the re-markably fertile imagination that must somehow be responsible for the world's menagerie of mythical creatures and demons. Since the monsters themselves have never existed, they had to be "in-vented," either deliberately or inadvertently (the way languages were invented). They are expensive creations, and the R & D re-quired for the task had to be generated by something that could pay for itself. I've left the hypothesis quite unspecific for the moment, but more constrained forms of it are available, and they have the great advantage of having testable consequences. We can start scouring the world's mythology for patterns that would be pre-dicted by some versions of the hypothesis but not others.

And we don't have to restrict ourselves to the human species. Ex-periments along the lines of Skinner's provocation of superstition in the pigeon might begin to uncover the biases and fault lines in ape memory mechanisms, in much the way Niko Tinbergen's ex-periments with gulls (1948, 1959) famously showed their percep-tual biases. The adult female gull has an orange spot on her beak, at which her chicks instinctually peck, to stimulate the female to regurgitate and feed them. Tinbergen showed that chicks would peck even more readily at exaggerated cardboard models of the orange spot, so-called *supernormal stimuli*. Pascal Boyer (2001) notes that, over the eons, human beings have discovered and exploited their own supernormal stimuli:

> There is no human society without some musical tradition. Al-though the traditions are very different, some principles can be found everywhere. For instance, musical sounds are always closer to pure sound than to noise. . . . To exaggerate a little, what you get from musical sounds are super-vowels (the pure frequen-cies as opposed to the mixed ones that define ordinary vowels) and pure consonants (produced by rhythmic instruments and the attack of most instruments). These properties make music an intensified form of sound-experience from which the cortex

receives purified and therefore intense doses of what usually acti-
vates it. . . . This phenomenon is not unique to music. Humans
everywhere also fill their environments with artifacts that over-
stimulate their visual cortex, for instance by providing pure satu-
rated color instead of the dull browns and greens of their familiar
environments. . . . In the same way, our visual system is sensitive
to symmetries in objects. Bilateral symmetry in particular is
quite important; when two sides of an animal or person look the
same it means that they are facing you, a relevant feature of interaction with people but also with prey and predators. Again,
you cannot find a human group where people do not produce visual gadgets with such symmetrical arrangements, from the simplest makeup or hairdressing techniques to textile patterns and
interior decoration. [pp. 132–33]

Why don't other species have art? Once again, the answer that
suggests itself—which does not mean that it is proven but only that
it may well be provable—is that, lacking language, they lack the
tools for creating surrogate stimulus combinations and hence they
lack the perspective that permits exploration of the combinatorics
of their own senses.[1] Using acute observation and trial and error,
Tinbergen cleverly devised the supernormal stimuli that enticed his
birds (and other animals) into a host of bizarre behaviors. No doubt
animals do on occasion trap themselves by inadvertently discovering a supernormal stimulus in nature and letting it do its thing on
them, but what would they do next? Do it again if it felt good, but
the generation of diversity on which true design exploration depends would probably not be possible for them.

To sum up the story so far: The memorable nymphs and fairies
and goblins and demons that crowd the mythologies of every people are the imaginative offspring of a hyperactive habit of finding
agency wherever anything puzzles or frightens us. This mindlessly
generates a vast overpopulation of agent-ideas, most of which are
too stupid to hold our attention for an instant; only a well-designed

few make it through the rehearsal tournament, mutating and improving as they go. The ones that get shared and remembered are the souped-up winners of billions of competitions for rehearsal time in the brains of our ancestors. This is not a new idea, of course, just a clarification and extension of an idea that has been around for generations. As Darwin himself surmised:

> . . . the belief in unseen or spiritual agencies . . . seems to be almost universal . . . nor is it difficult to comprehend how it arose. As soon as the important faculties of the imagination, wonder, and curiosity, together with some power of reasoning, had become partially developed, man would naturally have craved to understand what was passing around him, and have vaguely speculated on his own existence. [1886, p. 65]

So far so good, but what we have accounted for is superstition, not religion. Hunting for elves in the garden or the bogeyman under your bed is not (yet) having a religion.

What is missing? For one thing, belief! For, although Darwin speaks of *belief* in spiritual entities, we have not yet provided an account that secures anything so strong as that. Nothing has yet been said about having to *believe* the hobbyhorse idea that keeps recycling through your mind; it may be a hunch, or a wonder, or even an obsessively *disbelieved* little nugget of paranoia—or just a captivating morsel of story line. Nobody has ever believed in Cinderella or Little Red Riding Hood, but their fairy tales have been quite faithfully transmitted (with mutations) over many generations. Many fairy tales make up for not being true stories by having a moral, which gives them an apparent value—to the tellers and hearers—that makes up for their not being *information* about the wide world. Others conspicuously lack a moral—just what does "Goldilocks" teach our children: not to invade strangers' houses?—and must persist in the transmission tournament for less obvious reasons. As is usual in evolutionary circumstances, a gradual ramp of intermediate states of mind is there to be traversed, from shuddering doubt (are there *really* wicked witches in the woods?) and neutral fascination (a flying carpet—just imagine!) through nagging uncertainty (unicorns? well, I've never seen one) and on to robust conviction (Satan is as real as that horse over there). Fascination is enough to power rehearsal and replication. Almost everybody has a good strong copy of the idea of unicorns, though few people believe in them; but hardly anybody has the idea of pudus, which have the distinct advantage of being real (you can look it up). There is a lot more to religion than a fascination with counterintuitive agentlike entities.

2 Gods as interested parties

Why the gods above me
Who must be in the know
Think so little of me
They allow you to go . . .
—Cole Porter, "Every Time We Say Goodbye"

Ancestor worship must be an appealing idea to those who are about to become ancestors. —Steven Pinker, *How the Mind Works*

Whereas other species make limited use of the intentional stance—for anticipating the moves of predator and prey, plus a little bullying and bluffing—we human beings are obsessed about our personal relations with others: worrying about our reputations, our unfulfilled promises and obligations, and reviewing our affections and loyalties. Unlike other species, which have to worry all the time about lurking predators and dwindling food sources, we human beings have largely traded in these pressing concerns for others. The price our species has paid for the security of living in large groups of interacting communicators with different agendas is having to keep track of these complex agendas and shifting relationships. Whom can I trust? Who trusts me? Who are my rivals and my friends? To whom do I owe debts, and whose debts to me should I forgive or collect? The human world is teeming with such *strategic information*, to use Pascal Boyer's term, and what matters most about it (as in a card game) is this: "In social interaction, we presume that other people's access to strategic information is neither perfect nor automatic" (2001, p. 155). Does *she know* that *I know* that *she wants* to leave her husband? Does *anybody* know that I stole that pig? All the plots of all the great sagas and tragedies and novels, but also all the situation comedies and comic books, hinge on the tensions and complexities that arise because agents in the world don't all share the same strategic information.

How do people deal with all this complexity?[2] Sometimes when people are learning a new card game they are advised by their teacher to lay all their cards faceup on the table, so everybody can see what the others are holding. This is an excellent way of teaching the tactics of the game. It provides a temporary crutch for the imagination—you actually get to see what each person would normally be hiding, so you get to base your reasoning on the facts. You don't have to keep track of them in your head, since you can just look down on the table whenever you need a reminder. This helps you build up skill in visualizing where the cards must be when they are hidden. What works at the card table can't be done in real life. We can't get people to divulge all their secrets during a practice session of life, but we can get practice "off line" by telling and listening to stories, narrated by an agent who sees all the cards of the fictional or historic characters.

What if there really were agents who had access to all the strategic information! What an idea! It is easy enough to see that such a being—in Boyer's terms, a "full-access agent"—would be an attention-grabbing concoction, but aside from that, what good would it be? Why would it be any more important to people than any other fantasy? Well, it might help people simplify the thinking that has to be done to figure out what to do next. A survey of the world's religions shows that almost always the full-access agents turn out to be ancestors, gone but not at all forgotten. As the memory of Father is burnished and elaborated in many retellings to children and grandchildren and their grandchildren, his ghost may acquire many exotic properties, but at the heart of his image is his virtuosity in the strategic-information department. Remember how your mother and father often seemed to know just what you were thinking, just what mischief you were trying to hide? Ancestors are like that, only more so: you can't hide from them, not even your secret thoughts, and nobody else can either. Now you can reframe your puzzlement about what to do next: what would my ancestors want me to do in my current situation? You may not be able to tell what these vividly imagined agents would want you to do, but, whatever it is, it's what you should do.

Why, though, do we human beings so consistently focus our fantasies on *our ancestors*? Nietzsche, Freud, and many other theorists of culture have articulated elaborate conjectures about the subliminal motivations and memories that arose from mythic struggles deep in our human past, and there may be substantial gold to be refined from this lode of speculation once we re-examine it with an eye to testable hypotheses of evolutionary psychology, but in the meantime, we can more confidently identify the basic mental disposition that sets up this bias, for it is considerably older than our species. Mammals and birds, unlike most other animals, often devote considerable parental attention to their young, but there is wide variation in this: *precocial* species are those in which the young hit the ground running, as the saying goes, whereas *altricial* species have young that require prolonged parental care *and training*. This training period provides a host of opportunities for information transmission from parent to offspring that bypasses the genes entirely.

Biologists are often accused of *gene centrism*—thinking that everything in biology is explained by the action of genes. And some biologists do indeed go overboard in their infatuation with genes. They should be reminded that Mother Nature is not a gene centrist! That is, the process of natural selection itself doesn't require that all valuable information move "through the germ line" (via the genes). On the contrary, if the burden can be reliably taken over by continuities in the external world, that is fine with Mother Nature—it takes a load off the genome. Consider the various continuities *relied on* by natural selection: those supplied by the fundamental laws of physics (gravity, etc.) and those supplied by the long-term stabilities of environment that can be safely "expected" to persevere (salinity of the ocean, composition of the atmosphere, colors of things that can be used as triggers, etc.). To say that natural selection relies on these regularities means just this: it generates mechanisms that are tuned to work well in environments that exhibit those regularities. The design of these mechanisms *presupposes* these regularities in the same way that the design of a Mars rover presupposes the planet's gravity, the solidity and temperature range of its surface, and so forth. (It is not designed to operate in the Everglades, for instance.) Then there are the regularities that can be transmitted from generation to generation by social learning. These are a special case of reliable environmental regularities; they take on further importance since they are themselves subject to natural selection, directly and indirectly. *Two* information superhighways have been improved and enlarged over the eons. The *genetic* informational pathways have themselves been subject to incessant refinement over billions of years, with optimization of chromosome design and invention and improvement of proofreading enzymes and so forth, with the effect that high-fidelity, high-bandwidth transmission of genetic information has been achieved. The parent-child *instructional* pathway has also been optimized by a recursive or iterative process of enhancement. As Avital and Jablonka (2000) note, "The evolution of the transmission of mechanisms of transmission is of central importance for the evolution of learning and behaviour" (p. 132).

Among the adaptations for improving the bandwidth and fidelity of parent-offspring transmission is *imprinting,* in which the newborn has a readily triggered and powerful instinctual urge to approach and stay close to *and attend to* the first large moving thing it sees. In mammals the urge to find and cling to the nipple is hard-wired by the genes, and it has the side effect, opportunistically exploited by further adaptations, of keeping the young where they can watch Mother when they are not feeding. Human infants are no exception to the mammalian rules. Meanwhile, going in the other direction, parents have been genetically designed to attend to infants. Whereas gull chicks are irresistibly drawn to an orange spot, human beings are irresistibly captivated by the special proportions of a "baby face." It brings out the "Aw, isn't she *cunning!*" response in the steeliest curmudgeons. As Konrad Lorenz (1950) and others have argued, the correlation between an infant's facial appearance and an adult's nurturing response is no accident. It is not that baby faces are somehow *intrinsically darling* (what on earth could that mean?) but that evolution hit upon facial proportions as the signal to trigger parental responses, and this has been refined and intensified over the eons in many lineages. We don't love babies and puppies *because they're cute*. It's the other way around: we *see them as cute* because evolution has designed us to love things that look like that. So strong is the correlation that measurements of fossils of newborn dinosaurs have been used to support the radical hypothesis that some dinosaur species were altricial (Hopson, 1977; Horner, 1984). Stephen Jay Gould's classic analysis (1980) of the gradual juvenilization over the years of Mickey Mouse's features provides an elegant demonstration of the way cultural evolution can parallel genetic evolution, homing in on what human beings instinctually prefer.

But even more potent than the bias in adults to respond parentally to baby-faced young is the bias in those young to respond with obedience to parental injunction—a trait observable in bear cubs as well as human babies. The free-floating rationale is not far to seek: *it is in the genetic interests* of parents (but not necessarily other conspecifics!) to inform—not misinform—their young, so it is efficient (and relatively safe) to *trust* one's parents. (Sterelny, 2003, has particularly acute observations on the trade-offs between trust and suspicion in the evolutionary arms races of cognition.) Once the information superhighway between parent and child is established by genetic evolution, it is ready to be used—or abused—by any agents with agendas of their own, *or by any memes that happen to have features that benefit from the biases built into the highway.*3

One's parents—or whoever are hard to distinguish from one's parents—have something approaching a dedicated hotline to acceptance, not as potent as hypnotic suggestion, but sometimes close to it. Many years ago, my five-year-old daughter, attempting to imitate the gymnast Nadia Comaneci's performance on the horizontal bar, tipped over the piano stool and painfully crushed two of her fingertips. How was I going to calm down this terrified child so I could safely drive her to the emergency room? Inspiration struck: I held my own hand near her throbbing little hand and sternly ordered: "Look, Andrea! I'm going to teach you a secret! You can push the pain into *my* hand with your mind. Go ahead, *push! Push!*" She tried—and it worked! She'd "pushed the pain" into Daddy's hand. Her relief (and fascination) were instantaneous. The effect lasted only for minutes, but with a few further administrations of impromptu hypnotic analgesia along the way, I got her to the emergency room, where they could give her the further treatment she needed. (Try it with your own child, if the occasion arises. You may be similarly lucky.) I was exploiting her instincts—though the rationale didn't occur to me until years later, when I was reflecting on it. (This raises an interesting empirical question: would my attempt at instant hypnosis have worked as effectively on some other five-year-old, who hadn't imprinted on me as an authority figure? And if imprinting is implicated, how young must a child be to imprint so effectively on a parent? Our daughter was three months old when we adopted her.)

"Natural selection builds child brains with a tendency to believe whatever their parents and tribal elders tell them" (Dawkins, 2004a, p. 12). It is not surprising, then, to find religious leaders in every part of the world hitting upon the extra authority provided them by their taking on the title "Father"—but this is to get ahead of our story. We're still at the stage where, Boyer claims, our ancestors were unwittingly summoning up fantasies about *their* ancestors in order to relieve some of their quandaries about what to do next. An important feature of Boyer's hypothesis is that these imagined full-access agents are *not* typically deemed to be omniscient; if you lose your knife, you don't automatically suppose that they would know the whereabouts of your knife *unless somebody stole it from you or you dropped it in an incriminating place during a tryst*—unless, that is, it was *strategic* information. And the ancestors know all the strategic information, because they are interested in it. What you and your kin do is of concern to them for the same reason it is of concern to your parents, and for the same reason it matters to you what your children do and how they are perceived in the community. Boyer's suggestion is that the idea of omniscience—a god who knows absolutely everything about everything, including where your car keys are, the largest prime number smaller than a quadrillion, and the number of grains of sand on that beach—is a later wrinkle, a bit of sophistication or intellectual tidying-up much more recently adopted by theologians. There is some experimental evidence in support of this hypothesis. People have been taught since childhood, and hence will avow, that God knows *everything*, but they don't rely on this when reasoning un-self-consciously about God. The root idea, the one that people actually use when they are not worrying about "theological correctness" (Barrett, 2000), is that the ancestors or the gods *know the things that matter the most:* the secret longings and schemes and worries and pangs of guilt. Gods know where all the bodies are buried, as the saying goes.

3 Getting the gods to speak to us

Nothing is more difficult, and therefore more precious, than to be able to decide. —Napoleon Bonaparte

But what good to us is the gods' knowledge if we can't get it from them? How could one communicate with the gods? Our ancestors (while they were alive!) stumbled on an extremely ingenious solution: *divination*. We all know how hard it is to make the major decisions of life: should I hang tough or admit my transgression, should I move or stay in my present position, should I go to war or not, should I follow my heart or my head? We still haven't figured out any satisfactory systematic way of deciding these things. Anything that can relieve the burden of figuring out how to make these hard calls is bound to be an attractive idea. Consider flipping a coin, for instance. Why do we do it? To take away the burden of having to *find a reason* for choosing A over B. We like to have reasons for what we do, but sometimes nothing sufficiently persuasive comes to mind, and we recognize that we have to decide soon, so we concoct a little gadget, an external thing that will make the decision for us. But if the decision is about something momentous, like whether to go to war, or marry, or confess, anything like flipping a coin would just be too, well, flippant. In such a case, choosing *for no good reason* would be too obviously a sign of incompetence, and, besides, if the decision is really that important, once the coin has landed you'll have to confront the further choice: should you honor your just-avowed commitment to be bound by the flip of the coin, or should you reconsider? Faced with such quandaries, we recognize the need for some treatment stronger than a coin flip. Something more ceremonial, more impressive, like divination, which not only tells you what to do, but gives you a reason (if you squint just right and use your imagination). Scholars have uncovered a comically variegated profusion of ancient ways of delegating important decisions to uncontrollable externalities. Instead of flipping a coin, you can flip arrows (belomancy) or rods (rhabdomancy) or bones or cards (sortilege), and instead of looking at tea leaves (tasseography), you can examine the livers of sacrificed animals (hepatoscopy) or other entrails (haruspicy) or melted wax poured into water (ceroscopy). Then there is moleosophy (divination by blemishes), myomancy (divination by rodent behavior), nephomancy (divination by clouds), and of course the old favorites, numerology and astrology, among dozens of others.[4]

One of the more plausible arguments made by Julian Jaynes in his brilliant but quirky and unreliable book, *The Origins of Consciousness in the Breakdown of the Bicameral Mind* (1976), was that this riotous explosion of different ways of passing the buck to an external deciding-gadget was a manifestation of human beings' growing difficulties with self-control, as human groups became larger and more complicated (chapter 4, "A Change of Mind in Mesopotamia," pp. 223–54). And as Palmer and Steadman have more recently noted, "The most important effect of divination is that it reduces responsibility in decision-making, and thereby reduces the acrimony that can result from bad decisions" (2004, p. 145). The free-floating rationale is obvious enough: if you're going to pass the buck, pass it to something that can't duck the responsibility in turn, and that can be held responsible if things don't go well. And as usual with adaptations, you don't have to understand the rationale to benefit from it. Divination—what Jaynes called "exopsychic methods of thought or decision-making" (p. 245)—could have risen in popularity simply because those who happened to do it liked the results enough to do it again, and again, and then others began to copy them, and it became the thing to do even though nobody really knew why.

Jaynes noted (p. 240) that the very idea of randomness or chance is of quite recent origin: in earlier times, there was no way of even *suspecting* that some event was utterly random; everything was presumed to mean something, if only we knew what. Deliberately opting for a meaningless choice just to get *some choice or other* made, so one can get on with one's life, is probably a much later sophistication, even if that is the rationale that explains why it was actually useful to people. In the absence of that sophistication, it was important to believe that *somebody somewhere who knows what's right is telling you*. Like Dumbo's magic feather, some crutches for the soul work only if you believe that they do.[5]

But what does it mean to say that such a method *works?* Only that it actually helps people think about their strategic predicaments and then make timely decisions—even if the decisions themselves are not any better informed by the process. This is not nothing. In fact, it could be a tremendous boost under various circumstances. Suppose that people facing difficult decisions typically have all the information they need to make well-grounded decisions, but just don't realize that they do, or just don't trust their own judgment as much as they ought to. All they need to get them out of their funk and stiffen their spine for resolute action is . . . a little help from their friends, their imagined ancestors hovering invisibly nearby and telling them what to do. (Such a psychological asset would be jeopardized by skeptics going around pooh-poohing the integrity of divination, of course, and probably that recognition— even if subliminal and unarticulated—has always motivated hostility toward skeptics. Shh. Don't break the spell; these people need this crutch to keep their act together.)

Even if people are not, in general, capable of making good decisions on the information that they have, it may *seem to them* that divination helps them think about their strategic predicaments, and this may provide the motivation to cling to the practice. For reasons they can't fathom, divination provides relief and makes them feel good—rather like tobacco. And note that none of this is genetic transmission. We're talking about a culturally transmitted practice of divination, not an instinct. We don't have to settle the empirical question *now* of whether divination memes are mutualist memes that actually enhance the fitness of their hosts, or parasite memes that they'd be better off without. Eventually, it would be good to get an evidence-based answer to this question, but for the time being it is the questions I am interested in. Notice, too, that this leaves wide open the possibility that divination (under specific circumstances, to be discovered and confirmed) is a mutualist meme because it's *true*—because there *is* a God who knows what is in everyone's heart and on special occasions tells people what to do. After all, the reason why water is deemed essential to life in every human culture is that it *is* essential to life. For the moment, though, my point is just that divination, which appears just about everywhere in human culture (including, of course, among the astrology-seekers and numerologists who still inhabit our high-tech Western culture), could be understood as a natural phenomenon, paying for itself in the biological coin of replication, whether or not it is actually a source of reliable information, strategic or otherwise.

4 Shamans as hypnotists

Anyone who goes to a psychiatrist should have his head examined.
   —Samuel Goldwyn

Divination is one genus of rituals found throughout the world; healing rituals conducted by local shamans (or "witch doctors") are another. How did it arise? In *Guns, Germs, and Steel* (1997), Jared Diamond showed that, to a first approximation, in every culture on every continent, human exploration over the centuries has discovered *all* the local edible plants and animals, including many that require elaborate preparation to make them nonpoisonous. Moreover, they have domesticated whatever local species have been amenable to domestication. We have had the time, intelligence, and curiosity to have made a near-exhaustive search of the possibilities— something that can now be proved by high-tech methods of genetic analysis of domesticated species and their closest wild relatives. It stands to reason, then, that we should also have done an excellent job of uncovering most if not all the locally available medicinal herbs, even those requiring elaborate refinement and preparation. So powerful and reliable have these search procedures proved to be that pharmaceutical companies have in recent years invested in anthropological research, energetically acquiring—by theft, in some cases—the fruits of this "primitive" R & D by the indigenous populations in every rain forest and remote island. This eager appropriation of the "intellectual property rights" and "trade secrets" of economically naïve peoples is, however deplorable, an excellent instance of the *cui bono?* reasoning of evolutionary biology. R & D is expensive and time-consuming; whatever information has stood the test of time, replicating through the ages, must have *paid for itself somehow,* so it is *probably* worth plagiarizing! (*Cui bono?* It may have paid for itself by helping a long line of tricksters dupe their clients, so we mustn't assume the payment was a benefit *for all parties.*)

That people take herbs to alleviate their symptoms or even cure their conditions is not puzzling or surprising, but why all the accompanying (and often horrifying) rituals? Anthropologist James McClenon (2002) has examined the patterns in rituals of healers all over the world and finds that they strongly support the hypothesis that what people have discovered, over and over again, is the placebo effect—more specifically, the power of hypnotism, often aided by ingestion or inhalation of hallucinogens or other mind-altering substances (see also Schumaker, 1990). Ritual healing, McClenon argues, is ubiquitous because it actually works—not perfectly, of course, but much better than the Western medical establishment has typically been willing to grant. Indeed, there is a convergence: the ailments that people go to—and pay—shamans to alleviate are those that are particularly hospitable to placebo-effect treatment: psychological stress and its attendant symptoms, as well as the ordeals of childbirth, to name perhaps the most interesting case.

Childbirth in *Homo sapiens* is a particularly stressful event, and of course the timing of its arrival—unlike the traumas of accidents and hostility—can be anticipated quite accurately, making it an ideal occasion for elaborate ceremonies requiring considerable preparation time. Since infant and maternal mortality rates in childbirth were presumably as high in pretechnological days as they are now in nontechnological cultures, there has been plenty of room for a strong selection pressure for the coevolution of any treatment (culturally transmitted) *and susceptibility to treatment* (genetically transmitted) that could improve the odds. Just as lactose tolerance has evolved in peoples who had the culture of dairy-herding, hypnotizability could have evolved in peoples who had the culture of healing rituals.

> I hypothesize that shamanic rituals constitute hypnotic inductions, that shamanic performances provide suggestions, that client responses are equivalent to responses produced by hypnosis, and that responses to shamanic treatment are correlated with patient hypnotizability. [McClenon, 2002, p. 79]

These hypotheses are eminently testable, and, McClenon argues, they plausibly provide sources for some of the features (rituals and beliefs) to be found just about everywhere in religions. Interestingly, there is wide variation in hypnotizability, with about 15 percent of human populations exhibiting strong hypnotizability, and there is apparently a genetic component, which is not (to my knowledge) well studied yet. Shamans tend to run in families, according to a wealth of anthropological evidence, but this could, of course, be due entirely to vertical *cultural* transmission (of the shamanic memes from parent to child).

But why should human beings be susceptible to the placebo effect in the first place? Is this a unique human adaptation (depending on language and culture), or are related effects discernible in other species? This is a topic of current research and controversy. One of the most ingenious hypotheses under discussion is Nicholas

Humphrey's (2002) "economic resource management" hypothesis. The body has many resources to cure its own ailments: pain to discourage activity that can further damage an injury, fevers to combat infection, vomiting to rid the digestive system of toxins, and immune responses, to mention the most powerful. These are all effective but costly; overuse, or premature use, by the body could actually end up harming the body more than helping. (Full-scale immune responses are particularly costly, and only the healthiest animals can maintain a fully equipped army of antibodies.) When should a body *spare no expense* in hopes of a quick cure? Only when it is safe to do so, or when help is just around the corner. Otherwise, it might be more prudent for the body to be stingy with its costly self-treatments. The placebo effect, according to this hypothesis, is a releasing trigger, telling the body to pull out all the stops because there is *hope*. In other species, the hope variable is presumably tuned to whatever information the animal can glean from its current surroundings (is it safe in its den, or in the middle of its herd, and is there plenty of food around?); in us, the hope variable can be manipulated by authoritative figures. These are questions worth further investigation.

In chapter 3, I briefly introduced the hypothesis that our brains might have evolved a "god center" but noted that it would be better for the time being to consider it a *whatsis* center that had later been adapted or exploited by religious elaborations of one sort or another. Now we have a plausible candidate for filling in the blank: the hypnotizability-enabler. Moreover, in his recent book, *The God Gene*, the neurobiologist and geneticist Dean Hamer (2004) claims to have found a gene that could be harnessed for this role. The VMAT2 gene is one of many that provide recipes for the proteins that do the work in the brain helping to create and regulate the molecules that carry the signals that somehow combine to compose all our thought and behavior. VMAT2 makes a protein that transports a monoamine. (The mood-changing drug Prozac also adjusts the activity of monoamines, such as dopamine and serotonin, but there are many other psychoactive or mood-changing drugs developed in recent years that work by enhancing or suppressing the activity of other neuromodulators and neurotransmitters.) The VMAT2 gene is polymorphic in human beings, meaning that there are different mutations of it in different people. The VMAT2 gene variants are well-placed, then, to account for differences in people's emotional or cognitive responses to the same stimuli, and could explain why some people are relatively immune to hypnotic induction whereas others are readily put into a trance. None of this is close to proven yet. Other polymorphic genes have the same potential and Hamer's development of his hypothesis is marked by more enthusiasm than subtlety, a foible that may repel researchers who would otherwise take it seriously. Still, *something like* his hypothesis (but probably much more complicated) is a good bet for confirmation in the near future, as the roles of proteins and their gene recipes are further analyzed.

Part of what is tantalizing about this research avenue is how *non*reductionistic it is! McClenon and Hamer have worked independently of each other, so far as I know. Neither mentions the other, in any case, and neither is treated by Boyer or Atran or the other anthropologists. This is not surprising. The collaboration between geneticists and neurobiologists on the one hand and anthropologists and archeologists and historians on the other, pioneered by Luigi Luca Cavalli-Sforza and his colleagues, is a recent and spotty trend. False starts and disappointments are bound to outnumber triumphs in the early days of such interdisciplinary work, and I make no promises about the prospects of the specific hypotheses of either McClenon or Hamer. They make a vivid and accessible example of the possibilities in store, however. Recall Dawkins's point quoted in chapter 3: "If neuroscientists find a 'god center' in the brain, Darwinian scientists like me want to know why the god center evolved. Why did those of our ancestors who had a genetic tendency to grow a god center survive better than rivals who did not?" (2004b, p. 14). We now have one eventually testable answer to

Dawkins's question, and it invokes not just biochemical facts but the whole world of cultural anthropology.[6] Why did those with the genetic tendency survive? Because they, unlike those who lacked the gene, had health insurance! In the days before modern medicine, shamanic healing was your only recourse if you fell ill. If you were constitutionally impervious to the ministrations that the shamans had patiently refined over the centuries (cultural evolution), you had no health-care provider to turn to. If the shamans had not existed, there would have been no selection advantage to having this variant gene, but their accumulated memes, their culture of shamanic healing, could have created a strong ridge of selection pressure in the adaptive landscape that would not otherwise have been there.

This still doesn't get us to organized religion, but it does get us to what I am going to call *folk religion,* the sorts of religion that have no written creeds, no theologians, no hierarchy of officials.[7] Before any of the great organized religions existed, there were folk religions, and these provided the cultural environment from which organized religions could emerge. Folk religions have rituals, stories about gods or supernatural ancestors, prohibited and obligatory practices. Like folk tales, the sayings of folk religion are of such distributed authorship that it is better to say that they have no authors at all, not that their authors are unknown. Like folk music,[8] the rituals and songs of folk religion have no composers, and their taboos and other moral injunctions have no legislators. Conscious, deliberate authorship comes later, after the designs of the basic cultural items have been honed and polished for many generations, without foresight, without intent, by nothing but the process of differential replication during cultural transmission. Is all this possible? Of course. Language is a stupendously intricate and well-designed cultural artifact, and no individual human designers get to take credit for it. And just as some of the features of *written* languages are clearly vestigial traces of their purely oral ancestors,[9] some of the features of *organized* religion will turn out to be vestigial traces of the folk religions from which they are descended. By vestigial traces,

I mean this: the preservation over many generations of a folk religion—its self-replication in the face of inexorable competition—demands adaptations that are peculiar to an oral tradition and that are no longer strictly necessary (from a reverse-engineering point of view) but that persist simply because they haven't yet been sufficiently costly to be selected against.

5 Memory-engineering devices in oral cultures

The total corpus of Baktaman knowledge is stored in 183 Baktaman minds, aided only by a modest assemblage of cryptic concrete symbols (the meanings of which depend on the associations built up around them in the consciousness of a few seniors) and by limited, suspicious communication with the members of a few surrounding communities.

—Fredrik Barth, *Ritual and Knowledge Among the Baktaman of New Guinea*

Humans, it appears, are the only animals that spontaneously engage in creative, rhythmic bodily coordination to enhance possibilities for cooperation (e.g., singing and swaying when they work together).

—Scott Atran, *In Gods We Trust*

Every folk religion has rituals. To an evolutionist, rituals stand out like peacocks in a sunlit glade. They are usually stunningly expensive: they often involve the deliberate destruction of valuable food and other property—to say nothing of human sacrifices—are often physically taxing or even injurious to the participants, and typically require impressive preparation time and effort. *Cui bono?* Who or what is the beneficiary of all this extravagant outlay? We have already seen two ways rituals *might* pay for themselves, as psychologically necessary features of divination techniques, or hypnotic induction procedures in shamanic healing.[10] Once they were established on the scene for these purposes, they would be available to be adapted—*exapted*, as the late Stephen Jay Gould would say—for other uses. But there are other possibilities to explore.

Anthropologists and historians of religion have theorized about the meaning and function of religious ritual for generations, usually from blinkered perspectives that ignore the evolutionary background. Before we look at speculations about rituals as symbolic expressions of one deep need or belief or another, we should consider the case that can be made for rituals as memory-enhancement processes, designed by cultural evolution (and not by any conscious designers!) to improve the copying fidelity of the very process of meme transmission they ensure. One of the clearest lessons of evolutionary biology is that early extinction lies in the future of any lineage in which the copying machinery breaks down, or even just degrades a little. Without high-fidelity copying, any design improvements that happen to occur in a lineage will tend to be frittered away almost immediately. Hard-won gains accumulated over many generations can be lost in a few faulty replications, the precious fruits of R & D evaporating overnight. So we can be sure that would-be religious traditions that have no good ways of preserving their designs reliably over the centuries are doomed to oblivion.

We can observe today the birth and swift death of cults, as the early adherents lose faith or lose interest and drift away, leaving hardly a trace after a few years. Even when members of such a group fervently want to keep it going, their desires will be thwarted unless they avail themselves of the technologies of replication. Today, writing (not to mention videotape and other high-tech recording media) provides the obvious information highway to use. And from the earliest days of writing, there has been a keen appreciation of the need not only to protect the sacred documents from damage and decay, but to copy them over and over, minimizing the risk of loss by ensuring that multiple copies were distributed around. For many centuries before the invention of movable type, which made possible for the first time the mass production of identical copies, roomfuls of scribes, shoulder to shoulder at their writing desks, took dictation from a reader and thus turned one frail and dog-eared copy into dozens of fresh new copies—a copy machine made of people. Since the originals from which the copies were made have mostly turned to dust in the meantime, without the efforts of these scribes we would have no reliable texts for any of the literature of antiquity, sacred or secular, no Old Testament, no Homer, no Plato and Aristotle, no *Gilgamesh*. The earliest known copies of Plato's dialogues still in existence, for instance, were created centuries after his death, and even the Dead Sea Scrolls and the Nag Hammadi gospels (Pagels, 1979) are copies of texts that were composed hundreds of years earlier.

A text inked on papyrus or parchment is like the hard spore of a plant that may lie undamaged in the sand for centuries before finding itself in suitable conditions to shed its armor and sprout. In oral traditions, in contrast, the vehicle—a spoken verse or sung refrain—lasts for only a few seconds, and must enter some ears—as many as possible—and imprint itself firmly in as many brains as possible, if it is to escape oblivion. Getting registered in a brain—getting heard and noticed above the competition—is less than half the battle. Getting rehearsed and rehearsed, either in the privacy of a single brain or in unison public repetition, is a life-or-death matter for an orally transmitted meme.[11]

If you want to brush up your memory of the order of worship in your church's Sunday service, or check to see whether one should stand or sit down during the closing benediction, there is almost certainly a text you can consult. The details are printed in the back of every hymnal, perhaps, or in the *Book of Common Prayer*, or, if not there, at least in texts that are readily available to the priest or minister or rabbi or imam. Nobody has to memorize every line of every invocation, every prayer, every detail of the costumes, music, manipulation of sacred objects, and so forth, since they are all written down in one official record or another. But rituals are not by any means restricted to literate cultures. In fact, the religious rituals of nonliterate societies are often more detailed, typically much more demanding physically, and just plain longer in duration than the rituals of organized religions. Moreover, the shamans don't go to official shaman-seminaries, and there is no Council of Bishops or Ayatollahs to maintain quality control. How do the members of these religions keep all the details in memory over the generations?

A simple answer is: They don't! They can't! And it is surprisingly hard to prove otherwise. Whereas members of a nonliterate culture may be well-nigh unanimous in their conviction that their rituals and creeds have been perfectly preserved by them over "hundreds" or "thousands" of generations (a thousand years is only about fifty generations), why should we believe them? Is there any evidence that supports their traditional conviction? There is a little.

> Much of the excitement that accompanied scholars' discovery of the Nambudiri ritual tradition turned on the fact that although texts delineating Vedic rituals exist, the Nambudiri have not used them. Exclusively by non-literate means, they have sustained this elaborate ritual tradition with astonishing fidelity (as gauged by the centuries-old *Śrauta Sūtras*). [Lawson and McCauley, 2002, p. 153]

So at first it appears that the Nambudiri are perhaps a uniquely lucky oral culture, having some evidence in support of their conviction that they have preserved their rituals intact. If it were not for the Vedic texts, presumably unknown to them and never consulted over the years, there would be no fixed yardstick against which to measure their confidence in the antiquity of their traditions. But, alas, the story is too good to be entirely true. The Nambudiri tradition may be oral, but they are not illiterate (some of their priests teach engineering, for instance), and it is hard to believe they have kept themselves entirely isolated from the Vedic texts. "It is known that during their six-month initiation period of the training, preparation and rehearsals leading up to the actual event, use is made of notebooks, prepared by the senior AcAryas who have already taken part in previous rituals. . . ."[12] So the Nambudiri are not really an independent benchmark of how accurate oral transmission can be.

Compare the problem here with the ongoing research on the evolution of languages. Using complex and sophisticated probabilistic analyses, linguists can deduce features of extinct *oral* languages whose last speakers have been dead for millennia! How can this be done with no tape recordings to consult and no texts in the language they are extrapolating? The linguists make heavy use of the enormous corpus of textual data in other, later languages, tracing linguistic shifts from Attic Greek to Hellenistic Greek, and from Latin to the Romance languages, and so forth. Finding common patterns in these shifts, they have been able to extrapolate back with some confidence to what languages must have been like before writing came along to fossilize some of them for later ages to study. They have been able to extract regularities of pronunciation shift and grammatical shift, and juxtapose them on patterns of stability, to arrive at highly educated and cross-confirmed guesses about how, say, Indo-European words were pronounced long before there were written languages to preserve the clues like fossil insects in amber.[13]

If we tried to do the same extrapolation trick with religious beliefs, we would first have to establish benchmarks for stability and shift in them, and so far this has not proved feasible. What little we know about early religions is almost entirely dependent on surviving texts. Pagels (1979) offers a fascinating perspective on the Gnostic Gospels, for instance, early competitors for inclusion in the canon of Christian texts, thanks to the fortuitous survival of written texts that have been passed on as translations of copies of copies . . . of the originals.

We cannot, then, just take it on faith that nonliterate religious traditions still extant in the world are as ancient as advertised. And we already know that in some such religions there is *not* a tradition of obsessive preservation of ancient creed. Fredrik Barth, for instance, found lots of evidence of innovation among the Baktamans, and as Lawson and McCauley (2002, p. 83) dryly note, "Perfect fidelity to past practice is not an unwavering ideal for the Bakta-mans." So, whereas we can be quite sure that people in oral tradi-tions have had religion of one sort or another for thousands of years, we shouldn't ignore the possibility that the religion we see (and record) today may consist of elements that have been invented or reinvented quite recently.

People run and jump and throw stones pretty much the same way everywhere, and this regularity is explained by the physical properties of human limbs and musculature and the uniformity of wind resistance around the globe, not a tradition somehow passed down from generation to generation. On the other hand, where no such constraints ensure reinvention, items of culture will be able to wander swiftly, widely, and unrecognizably in the absence of mecha-nisms of copying fidelity. Different strokes for different folks.[14] And wherever this wandering transmission occurs, there will auto-matically be selection for mechanisms that enhance copying fi-delity whenever they arise, *whether or not people care,* since any such mechanisms will tend to persist longer in the cultural medium than alternative (and no less costly) mechanisms that get them-selves copied indifferently.

One of the best ways of ensuring copying fidelity over many replications is the "majority rule" strategy that is the basis for the uncannily reliable behavior of computers. It was the great mathe-matician John von Neumann who saw a way of applying this trick in the real world of engineering, so that Alan Turing's imaginary computing machine could become a reality, permitting us to manu-facture highly reliable computers out of unavoidably unreliable parts. Practically perfect transmission of trillions of bits is routinely executed by even the cheapest computers these days, thanks to "von Neumann multiplexing," but this trick has been invented and reinvented over the centuries in many variations. In the days before radio communication and GPS satellites, navigators used to take not one or two but three chronometers aboard their ships on long voyages. If you have just one chronometer and it starts running slow or fast, you'll never know it is in error. If you bring two and they eventually begin to disagree, you won't know whether one is running slow or the other is running fast. If you bring three, you can be quite sure that the odd one out is the one in which the error is accumulating, since otherwise the two that are still in agreement would have to be going bad in exactly the same way, an unlikely co-incidence under most circumstances.

Long before it was consciously invented or discovered, this Good Trick was already embodied as an adaptation of memes. It can be seen at work in any oral tradition, religious or secular, in which people act in unison—praying or singing or dancing, for instance. Not everybody will remember the words or the melody or the next step, but most will, and those who are out of step will quickly correct themselves to join the throng, preserving the traditions much more reliably than any of them could do on their own. It doesn't depend on virtuoso memorizers scattered among them; nobody needs to be better than average. It is mathematically provable that such "multiplexing" schemes can overcome the "weakest link" phenomenon, and make a mesh that is much *stronger* than its weakest links. It is no accident that religions all have occasions on which the adherents come together to act in public unison in rituals. Any religion without such occasions would already be extinct.[15]

A public ritual is a great way of preserving content with high fidelity, but why are people so eager to participate in rituals in the first place? Since we are presuming that they are *not* intent on preserving the fidelity of their meme-copying by constituting a sort of social computer-memory, what motivates them to join in? Here there are currently a welter of conflicting hypotheses that will take some time and research to resolve, an embarrassment of riches in need of culling.[16] Consider what we can call the shamanic-advertising hypothesis. Shamans the world over conduct much of their medicine in public ceremonies, and they are adept at getting the local people not just to watch while they induce a trance in themselves or their clients but to participate, with drumming, singing, chanting, and dancing. In his classic *Witchcraft, Oracles and Magic Among the Azande* (1937), the anthropologist Edward Evans-Pritchard vividly describes these proceedings, observing how the shaman cleverly enlists the crowd of knowing onlookers, turning them into shills, in effect, to impress the uninitiated, for whom this ceremonial demonstration is a novel spectacle.

> It may be supposed, indeed, that attendance at them has an important formative influence on the growth of witchcraft beliefs in the minds of children, for children make a point of attending them and taking part in them as spectators and chorus. This is the first occasion on which they demonstrate their belief, and it is more dramatically and more publically affirmed at these séances than in any other situation. [Evans-Pritchard, 1937 (1976 abridged edition, pp. 70–71)]

Innate curiosity, stimulated by music and rhythmic dancing and other forms of "sensory pageantry" (Lawson and McCauley, 2002), could probably account for the initial motivation to join the chorus—especially if we have an evolved innate desire to belong, to join with the others, especially the elders, as many have recently argued. (This will be a topic in the next chapter.) Then there are the phenomena of "mass hypnosis" and "mob hysteria," still poorly understood but undeniably potent effects observable when people are brought together in crowds and given something exciting to react to. Once people find themselves in the chorus, other motivations can take over. Anything that makes the cost of nonparticipation steep will do the trick, and if community members get the idea of encouraging other members not only to participate but to inflict costs on those who shirk their responsibility to participate, the phenomenon can become self-sustaining (Boyd and Richerson, 1992).

Doesn't there have to be some*one* to prime the pump? How would this initially get started unless there were some people, some agents, who *wanted* to start a ritual tradition? As usual, this hunch betrays a failure of evolutionary imagination. It is of course possible—and in some instances surely *likely* or even *proven*—that some community leader or other agent set out to design a ritual to serve a particular purpose, but we have seen that such an author is not strictly necessary. Even elaborate and expensive rituals of public rehearsal could *emerge* out of earlier practices and habits without conscious design.[17]

Public rehearsal is a key process of memory enhancement, but it is not enough. We also have to look at the features of what is rehearsed, for these can themselves be designed to be more and more memory-friendly. A key innovation is breaking down the material to be transmitted into something like an alphabet, a smallish repertoire of *norms of production*. In appendix A, I describe how the reliability of DNA replication itself depends on there being a finite code or ensemble of elements, an alphabet of sorts, such as A, C, G, T. This is a form of *digitization* that allows tiny fluctuations or variations in execution to be absorbed or wiped out in the next round. The design idea of digitization has been made famous in the computer age, but earlier applications of it can be seen in the ways in which religious rituals—like dances and poems and words themselves—can be broken down into easily recognizable elements fit for what Dan Sperber (2000) calls "triggered production" (see appendixes A and C). No two people may do their curtsy or salute or kowtow in exactly the same way, but each will be clearly recognizable *as* a curtsy or salute or kowtow by the rest of the group, which thereby absorbs the noise of the moment and transmits to the future only the essential skeleton, the spelling out of the moves. When the children watch their elders doing the moves, whether in a secular folk dance or a folk-religious ceremony (and that distinction will be quite arbitrary or nonexistent in some cultures), they learn an alphabet of behaviors, and they may vie with one another to see who can do the most dashing *A-move* or the curliest *B-move* or the loudest *C-chant*, but they all agree on what the moves are, and therein lies a huge compression of the information that must be transmitted. This kind of compression can be accurately measured on your home computer by comparing a bitmap of a page of text (which makes no distinction between alphabet characters and smudges or inkblots, laboriously representing every dot) and a text file of the same page, which will be orders of magnitude smaller.

To speak of an "alphabet" as composed of a "canonical" set of things to remember is to be doubly anachronistic, using later technology (*written* language and the conscious and deliberate elevation of a restricted *canon* of prescribed beliefs and texts) to analyze the design strengths of earlier innovations in transmission methods that had no authors. These were further enhanced by the use of *rhythm* and *rhyme*—to commit a further anachronism, since these "technical" terms were surely invented long after the effectiveness of the properties was "recognized" by the blind watchmaker of cultural selection. Rhythm and rhyme and musical pitch all provided additional bolstering (Rubin, 1995), turning unmemorable strings of words into *sound bites* (let's wallow in anachronism, while we're at it).

A somewhat less obvious design feature was the inclusion of *incomprehensible* elements! Why would this help transmission? By obliging the transmitters to fall back on "direct quotation" in circumstances where they might otherwise be tempted to use "indirect quotation" and just transmit the *gist* of the occasion "in their own words"—a dangerous source of mutation. The underlying idea is familiar enough to us all in the (usually despised, but effective) pedagogical method: rote learning. "Don't try to understand these formulas! Just *memorize* them!" If you are simply unable to understand the formulas, or some aspect of them, you don't need the admonition; you have no recourse but memorization, and that reinforces the reliance on strict rehearsal and the error-correcting genius of alphabets. The admonition, however, may well be there as well, as yet another memory-enhancing feature: *Say the formula exactly! Your life depends on it!* (If you don't say the magic word just right, the door won't open. The devil will get you if you misspeak.) To repeat the refrain that should be familiar by now: nobody had to *understand* these rationales, or even *want* to improve the copying fidelity of the rituals in which they participated; it is rather that any rituals that just happened to be favored by these features would have a powerful replicative advantage over competing rituals that lacked them.

Note that, so far, the adaptations that we have uncovered as likely contributors to the survival of religions have been neutral on the subject of whether or not *we* are beneficiaries. They are features of the medium, not the message, designed to ensure the transmission fidelity—a requirement of evolution—while almost entirely neutral with regard to whether what is transmitted is good (a mutualist), bad (a parasite), or neutral (a commensal). To be sure, we hypothesized that the evolution of shamanic healing rituals was probably a benign or mutualist development, not just a *bad habit* for which our ancestors suckered, and there is a good chance that divination actually helped (and didn't just *seem* to help) our ancestors make up their minds when they needed to, but these are still open empirical questions on which we could revise our opinion without collapse of the theory if the evidence warranted. And no one should object, at this point, that we haven't begun talking about all the good that religion does. We haven't had to address that issue yet, which is as it should be. We should exhaust our minimalist options in order to *lay the foundations* for a proper consideration of that question.

\\\

Chapter 5 The obvious expensiveness of folk religion, a challenge to biology, can be accounted for by hypotheses that are not yet confirmed but testable. Probably the excess population of imaginary agents generated by the HADD yielded candidates to press into service as decision aids, in divination, or as shaman's accomplices, in health maintenance, for instance. These co-opted or exapted mental constructs were then subjected to extensive design revision under the selective pressure for reproductive prowess.

\\\

Chapter 6 As human culture grew and people became more reflective, folk religion became transformed into organized religion; the free-floating rationales of the earlier designs were supplemented and sometimes replaced by carefully crafted reasons as religions became domesticated.

CHAPTER SIX

|||\\\|||

The Evolution of Stewardship

1 The music of religion

It don't mean a thing if it ain't got that swing. —Duke Ellington

The central claim of this chapter is that folk religion turned into organized religion in much the same way folk music spawned what we might call organized music: professional musicians and composers, written representations and rules, concert halls, critics, agents, and the rest. In both cases the shift happened for many reasons but largely because, as people became more and more reflective about both their practices and their reactions, they could then become more and more inventive in their explorations of the space of possibilities. Both music and religion gradually became more "artful" or sophisticated, more elaborate, more of a production. Not necessarily better in any absolute sense, but better able to respond to increasingly complicated demands from populations that were biologically pretty much the same as their distant ancestors but culturally enlarged, both equipped and encumbered.

There is artifice in the design and execution of religious practices, as anyone knows who has ever suffered through an ineptly conducted religious ceremony. A stammering and prosaic minister and boring liturgy, shaky singing from the choir, people forgetting when to stand and what to say and do—such a flawed performance can drive away even the best-intentioned congregants. More artfully celebrated occasions can raise the congregation to sublime ecstasy. We can analyze the artifice in religious texts and ceremonies just as we can analyze the artifice in literature, music, dance, architecture, and other arts. A good professor of music theory can take apart a Mozart symphony or a Bach cantata and show you how the various design features work to achieve their "magic," but some people prefer not to delve into these matters, for the same reason that they don't want stage magic tricks explained: for them, explanation diminishes the "wonder." Maybe so, but compare the uncomprehending awe with which the musically uneducated confront a symphony to the equally superficial appreciation of someone at a soccer match who doesn't know the rules or the fine points of the game, and just sees lots of kicking the ball back and forth and vigorous running around. "Great action!" they may sincerely exclaim, but they're missing most of the excellence on offer. Mozart and Bach—and Manchester United—deserve better. The designs and techniques of religion can also be studied with the same detached curiosity, with valuable results.

Consider adopting the same inquisitive attitude to religion, especially to your own religion. It is a finely tuned amalgam of brilliant plays and stratagems, capable of holding people enthralled and loyal for their entire lives, lifting them out of their selfishness and mundane ways in much the way music often does, but even more so. Understanding how it works is as much a preamble to better appreciating it or making it work better as it is to trying to dismantle it. And the analysis I am urging is, after all, just the continuation of the reflective process that has brought religion to the state it is now in. Every minister in every faith is like a jazz musician, keeping traditions alive by playing the beloved standards the way they are supposed to be played, but also incessantly gauging and deciding, slowing the pace or speeding up, deleting or adding another phrase to a prayer, mixing familiarity and novelty in just the right proportions to grab the minds and hearts of the listeners in attendance. The best performances are not just *like* good music; they are a kind of music. Listen to the recorded sermons of the Reverend C. L. Franklin (Aretha Franklin's father, and famous among gospel preachers before she recorded any hits), or the white Baptist preacher Brother John Sherfey, for example.[1]

Such performer-composers are not just vocalists; their instrument is the congregation, and they play it with the passionate but knowledgeable artistry of a violinist entrusted with a Stradivarius. In addition to the immediate effects today—a smile or "Amen!" or "Hallelujah!"—and short-term effects—returning to church next Sunday, putting another dollar in the collection plate—there are long-term effects. By choosing which passages of Scripture will be replicated this week, the minister shapes not just the order of worship but the minds of the worshipers. Unless you are a remarkable and rare scholar, you carry around in your personal memory only a fraction of the holy texts of your faith—those that you have heard over and over again since your childhood, sometimes intoning them in unison with the congregation, whether or not you have deliberately committed any of them to memory. Just as the Latin minds of ancient Rome gave way to French and Italian and Spanish minds, Christian minds today are quite unlike the minds of the earliest Christians. The major religions of today are as different from their ancestral versions as today's music is different from the music of ancient Greece and Rome. The changes that have been established are far from random. They have tracked the restless curiosity and changing needs of our encultured species.

The human capacity for reflection yields an ability to notice and evaluate patterns in our own behavior ("Why do I keep falling ̂ *that?*"; "It seemed like a good idea at the time, but wh ̀ enhances our ability to represent future prospec ties, which in turn threatens the stability of any il practices that cannot survive such skeptical attentio start "catching on," a system that has "worked" for generations can implode overnight. Traditions can erode more swiftly than stone walls and slate roofs, and preventive maintenance of an institution's creeds and practices can become a full-time occupation for professionals. But not all institutions get, or require, such maintenance.

2 Folk religion as practical know-how

Among the Nuer it is particularly auspicious to sacrifice a bull, but since bulls are particularly valuable, a cucumber will do just fine most of the time. —E. Thomas Lawson and Robert N. McCauley, *Bringing Ritual to Mind*

In the face of inevitable wear and tear, no designed thing persists for long without renewal and replication. The institutions and habits of human culture are just as bound by this principle, the second law of thermodynamics, as are the organisms, organs, and instincts of biology. But not all culturally transmitted practices need *stewardship*. Languages, for instance, don't require the services of usage police and grammarians—though in European languages they have long had a surfeit of these self-appointed protectors of integrity. One of the main claims of the previous chapter is that folk religions are like languages in this regard: they can pretty much take care of themselves. The rituals that persist are those that are *self*-perpetuating, whether or not anybody devotes serious effort to the goal of maintaining them. Memes could acquire new tricks—adaptations—that could help them secure this longevity of their lineages whether or not anybody appreciated them. Thus the question of whether folk religions have ever provided a clear benefit to people—whether the memes that compose them are mutualist memes, not commensals or parasites—could be left unanswered for the time being. The benefits of folk religion may seem obvious—as obvious as the benefits of language—but we need to remind ourlves that a benefit to *human genetic fitness* is not the same thing as ʾnefit to *human happiness* or *human welfare*. What makes us happy may not make us more prolific, which is all that matters to genes.

Even language should be viewed with as much neutrality as we can muster. Perhaps language is just a bad habit that happened to spread! How on earth could that be? Like this: Once language began to be the fad among our ancestors, those who didn't swiftly catch on to language were pretty much left out of the mating game. Chat or go childless. (This would be the *sexual-selection* theory of language: glibness as the peacock's tail for *Homo sapiens*. According to this theory, it might be true that if none of us had ever had language we'd all have done better in the offspring department, but once the costly handicap of language caught on among the females, males without it tended to die without offspring, so they couldn't afford not to make the investment, however difficult it made their lives.) Unlike tail feathers, which you have to grow with whatever equipment your parents endowed you with, languages spread horizontally or culturally, so we need to consider them as interactors in the drama as well, with their own prospects for reproduction. On this theory, the reason we love speaking is like the reason that mice infected with *Toxoplasma gondii* love to taunt cats—languages have enslaved our poor brains and made us eager accomplices in their own propagation!

That's a far-fetched hypothesis, since language's contributions to *genetic fitness* are all too obvious. There are now over six billion of us crowding up the planet and monopolizing its resources, while our nearest kin, the languageless bonobos, chimpanzees, orangutans, and gorillas, are all threatened with extinction. Setting aside the hypotheses that our running ability or hairlessness is the secret of our success, we can be quite confident that the memes of language have been fitness-enhancing mutualists, not parasites. Nevertheless, framing the hypothesis reminds us that genetic evolution doesn't foster happiness or well-being directly; it cares only about the number of our offspring that survive to make grand-offspring and so on. Folk religion may well have played an important role in the propagation of *Homo sapiens,* but we don't know that yet. The fact that, so far as we know, all human populations have had some version of it doesn't establish that. All known human populations have also had the common cold, which—so far as we know—is no mutualist.

How long could folk religion be carried along by our ancestors before reflection began to transform it? We may get some perspective on this by looking at other species. It is obvious that birds don't need to understand the principles of aerodynamics that dictate the shapes of their wings. It is less obvious—but still true—that birds can be uncomprehending participants in such elaborate rituals as *leks*—the mating meeting places sometimes called "nature's nightclubs"—where females of a local population of a species gather to observe the competitive performances by the males, who strut their stuff. The rationale for leks, which are also found in some mammals, fish, and even insects, is clear: leks *pay for themselves* as efficient methods of mate selection under specifiable conditions. But the animals that participate in leks don't need to have any understanding of why they do what they do. The males show up and show off, and the females pay attention and let their choices be guided by the "dictates of their hearts," which, unbeknownst to them, have been shaped by natural selection over many generations.[2]

Could our proclivity for participating in religious rituals have a similar explanation? The fact that our rituals are passed on through culture, not genes, doesn't rule out this prospect at all. We know that specific languages are passed on through culture, not genes, but there has also been genetic evolution that has tuned our brains for ever more adept acquisition and use of language.[3] Our brains have evolved to become more effective word processors, and they may also have evolved to become more effective implementers of the culturally transmitted habits of folk religions. We have already seen how hypnotizability could be the talent for which the *whatsis* center imagined in chapter 3 has been shaped. Sensitivity to ritual (and music) could be part of that package.

There is really no reason to suppose that animals have a clue about why they do what they instinctually do, and human beings are no exception; the deeper purposes of our "instincts" are seldom transparent to us. The difference between us and other species is that we are the only species that cares about this ignorance! Unlike other species, we feel a general need to understand, so even though nobody *had* to understand or intend any of the design innovations that created folk religions, we should recognize that people, being naturally curious and reflective, and having language in which to frame and reframe their wonders, would have been likely—unlike the birds—to *ask themselves* what these rituals were all about. Not everybody. The itch of curiosity is not strong in some people, apparently. Judging by the variation observable around us today, it is a fair bet that only a small minority of our ancestors ever had the time or inclination to question the activities they found themselves engaging in with their kinfolk and their neighbors.

Our hunter-gatherer ancestors in Paleolithic times may well have lived a relatively easy life, with abundant food and leisure time (Sahlins, 1972), compared with the hard work that was required to scratch out a living once agriculture was invented, more than ten thousand years ago, and populations grew explosively. From the beginning of this, the Neolithic period, until very recently indeed on the biological timescale—the last two hundred generations—life for just about all our ancestors was, as Hobbes famously said, nasty, brutish, and short, with few brief pockets of spare time in which to get . . . *theoretical.* So it is probably safe to imagine that pragmatism compressed their horizons. Among the gems of folk wisdom found around the world is the idea that a *little* knowledge can be a dangerous thing. A corollary not often noted is that sometimes it might therefore be safer to substitute a potent myth for incomplete knowledge. As the anthropologist Roy Rappaport put it in his last book:

. . . in a world where the processes governing its physical elements are in some degree unknown and in even larger degree unpredictable, empirical knowledge of such processes cannot replace respect for their more or less mysterious integrity, and it may be more adaptive—that is, adaptively true—to drape such processes in supernatural veils than to expose them to the misunderstandings that may be encouraged by empirically accurate but incomplete naturalistic understanding. [1999, p. 452]

The practical demands of coming up with a way of putting together all the puzzling bits and pieces of life on the fly are not the same as the practical demands of science, and as Dunbar (2004, p. 171) observes, "The law of diminishing returns means that there will always be a point after which it is just not worth investing more time and effort into figuring out the underlying reality. In traditional societies, anything that does the trick will do."

So we can expect that our ancestors, no matter how curious they were by temperament, did more or less what we all still do today: rely on "what everyone knows." *Most* of what you (think you) know you just accept on faith. By this I do *not* mean the faith of religious belief, but something much simpler: the practical, always revisable policy of simply trusting the first thing that comes to your mind without obsessing over why it does so. What are the odds that "everybody" is *just wrong* to think that yawning is harmless or that you should wash your hands after going to the bathroom? (Remember those "good healthy tans" we used to covet?) Unless somebody publishes a study that surprises us all, we take for granted that the common lore we get from our elders and others is correct. And we are wise to do so; we need huge amounts of common knowledge to guide our way through life, and there is no time to sort through all of it, testing every item for soundness.4 And so, in a tribal society in which "everyone knows" that you need to sacrifice a goat in order to have a healthy baby, you make sure that you sacrifice a goat. Better safe than sorry.

This feature marks a profound difference between folk religion and organized religion: those who practice a folk religion *don't think of themselves as practicing a religion at all.* Their "religious" practices are a seamless part of their practical lives, alongside their hunting and gathering or tilling and harvesting. And one way to tell that they really believe in the deities to which they make their sacrifices is that they aren't forever talking about how much they believe in their deities—any more than you and I go around assuring each other that we believe in germs and atoms. Where there is no ambient doubt to speak of, there is no need to speak of faith.

Most of us know of atoms and germs only by hearsay, and would be embarrassingly unable to give a good answer if a Martian anthropologist asked us how we knew that there are such things—since you can't see them or hear them or taste them or feel them. If pressed, most of us would probably concoct some seriously mistaken lore about these invisible (but important!) things. We're not the experts—we just go along with "what everybody knows," which is just what the tribal people do. It happens that their experts have got it wrong.[5] Many anthropologists have observed that when they ask their native informants about "theological" details—their gods' whereabouts, specific history, and methods of acting in the world—their informants find the whole inquiry puzzling. Why should they be expected to know or care anything about *that*? Given this widely reported reaction, we should not dismiss the corrosive hypothesis that many of the truly exotic and arguably incoherent doctrines that have been unearthed by anthropologists over the years are artifacts of inquiry, not pre-existing creeds. It is possible that persistent questioning by anthropologists has composed a sort of innocently collaborative fiction, newly minted and crystallized dogmas generated when questioner and informant talk past each other until a mutually agreed-upon story results. The informants deeply believe in their gods—"Everybody knows they exist!"—but they may never before have thought about these details (maybe nobody in the culture has!), which would explain why their convictions are vague and indeterminate. Obliged to elaborate, they elaborate, taking their cues from the questions posed.[6]

In the next chapter, we will look at some striking implications of these methodological issues, once we have sketched more of an account to serve as our test bed. For the moment, it may help if you try to put yourself in the shoes of an anthropologist's informant. Now that the modern world with its particular complexities is descending on tribal people, they have to make wholesale revisions in their views of nature, and, not surprisingly, this prospect is daunting to them. I daresay if Martians arrived with marvelous technology that struck us as "impossible" and told us that we had to abandon our germs and atoms and get with their program, only the most nimble-minded of our scientists would make the transition swiftly and gladly. The rest of us would cling to our dear old atoms and germs as long as we could, matter-of-factly telling our children about how water is made of hydrogen and oxygen atoms—at least that's what we've always been told—and warning them about germs, just to stay on the safe side. What looms large in every person's life is the problem of *what to do now,* and there are few discomforts more stressful than the quandary of not knowing what to do, or what to think about, when baffling novelty strikes. At times like that, we all seek refuge in the familiar. The tried-and-true may not be true, but at least it is tried, so it gives us something to do that we know how to do. And usually it will work pretty well, about as well as it ever did in any case.

3 Creeping reflection and the birth of secrecy in religion

You can fool all the people some of the time, and some of the people all the time, but you cannot fool all the people all the time.
—Abraham Lincoln

Those to whom his word was revealed were always alone in some remote place, like Moses. There wasn't anyone else around when Mohammed got the word, either. Mormon Joseph Smith and Christian Scientist, Mary Baker Eddy, had exclusive audiences with God. We have to trust them as reporters—and you know how reporters are. They'll do anything for a story. —Andy Rooney, *Sincerely, Andy Rooney*

Everyday folk physics and folk biology and folk psychology work very well as a rule, and so does folk religion, but occasional doubts surface. The exploratory reflections of human beings have a way of snowballing into waves of doubt, and if these threaten our equanimity, we can be expected to seize upon any responses that happen to shore up the consensus or damp the challenge. When curiosity stubs its toe on an unexpected event, something has to give: "what everybody knows" has a counterexample, and either the doubt blossoms into a discovery, which leads to the abandonment or extinction of a dubious bit of local lore, or the dubious item secures itself with an ad hoc repair of one sort or another, or it allies itself with other items that have in one way or another put themselves out of the reach of gnawing skepticism.[7]

This winnowing has the effect of sequestering a special subset of cultural items behind the veil of *systematic* invulnerability to disproof—a pattern found just about everywhere in human societies. As many have urged (see, e.g., Rappaport, 1979; Palmer and Steadman, 2004), this division into the propositions that are *designed* to be immune to disconfirmation and all the rest looks like a hypothetical joint at which we could well carve nature. Right here, they suggest, is where (proto-)science and (proto-)religion part company. Not that the two types of lore aren't often thoroughly mixed together in many cultures. Detailed natural history of the local region, with the habits and properties of all the different species acutely observed, is typically intermingled with myths and rituals involving these species—which deities inform which birds, which sacrifices need to be offered before hunting which prey, and so forth. The dividing line may, moreover, be blurred in practice, with one father telling his son how the starling gives an alarm call to its kin that is overheard by the wild boar whereas another father tells his son that he doesn't know how the boar learns from the starling—perhaps a god carries the message—and this son may tell his own son a story about a god who protects starlings and boars but not antelopes.

Would-be scientists know temptation: whenever your favorite theory yields a prediction that turns out wrong, why not let your hypothesis metamorphose a little into one that is conveniently untestable under just those conditions? Scientists are supposed to be leery of these migrations away from refutation, but it's a hard lesson to learn. Sticking to your hypothesis and letting the facts decide is an unnatural act, and you have to brace yourself to perform it. Shamans have a different agenda: they're trying to heal and advise people in real time, and can gratefully hide behind mystery when the unexpected happens. (A cartoon shows a witch doctor standing dejectedly over the body of his late patient and saying to the grieving widow, "There is so much that we still don't know!")

The postulation of invisible, undetectable effects that (unlike atoms and germs) are *systematically* immune to confirmation or disconfirmation is so common in religions that such effects are sometimes taken as definitive. No religion lacks them, and anything that lacks them is not really a religion, however much it is like a religion in other regards. For instance, elaborate sacrifices to gods are everywhere to be found, and of course nowhere do the gods emerge from invisibility and sit down to eat the beautiful roast pork or drink the wine. Rather, the wine is poured into the ground or onto the fire, where the gods may enjoy it in unobservable privacy, and the partaking of the food is accomplished by either burning it to ashes or delegating it to the shamans, who get to eat it as part of their official duties as representatives of the gods. As Dana Carvey's Church Lady would exclaim, "How convenient!" As usual, we don't *have* to implicate the shamans, individually or even as a diffuse group of conspirators, in the devising of this rationale, since it could just emerge by the differential replication of rites, but the shamans would have to be pretty dense not to appreciate this adaptation, and even appreciate the need for deflecting attention from it. In some cultures, a more egalitarian convenience has emerged: *everybody* gets to eat the food that has somehow also been invisibly and nondestructively eaten by the gods. The gods can have their cake and we can eat it too. Isn't the transparency of these all-too-convenient arrangements risky? Yes, so it is almost always protected by a second veil: *These are mysteries beyond all comprehension! Don't even try to understand them!* And as often as not, a third veil is provided: *it is forbidden* to ask too many questions about all these mysteries!

What about the shamans themselves? Is their own inquisitiveness blunted by these taboos? Not always, obviously. Like every conscientious worker, shamans can be expected to notice or suspect shortcomings in their own performance and then experiment with alternative methods: "I'm losing customers to that other shaman; what is he doing that I'm not doing? Is there a better way to do the healing rituals?" A familiar folk idea about hypnosis is that the hypnotist somehow *disables* the subject's sentries, the skeptical defense mechanisms, whatever they are, that inspect all incoming material for credibility. (Perhaps he puts the guards to sleep!) A better idea is that the hypnotist doesn't disable the sentries but, rather, *co-opts* them, turning them into allies, getting them to vouch for the hypnotist, in effect. One way to do that is to throw them some little facts ("You are getting sleepy, your eyelids feel heavy . . .") that they can check for accuracy and readily confirm. If it isn't obvious to the subject that the hypnotist would know these facts, this creates a mild illusion of unexpected authority ("How did he know *that*?"), and then the hypnotist, armed with the blessing of the sentries, can go to town.

This bit of more or less secret folk wisdom gets some support from experiments: the success a hypnotist has on a subject is significantly affected by whether the subject is told in advance that the hypnotist is a novice or an expert (Small and Kramer, 1969; Coe et al., 1970; Balaschak et al., 1972), and this tactic has been discovered and exploited again and again by shamans. Everywhere, they are assiduous, discreet gatherers of little-known facts about the individuals who may become their clients, but they don't stop there. There are other ways of demonstrating unexpected mastery. As McClenon (2002) notes, the ritual of walking unscathed on a bed of hot coals has been observed around the world—in India, China, Japan, Singapore, Polynesia, Sri Lanka, Greece, and Bulgaria, for instance. Two other widespread practices by shamans are sleight-of-hand moves such as the concealment of animal entrails that can then be miraculously "removed" from the afflicted person's torso in "psychic surgery," and the trick of being bound hand and foot and then somehow causing the tent to shake noisily. In the huge Design Space of possibilities, these three seem to be the most accessible ways of creating astonishing "supernatural" effects to impress one's clients, since they have been rediscovered again and again. "The close equivalences among cultures seem more than coincidental: shamans may use similar forms of conjuring without any formal training and without having had contact with others who use the same strategies," McClenon asserts, so any " 'diffusion explanation' seems implausible" (p. 149).

One of the most interesting facts about these unmistakable acts of deceit is that the practitioners, when pressed by inquiring anthropologists, exhibit a range of responses. Sometimes we get a candid admission that they are knowingly using the tricks of stage magic to gull their clients, and sometimes they defend this as the sort of "sacred dishonesty" (for the cause) of which the theologian Paul Tillich speaks (see appendix B). And sometimes, more interestingly, a sort of holy fog of incomprehension and mystery swiftly descends on the responder to protect him or her from any further corrosive inquiries. These shamans are not *quite* con men—not all of them, at any rate—and yet they know that the effects they achieve are trade secrets that must not be revealed to the uninitiated for fear of diminishing their effects. Every good doctor knows that a few simple tricks of self-presentation that compose a good "bedside manner" can make a huge difference.[8] It isn't really dishonest, is it? Every priest and minister, every imam and rabbi, every guru knows the same thing, and the same gradation from knowingness to innocence can be found today in the practices of revival preachers, as vividly revealed in *Marjoe*, the Oscar-winning 1972 documentary film that followed Marjoe Gortner, a charismatic young evangelical preacher who lost his faith but made a comeback as a preacher in order to reveal the tricks of the trade. In this disturbing and unforgettable film, he shows how he makes people faint when he does the laying on of hands, how he rouses them to passionate declarations of their love for Jesus, how he gets them to empty their wallets into the collection basket.[9]

4 The domestication of religions

When a race of plants is once pretty well established, the seed-raisers do not pick out the best plants, but merely go over their seed-beds, and pull up the "rogues," as they call the plants that deviate from the proper standard. —Charles Darwin, *On the Origin of Species*

We now begin to see that what we call Christianity—and what we identify as Christian tradition—actually represents only a small selection of specific sources, chosen from among dozens of others. Who made that selection, and for what reasons? Why were these other writings excluded and banned as "heresy"? What made them so dangerous?
—Elaine Pagels, *The Gnostic Gospels*

Folk religions emerge out of the daily lives of people living in small groups, and share common features the world over. How and when did these metamorphose into organized religions? There is a general consensus among researchers that the big shift responsible was the emergence of agriculture and the larger settlements that made both possible and necessary. Researchers disagree. [1]

on what to emphasize in this major transition. The creation of non-portable food stockpiles, and the resultant shift to fixed residence, permitted the emergence of an unprecedented division of labor (Seabright, 2004, is especially clear about this), and this in turn gave rise to *markets,* and opportunities for ever more specialized occupations. These new ways for people to interact created novel opportunities and novel needs. When you find that you have to deal on a daily basis with people *who are not your close kin,* the prospect of a few like-minded people forming a coalition that is quite different from an extended family must almost always present itself, and often be an attractive option. Boyer (2001) is not alone in arguing that the transition from folk religion to organized religion was primarily one of these market phenomena.

> Throughout history, guilds and other groups of craftsmen and specialists have tried to establish common prices and common standards and to stop non-guild members from delivering comparable services. By establishing a quasi monopoly, they make sure that all the custom comes their way. By maintaining common prices and common standards, they make it difficult for a particularly skilled or efficient member to undersell the others. So most people pay a small price for being members of a group that guarantees a minimal share of the market to each of its members. [p. 275]

The first step to such organization is the big one, but the next steps, from a guild of priests or shamans to what are, in effect, *firms* (and *franchises* and *brand names*), are an almost inevitable consequence of the growing self-consciousness and market savvy of those individuals who joined to form the guilds in the first place. *Cui bono?* When individuals start asking themselves how best to enhance and preserve the organizations they have created, they radically change the focus of the question, bringing new selective pressures into existence.

Darwin appreciated this, and used the transition from what he

called "unconscious" selection to "methodical" selection as a pedagogical bridge to explain his great idea of natural selection in the
opening chapter of his masterpiece. (*On the Origin of Species* is a
great read, by the way. Just as atheists often read "the Bible as literature" and come away deeply moved by the poetry and insight without being converted, creationists and others who cannot bring
themselves to believe in evolution can still be thrilled by reading
the founding document of modern evolutionary theory—whether
or not it changes their minds about evolution.)

> At the present time, eminent breeders try by methodical selec-
> tion, with a distinct object in view, to make a new strain or sub-
> breed, superior to any existing in the country. But for our
> purpose, a kind of Selection, which may be called Unconscious,
> and which results from every one trying to possess and breed
> from the best individual animals, is more important. Thus a man
> who intends keeping pointers naturally tries to get as good dogs
> as he can, and afterwards breeds from his own best dogs, but he
> has no wish or expectation of permanently altering the breed.
> Nevertheless I cannot doubt that this process, continued during
> centuries, would improve and modify any breed.... There is
> reason to believe that King Charles's spaniel has been uncon-
> sciously modified to a large extent since the time of that monarch.
> [pp. 34–35]

Domestication of both plants and animals occurred without any
farseeing intention or invention on the part of the stewards of the
seeds and studs. But what a stroke of good fortune for those lineages that became domesticated! All that remains of the ancestors
of today's grains are small scattered patches of wild-grass cousins,
and the nearest surviving relatives of all the domesticated animals
could be carried off in a few arks. How clever of wild sheep to have
acquired that most versatile adaptation, the shepherd! By forming a
symbiotic alliance with *Homo sapiens,* sheep could *outsource* their
chief survival tasks: food finding and predator avoidance. They

even got shelter and emergency medical care thrown in as a bonus. The price they paid—losing the freedom of mate selection and being slaughtered instead of being killed by predators (if that is a cost)—was a pittance compared with the gain in offspring survival it purchased. But of course it wasn't *their* cleverness that explains the good bargain. It was the blind, foresightless cleverness of Mother Nature, evolution, which ratified the free-floating rationale of this arrangement. Sheep and other domesticated animals are, in fact, significantly more stupid than their wild relatives—because they can be. Their brains are smaller (relative to body size and weight), and this is not just due to their having been bred for muscle mass (meat). Since both the domesticated animals and their domesticators have enjoyed huge population explosions (going from less than 1 percent of the terrestrial vertebrate biomass ten thousand years ago to over 98 percent today—see appendix B), there can be no doubt that this symbiosis was mutualistic—fitness-enhancing to both parties.

What I now want to suggest is that, alongside the domestication of animals and plants, there was a gradual process in which the wild (self-sustaining) memes of folk religion became thoroughly domesticated. They acquired stewards. Memes that are fortunate enough to have stewards, people who will work hard and use their intelligence to foster their propagation and protect them from their enemies, are relieved of much of the burden of keeping their own lineages going. In extreme cases, they no longer need to be particularly catchy, or appeal to our sensual instincts at all. The multiplication-table memes, for instance, to say nothing of the calculus memes, are hardly crowd-pleasers, and yet they are duly propagated by hardworking teachers—meme shepherds—whose responsibility it is to keep these lineages strong. The wild memes of language and folk religion, in other words, are like rats and squirrels, pigeons and cold viruses—magnificently adapted to living with us and exploiting us whether we like them or not. The domesticated memes, in contrast, depend on help from human guardians to keep going.

People have been poring over their religious practices and institutions for almost as long as they have been refining their agricultural practices and institutions, and these reflective examiners have all had agendas—individual or shared *conceptions* of what was valuable and why. Some have been wise and some foolish, some widely informed and some naïve, some pure and saintly, and some venal and vicious. Jared Diamond's hypothesis about the practically exhaustive search by our ancestors for domesticatable species in their neighborhoods (discussed in chapter 5) can be extended. Curious practitioners will also have uncovered whatever Good Tricks are in the nearest neighborhoods in the Design Space of possible religions. Diamond sees the transition from bands of fewer than a hundred people to tribes of hundreds to chiefdoms of thousands to states of over fifty thousand people as an inexorable march "from egalitarianism to kleptocracy," government by thieves. Speaking of chiefdoms, he remarks:

> At best, they do good by providing expensive services impossible to contract for on an individual basis. At worst, they function unabashedly as kleptocracies, transferring net wealth from commoners to upper classes. . . . Why do the commoners tolerate the transfer of the fruits of their hard labor to kleptocrats? This question, raised by political theorists from Plato to Marx, is raised anew by voters in every modern election. [1997, p. 276]

There are four ways, he suggests, that kleptocrats have tried to maintain their power: (1) disarm the populace and arm the elite, (2) make the masses happy by redistributing much of the tribute received, (3) use the monopoly of force to promote happiness, by maintaining public order and curbing violence, or (4) construct an ideology or religion justifying kleptocracy (p. 277).

How might a religion support a kleptocracy? By an alliance

between the political leader and the priests, of course, in which, first of all, the leader is declared to be divine, or descended from the gods, or, as Diamond puts it, at least having "a hotline to the gods."

> Besides justifying the transfer of wealth to kleptocrats, institutionalized religion brings two other important benefits to centralized societies. First, shared ideology or religion helps solve the problem of how unrelated individuals are to live together without killing each other—by providing them with a bond not based on kinship. Second, it gives people a motive, other than genetic self-interest, for sacrificing their lives on behalf of others. At the cost of a few society members who die in battle as soldiers, the whole society becomes much more effective at conquering other societies or resisting attacks. [p. 278]

So we find the same devices invented over and over again, in just about every religion, and many nonreligious organizations as well. None of this is new today—as Lord Acton said more than a century ago, "All power tends to corrupt; absolute power corrupts absolutely"—but it was new once upon a time, when our ancestors were first exploring design revisions to our most potent institutions.

For instance, *accepting inferior status to an invisible god* is a cunning stratagem, whether or not its cunning is consciously recognized by those who stumble upon it. Those who rely on it will thrive, wittingly or otherwise. As every subordinate knows, one's commands are more effective than they might otherwise be if one can accompany them with a threat to tell the bigger boss if disobedience ensues. (Variations on this stratagem are well known to Mafia underlings and used-car salesmen, among others—"I myself am not authorized to make such an offer, so I'll have to check with my boss. Excuse me for a minute.")

This helps to explain what is otherwise a bit of a puzzle. Any dictator depends on the fidelity of his immediate staff—in the simple sense that any two or three of them could easily overpower him (he can't go around with dagger drawn all his life). How do you, as a dictator, ensure that your immediate staff puts its fidelity to you above any thoughts they may very well have about replacing you? Putting the fear of a higher power in their heads is a pretty good move. There is often, no doubt, an unspoken détente between chief priest and king—each needs the other for his power, and together they need the gods above. Walter Burkert is particularly Machiavellian in his account of how this stratagem brings the institution of ritual praise in its wake, and notes some of its useful complexity:

> By the force of his verbal competence [the priest] not only rises to a superior level in imagination but succeeds in reversing the attention structure: it is the superior who is made to pay heed to the inferior's song or speech of praise. Praise is the recognized form of making noise in the presence of superiors; in a well-structured form, it tends to become music. Praise ascends to the heights like incense. Thus the tension between high and low is both stressed and relaxed, as the lower one establishes his place within a system he accepts emphatically. [1996, p. 91]

The gods will get you if you try to cross either one of us. We have already noted the role of rituals, both individual rehearsals and unison error-absorption sessions, in enhancing the fidelity of memetic transmission, and noted that these are enforced by making nonparticipation costly in one way or another. Moreover, as Joseph Bulbulia suggests, "It may be that religious rituals put on display the natural power of a religious community, an awesome show to potential defectors of what they are up against" (2004, p. 40). But what drives the community spirit in the first place? Is the project of keeping groups united mainly just a matter of kleptocrats' inventing ways of keeping their sheep? Or is there a more benign story to uncover?

\\\

Chapter 6 The transmission of religion has been attended by voluminous revision, often deliberate and foresighted, as people became stewards of the ideas that had entered them, domesticating them. Secrecy, deception, and systematic invulnerability to disconfirmation are some of the features that have emerged, and these have been designed by processes that were sensitive to new answers to the *cui bono?* question, as the stewards' motives entered the process.

\\\

Chapter 7 Why do people join groups? Is this simply a rational decision on their part, or are there relatively mindless forces of *group selection* at work? Though there is much to be said in favor of both of these proposals, they do not exhaust the plausible models that attempt to explain our readiness to form lasting allegiances.

CHAPTER SEVEN

||| \\\ |||

The Invention of Team Spirit

I A path paved with good intentions

And here comes the catch. Only a bad person needs to repent: only a good person can repent perfectly. The worse you are the more you need it and the less you can do it. The only person who could do it perfectly would be a perfect person—and he would not need it.

—C. S. Lewis, *Mere Christianity*

Every control system, whether it is an animal nervous system, a plant's system of growth and self-repair, or an engineered artifact such as an airplane-guidance system, is designed to protect *something*. And that something must include itself! (If it "dies" prematurely, it fails on its mission, whatever it is.) The "self-interest" that thus defines the evaluation machinery of all control systems can splinter, however, when a control system gets reflective. Our human reflectiveness opens up a rich field of opportunities for us to revise our aims, including our largest purposes. When you can start to *think about* the pros and cons of joining an existing coalition versus breaking away and trying to start a new one, or about how to deal with the problems of loyalty among your kin, or the need to change the power structure of your social environment, you create avenues by which to escape the default presumptions of your initial design.

Whenever an agent—an intentional system, in my terminology—makes a decision about the best course of action, all things considered, we can ask from whose perspective this optimality is being judged. A more or less standard default assumption, at least in the Western world, and especially among economists, is to treat each human agent as a sort of isolated and individualistic locus of well-being. What's in it for *me*? Rational *self*-interest. But although there has to be something in the role of the self—something that answers the *cui bono*? question for the decision-maker under examination—there is no necessity in this default treatment, common as it is. A self-as-ultimate-beneficiary can in principle be indefinitely distributed in space and time. I can care for others, or for a larger social structure, for instance. There is nothing that restricts me to a *me* as contrasted to an *us*.[1] I can still take my task to be looking out for Number One while including, under Number One, not just myself, and not just my family, but also Islam, or Oxfam, or the Chicago Bulls! The possibility, opened up by cultural evolution, of installing such novel perspectives in our brains is what gives our species, and only our species, the capacity for moral—and immoral—thinking.

Here is a well-known trajectory: You begin with a heartfelt desire to help other people and the conviction, however well or ill founded, that your guild or club or church is the coalition that can best serve to improve the welfare of others. If times are particularly tough, this conditional stewardship—I'm doing what's good for the guild because that will be good for everybody—may be displaced by the narrower concern for the integrity of the guild itself, and for good reason: if you believe that the institution in question is the best path to goodness, the goal of preserving it for future projects, still unimagined, can be the most rational higher goal you can define. It is a short step from this to losing track of or even forgetting the larger purpose and devoting yourself singlemindedly to furthering the interests of the institution, at whatever costs. A conditional or instrumental allegiance can thus become indistinguishable in practice from a commitment to something "good in itself." A further short step perverts this parochial *summum bonum* to the more selfish goal of doing whatever it takes to keep yourself at the helm of the institution ("Who better than I to lead us to triumph over our adversaries?").

We have all seen this happen many times, and may even have caught ourselves in the act of forgetting just why we wanted to be leaders in the first place. Such transitions bring conscious decision-making to bear on issues that had previously been tracked by the foresightless process of differential replication by natural selection (of memes, or of genes), and this creates new rivals as answers to the *cui bono?* question. What is good all things considered may not coincide with what is good for the institution, which may not be what makes life easiest for the institution's leader, but these different benchmarks have a way of being substituted for one another under the pressure of real-time reflective control. When this happens, the free-floating rationales that are blindly sculpted by earlier competitions can come to be augmented or even replaced by *represented* rationales, rationales that are not just anchored in individual minds, in diagrams and plans, and in conversations but *used*— argued over, reasoned about, agreed upon. People thus become conscious stewards of their memes, no longer taking their survival for granted the way we take our language for granted, but *taking on* the goal of fostering, protecting, enhancing, spreading the Word.[2]

Why do people want to be stewards of their religions? It is obvious, isn't it? They believe that this is the way to lead a moral life, a good life, and they sincerely want to be good. Are they right? Notice that this is *not* the question of whether religions have enhanced human biological fitness. Biological fitness and moral value are entirely different issues. I have postponed the fitness question until we could see that, although it is a good, empirical question, a question that we ought to try to answer, answering it will still leave wide open the question about whether we *ought* to be stewards of religion.

With that point firmly established, let us at last *consider*—not answer—the question of whether, in the end, folk religions, and the organized religions they have morphed into, have conferred fitness benefits on those who practice them. This question has preoccupied anthropologists and other researchers for centuries, often because they confused it with the question of the *ultimate* (moral) value of religion, and there is no dearth of familiar hypotheses to explore once we've cleared the decks. Two of the most plausible will receive further attention in later chapters, so for now I will just acknowledge them. Dunbar (2004) summarizes one of them well:

> It is surely no accident that almost every religion promises its adherents that they—and they alone—are the "chosen of god", guaranteed salvation no matter what, assured that the almighty (or whatever form the gods take) will assist them through their current difficulties if the right rituals and prayers are performed. This undoubtedly introduces a profound sense of comfort in times of adversity. [p. 191]

Notice that comfort, in and of itself, would not be a fitness booster unless it also provided (as it almost certainly does) the practical advantages of resolution and confidence, in both decision-making and action. *May the Force be with you!* When you are faced with the often terrifying uncertainty of a dangerous world, the belief that *somebody is watching over you* may well be a decisively effective morale booster, capable of turning people who would otherwise be disabled by fear and indecision into stalwart agents. This is a hypothesis about *individual* effectiveness in times of strife, and it may—or may not—be true.

An entirely distinct hypothesis is that participation in religion (in harrowing initiation rites, for instance) creates or strengthens bonds of trust that permit *groups* of individuals to act together much more effectively. Versions of this group-fitness hypothesis have been advanced by Boyer, Burkert, Wilson, and many others. It may or may not be true—indeed, both hypotheses could be true, and we should try to confirm or disconfirm them both if only for the light they will shed—no more—on the question of the moral value of religion.

2 The ant colony and the corporation

Religions exist primarily for people to achieve together what they cannot achieve alone. —David Sloan Wilson, *Darwin's Cathedral*

But what are the benefits; why do people want religion at all? They want it because religion is the only plausible source of certain rewards for which there is a general and inexhaustible demand.
 —Rodney Stark and Roger Finke, *Acts of Faith*

Why do people join groups? Because they want to—but why do they want to? For many reasons, including the obvious: for mutual protection and economic security, to promote efficiency of harvesting and other necessary activities, to accomplish large-scale projects that would otherwise be impossible. But the manifest utility of these group arrangements does not in itself explain how they ever came to pass, for there are barriers to overcome, in the form of mutual fear and hostility, and the always looming prospect of opportunistic defection or betrayal. Our inability to achieve truly global cooperation in spite of persuasive arguments demonstrating the benefits to be had, and in spite of many failed campaigns intended to create enabling institutions, shows that the limited cooperation and loyalty we do enjoy is a rare achievement. We have somehow managed to civilize ourselves to some degree, in ways no other species has even attempted, so far as we can tell. Other species often form populations that cluster together in herds or flocks or schools, and it is clear why these groupings, when they occur, are adaptive. But we are not grazing animals, for instance, and among the foraging (and predating) apes that are our nearest animal relatives, the largest stable groups are generally restricted to close kin, extended families into which newcomers are admitted only after a struggle and a test. (Among chimpanzees, the newcomers are always females emigrating from their home groups to find mates; any male that tried to join another group would be summarily killed.) There is no mystery about why we, like other apes, would have evolved a craving for the company of conspecifics, but that instinct for gregariousness has its limits.

It is remarkable that we have learned to be comfortable in the company of strangers, as Seabright (2004) puts it, and a perennially persuasive idea about religion is that it works to promote just such group cohesiveness, turning otherwise hapless populations of unrelated and mutually suspicious people into tightly knit families or even highly effective super-organisms, rather like ant colonies or beehives. The impressive solidarity achieved by many religious organizations is not in doubt, but could this *explain* the rise and continued existence of religions? Many have thought so, but just how could this work? Theorists of all persuasions agree that the R & D required to set up and maintain such a system has to be accomplished somehow, and there seem at first to be just two paths to choose between: the ant-colony route and the corporation route. Natural selection has shaped the design of ants over the eons, tooling the individual ant types into specialists that automatically coordinate their efforts so that a normally harmonious and vigorous colony results. There were no heroic individual ants who figured it out and implemented it. They didn't have to, since natural selection did all the trial and error for them, and there is not now and never was any individual ant—or council of ants—to play the role of governor. In contrast, it is precisely the rational choices of individual human beings that bring a corporation into existence: they design the structure, agree to incorporate, and then govern its activities. Individual rational agents, looking out for their own interests and doing their own individual cost-benefit analyses, make the decisions that shape, directly or indirectly, the features of the corporation.

Is the robustness of a religion, its ability to persevere and thrive in defiance of the second law of thermodynamics, like the robustness of an ant colony or a corporation? Is religion the product of blind evolutionary instinct or rational choice? Or is there some other possibility? (Might it be a gift from God, for instance?) The failure to ask—let alone answer—this question is the charge that has long been used to discredit the *functionalist* school of sociology initiated by Emil Durkheim. According to its critics, functionalists treated societies *as if* they were living things, maintaining their health and vigor by a host of adjustments in their organs, without showing how the R & D required to design and adjust these super-organisms was accomplished. This criticism is essentially the same criticism aimed by evolutionary biologists at the Gaia hypothesis of Lovelock (1979) and others. According to the Gaia hypothesis, Earth's biosphere is itself a sort of super-organism, maintaining its various balances in order to preserve life on Earth. A pretty idea, but, as Richard Dawkins succinctly puts it:

> For the analogy to apply strictly, there would have to have been a set of rival Gaias, presumably on different planets. Biospheres which did not develop efficient homeostatic regulations of their planetary atmospheres tended to go extinct.... In addition we would have to postulate some kind of reproduction, whereby successful planets spawned copies of their life forms on new planets. [1982, 1999, p. 236]

Gaia enthusiasts, if they want to be taken seriously, have to ask, and answer, the question of how the presumed homeostatic systems got designed and installed. Functionalists in the social sciences must assume the same burden.

Enter David Sloan Wilson (2002) and his "multi-level selection theory" to try to save the day for a brand of functionalism by grounding the design process in the same R & D algorithms that account for the rest of the biosphere. According to Wilson, the design innovations that work systematically to bind human groups together are the result of Darwinian descent with modification guided by differential replication of the most fit, at many levels, *including the group level*. In short, he accepts the challenge of showing that competition between rival groups led to the extinction of the ill-designed groups in failing competition with the better-designed groups, which were beneficiaries of free-floating rationales (to put it my way) that none of their members needed to understand. *Cui bono?* The fitness of the *group* must trump the individual fitness of its *members,* and if groups are going to be the ultimate beneficiaries, groups must be the competitors. Selection can go on at several levels at once, however, thanks to competitions at several levels.

Critics have long scoffed at the functionalists' invocation of something like mystical societal wisdom (like the imagined wisdom of Gaia), but Wilson is right to insist that there *need be* nothing mystical or even mysterious about Durkheimian group-friendly functions' getting installed by evolutionary processes—*if* he can demonstrate group-selection processes. The distributed wisdom of an ant colony, which really is a sort of super-organism, has been analyzed in depth and detail by evolutionary biologists, and there is no doubt that evolutionary processes *can* shape group adaptations under special conditions like those that prevail among the social insects. But people aren't ants, or very much like ants, and only the most regimented religious orders approach the fascistic lock step of the social insects. Human minds are hugely complex exploration devices, corrosive questioners of every detail of the world they encounter, so evolution had better add some remarkable bells and whistles to its adaptations for human groupishness if there is to be any chance of success by the group-selection route.

Wilson thinks that competition between religious groups, with differential survival and replication of some of those groups, can generate (and "pay for") the excellent design features we observe in religions. The opposite theoretical pole—the only alternative, or so it first appears—is occupied by the *rational choice theorists,* who have recently arisen to challenge the widespread presumption by social scientists that religion is some kind of lunacy. As Rodney

Stark and Roger Finke (2000) note with scorn, "For more than three centuries, the standard social scientific wisdom was that religious behavior must be irrational precisely because people do make sacrifices on behalf of their faith—since, obviously, no rational person would do such a thing" (p. 42), but as they insist:

> One need not be a religious person in order to grasp the underlying rationality of religious behavior, any more than one need be a criminal in order to impute rationality to many deviant acts (as the leading theories of crime and deviance do). . . . What we are saying is that religious behavior—to the degree that it occurs— is generally based on cost-benefit calculations and is therefore rational behavior in precisely the same sense that other human behavior is rational. [p. 36]

Religions are indeed like corporations, they claim: "Religious organizations are social enterprises whose purpose is to create, maintain, and supply religion to some set of individuals and to support and supervise their exchanges with a god or gods" (p. 103). Demand for the goods that religion has to offer is inelastic; in a free market of religious choice (as in the United States, with no state religion and many competing denominations) there is vigorous competition among denominations for market dominance—a straightforward application of "supply-side" economics. But as Wilson notes in a useful comparison between his theory and theirs, even if we were to grant that *now* it is rational for church members to make what are basically market decisions about which religion to invest in (an assumption we will soon examine), this doesn't answer the question about R & D:

> But how did the religion acquire its structure that adaptively constrains the choices of utility-maximizers in just the right way? We must explain the structure of the religion in addition to the behavior of individuals once the structure is in place. Were the bizarre customs consciously invented by rational actors attempting to

maximize their utilities? If so, why did they have the utility of maximizing the common good of their church? Must we really attribute all adaptive features of a religion to a psychological process of cost-benefit reasoning? Isn't a process of blind variation and selective retention possible? After all, thousands of religions are born and die without notice because they never attract more than a few members (Stark and Bainbridge, 1985). Perhaps the adaptive features of the few that survive are like random mutations rather than the product of rational choice. [p. 82]

Wilson is right to stress the alternative of a blind variation and selective retention process, but by clinging to his radical *group-selection* version he misses a better opportunity: the evolutionary design process that has given us religions involves the differential replication of *memes,* not *groups.*[3] Wilson briefly mentions this as an alternative, but dismisses it with hardly a glance, largely because he views its defining doctrine to be that religious features must be *dysfunctional.* He thinks the meme theory requires that all religious memes be (fitness-reducing) parasites, and seldom if ever fitness-neutral commensals or fitness-enhancing mutualists.[4] Here Wilson is led astray by a common misunderstanding: Richard Dawkins, who coined the term *meme,* is no friend of religion and has often likened memes—religious memes in particular—to viruses, stressing the capacity of memes to proliferate in spite of their deleterious effects on their human hosts. Although this jarring claim needs to be considered as a major possibility, we should not forget that the vast majority of memes, like the vast majority of bacterial and viral symbionts that inhabit our bodies, are neutral or even helpful (from the perspective of host fitness). Here, then, is my *mild memetic alternative* to Wilson's group-level hypothesis:

Memes that foster human group solidarity are particularly fit (as memes) in circumstances in which host survival (and hence host fitness) most directly depends on hosts' joining forces in groups. The success of such meme-infested groups is itself a potent broadcasting device, enhancing outgroup curiosity (and envy) and thus permitting linguistic, ethnic, and geographic boundaries to be more readily penetrated.

Like Wilson's more radical group-selection theory, this hypothesis can in principle account for the excellence of design encountered in religion *without postulating rational designers* (the religion-as-corporation route). And it can account for the fact that individual fitness is *apparently* subordinated to group fitness in religions. According to this theory, we don't need to postulate group-replication tournaments but only a cultural environment in which *ideas* compete. Ideas that encourage people to act together in groups (the way *Toxoplasma gondii* encourages rats to approach cats fearlessly) will spread more effectively as a result of this groupishness than ideas that do a less effective job of uniting their hosts into armies.[5]

Using the meme's-eye view, we can unite the two "opposite" poles of theory—ant colony versus corporation—and explain the R & D of human groupishness as a mixture of blind and foresighted processes, including intermediate selection processes of every flavor of knowingness. Since people are not like ants but really quite rational, they are unlikely to be encouraged to invest heavily in group activities unless they perceive (or think they perceive) benefits worth the investment. Hence the ideas that maximize groupishness will be those that appeal, just as Stark and Finke say, to "rewards for which there is a general and inexhaustible demand."

An unexpected bonus of this unified perspective is that it makes elbow room for an intermediate position on the status of religion that modifies one of the most troubling features of the rational choice model. Stark and Finke and the other rational choice theorists of religion like to portray themselves as defenders of those with religious faith, saying in effect: "They're not crazy, they're smart!" However, this deliberately cold-blooded rational analysis of the market for religious goods deeply offends many religious people.[6] They don't want to see themselves as cannily making a sound investment in the most effective purveyor of supernatural benefits. They want to see themselves as having set aside all such selfish considerations, as having relinquished their rational control to a higher power.

The meme theory accounts for this. According to this theory, the ultimate beneficiaries of religious adaptations are the memes themselves, but their proliferation (in competition with rival memes) depends on their ability to attract hosts *one way or another*. Once allegiance is captured, a host is turned into a rational servant, but the initial capture need not be—indeed, should not be—a rational choice by the host. Memes sometimes need to be gently inserted into their new homes, overcoming "rational" resistance by encouraging a certain passivity or receptivity in the host. William James, a memeticist ahead of his time, notes the importance of this feature for some religions, and usefully draws our attention to a secular counterpart: the music teacher who admonishes the student, "Stop trying and it will do itself!" (1902, p. 206). Just let go and clear your mind, and let that little information packet, that little habit-recipe, take over!

> One may say that the whole development of Christianity in inwardness has consisted in little more than the greater and greater emphasis attached to this crisis of self-surrender [pp. 210–11]. . . . Were we writing the story of the mind from the purely natural-history point of view, with no religious interest whatever, we should still have to write down man's liability to sudden and complete conversion as one of his most curious peculiarities. [p. 230]

It is worth recalling that the Arabic word *islam* means "submission." The idea that Muslims should put the proliferation of Islam ahead of their own interests is built right into the etymology of its name, and Islam is not alone. What is more important to devout Christians than their own well-being, than their own lives, if it comes to that? They will tell you: the Word. Spreading the Word of God is their *summum bonum,* and if they are called upon to forgo having children and grandchildren for the sake of spreading the Word, that is the command they will try hard to obey. They do not shrink from the idea that a meme has commandeered them and obtunded their reproductive instinct; they embrace it. And they declare that this is precisely what distinguishes them from *mere* animals; it gives them a value to pursue that transcends the genetic imperative that limits the decision horizon of all other species. In the pursuit of that value, however, they will be as rational as they can be. When they look out for Number One, Number One is the Word, not their own skin, let alone their selfish genes.

No ant can put itself in the service of a Word. It doesn't have language, or any culture to speak of. We language-users get not just one Word but many, however, and the many words compete for our attention, and in combination these can form coalitions that vie for our allegiance. This is where rational choice theory comes into its own. For, as we have seen, once people are turned into stewards of their own favorite memes, an arms race of would-be improvements ensues. All design work is *ultimately* a matter of trial and error, but a lot of it takes place "off line," in *representations* of decisions in the minds of people who consider them carefully before deciding for real on what they think will work best, given their limited information about the cruel world in which the designs must ultimately be tested. Thinking it through is quicker and cheaper than running the trials in the world and letting nature do the winnowing, but the human foresight that provides the extra speed is fallible and biased, so we often make mistakes. Memetic engineering, like genetic engineering, can spawn monsters if we're not careful, and if they escape the laboratory, they may proliferate in spite of our best efforts. We always need to remember Orgel's Second Rule: Evolution is cleverer than you are.

(Permit me to pause here for a moment and point out what we

have just done. The ardent anti-Darwinians in the humanities and social sciences have traditionally feared that an evolutionary approach would drown their cherished way of thinking—with its heroic authors and artists and inventors and other defenders and lovers of ideas. And so they have tended to declare, with desperate conviction but no evidence or argument, that human culture and human society can only be interpreted and never causally explained, using methods and presuppositions that *are completely incommensurable with,* or *untranslatable into,* the methods and presuppositions of the natural sciences. "You can't get here from there!" could be their motto. "The chasm is unbridgeable!" And yet we have just completed a sketchy but nonmiraculous and matter-of-fact stroll, all the way from blind, mechanical, robotic nature to the passionate defense and elaboration of the most exalted ideas known to humankind. The chasm was a figment of fearful imagination. We can do a better job of understanding ourselves as champions of ideas, and defenders of values, if we first see how we came to occupy such a special role.)

Once there are alternatives on offer in the "marketplace of ideas," bigger and better rivals compete for allegiance, including not just mutating religions but—eventually—secular institutions as well. Among the coalitions not based on genetic kinship that have thrived in recent human history are political parties, revolutionary groups, ethnic organizations, labor unions, sports teams, and, last but not least, the Mafia. The dynamics of group membership (entrance and exit conditions, loyalty and its enforcement by punishment or otherwise) have been intensively studied in recent years by evolutionary thinkers in a variety of disciplines: economics, political science, cognitive psychology, biology, and, of course, philosophy.[7] The results shed light on cooperation and altruism in secular as well as religious contexts, and this helps highlight the features that distinguish religious organizations from others.

3 The growth market in religion

Proposition 75: To the degree that religious economies are unregulated and competitive, overall levels of religious participation will be high. (Conversely, lacking competition, the dominant firm[s] will be too inefficient to sustain vigorous marketing efforts, and the result will be a low overall level of religious participation, with the average person minimizing and delaying payment of religious costs.)
  —Rodney Stark and Roger Finke, *Acts of Faith*

In every aspect of the religious life, American faith has met American culture—and American culture has triumphed.
  —Alan Wolfe, *The Transformation of American Religion*

We have a better product than soap or automobiles. We have eternal life.
  —Reverend Jim Bakker[8]

Why make great sacrifices in order to further the prospects of a religious organization? Why, for instance, might one choose loyalty to a religion when one is also, perhaps, a contributing member of a labor union, a political party, and a social club? These "why" questions start by being neutral between two quite different types of answers: they could be asking *why it is rational* to choose loyalty to a religion, or they could be asking *why it is natural* (somehow) for people to be drawn into a religion which then commands their loyalty. (Consider the question *Why do so many people fear heights?* One answer is: because it is rational to fear heights; you can fall and hurt yourself! Another is: we have evolved an instinctual caution triggered by the perception that we are exposed at a great height; in some people this anxiety is exaggerated beyond what is useful; *their* fear is natural—we can explain its existence without residual mystery—but *irrational*.) If we take a good hard look at the first answer regarding religion, as proposed by rational choice theory, it will help us see the forces and constraints that shape the alternatives.

Over the last two decades, Rodney Stark and his colleagues have done a remarkable job of articulating the rational choice answer, and they claim that, thanks to their efforts, "it now is impossible to do credible work in the social scientific study of religion based on the assumption that religiousness is a sign of stupidity, neurosis, poverty, ignorance, or false consciousness, or represents a flight from modernity" (Stark and Finke, 2000, p. 18). They concentrate on religion in the U.S.A., and their basic model is a straightforward application of economic theory:

> Indeed, having now had more than two centuries to develop under free market conditions, the American religious economy surpasses Adam Smith's wildest dreams about the creative forces of a free market (Moore, 1994). There are more than 1,500 separate religious "denominations" (Melton, 1998), many of them very sizable—24 have more than 1 million members each. Each of these bodies is entirely dependent on voluntary contributions, and American religious donations currently total more than $60 billion per year or more than $330 per person over age 18. These totals omit many contributions to church construction funds (new church construction amounted to $3 billion in 1993), as well as most donations to religious schools, hospitals, and foreign missions. In 1996, more than $2.3 billion was donated to support missionaries and a significant amount of this was spent on missionaries to Europe. [p. 223]

H. L. Mencken once opined: "The only really respectable Protestants are the Fundamentalists. Unfortunately, they are also palpable idiots." Many share that opinion, especially in academia, but not Stark and Finke. They are particularly eager to dispel the familiar idea that the more fundamentalist or evangelical the denomination is, the less rational it is:

> Among the more common suggestions as to why evangelical churches grow are repressed sexuality, divorce, urbanization,

racism, sexism, status anxieties, and rapid social change. Never do proponents of the old paradigm even explore possible religious explanations: for example, that people are drawn to the evangelical churches by a superior product. [p. 30]

People bear the heavy expenses of church membership, and the church in return contracts "to support and supervise their exchanges with a god or gods" (p. 103). Stark and Finke have worked this out carefully, and their driving premise is their Proposition 6, "In pursuit of rewards, humans will seek to utilize and manipulate the supernatural" (p. 90). Some people go it alone, but most think they need help, and that is what churches provide. (Do churches *actually* manipulate the supernatural? Are Stark and Finke committed to the claim that exchanges with a god or gods really occur? No, they are studiously agnostic—or so they claim—on this score. They often point out that it can be perfectly rational to invest in a stock that turns out to be worthless, after all.)

In a later book, *One True God: Historical Consequences of Monotheism* (2001), Stark takes on the role of memetic engineer, analyzing the pros and cons of doctrine as if he were an advertising consultant. "What sorts of Gods have the greatest appeal?" (p. 2). Here he distinguishes two strategies: *God as essence* (such as Tillich's God as the Ground of All Being, entirely nonanthropomorphic, not in time and space, abstract) and *God as conscious supernatural being* (a God who listens to and answers prayers in real time, for instance). "There is no more profound religious difference than that between faiths involving divine beings and those limited to divine essences," he says, and the latter he judges to be hopeless, because "only divine beings *do* anything" (p. 10). Supernatural conscious beings are much better sellers because "the supernatural is the only plausible source of many benefits we greatly desire" (p. 12).

People care about Gods because, if they exist, they are potential exchange partners possessed of immense resources. Furthermore, untold billions of people are certain that Gods do exist, precisely because they believe they have experienced long and satisfying exchange relations with them [p. 13]. . . . Because Gods are conscious beings, they are potential exchange partners because all beings are assumed to want something for which they might be induced to give something valuable. [p. 15][9]

He adds that a responsive, fatherly God "makes an extremely attractive exchange partner who can be counted on to maximize human benefits" (p. 21), and he even proposes that a God without a counterbalancing Satan is an unstable concept—"*irrational* and *perverse.*" Why? Because "one God of infinite scope must be responsible for *everything,* evil as well as good, and thus must be dangerously capricious, shifting intentions unpredictably and without reason" (p. 24). This is pretty much the same *raison d'être* that Jerry Siegel and Joe Shuster, the creators of *Superman,* appreciated when they invented kryptonite as something to counteract the Man of Steel: there is no drama possible—no defeats to overcome, no cliff-hangers— if your hero is too powerful! But, unlike the concept of kryptonite, these concepts of God and Satan have free-floating rationales, and are not the brainchildren of any particular authors:

> I do not mean to suggest that this portrait of the Gods is the product of conscious human "creation." No one sat down and decided, Let's believe in a supreme God, surround him/her with some subordinate beings, and postulate an inferior evil being on whom we can blame evil. Rather, this view tends to evolve over time because it is the most reasonable and satisfying conclusion from the available religious culture. [pp. 25–26]

Stark's footnote on this passage is not to be missed: "Nor am I prepared to deny that this evolution reflects progressive human discovery of the truth." Ah, that's the ticket! The story doesn't just get better; it *happens* to get closer to the truth. A lucky break? Maybe not. Wouldn't a really good God arrange things that way? Maybe, but the fact that dramatic considerations so conveniently dictate the details of the story does provide an explanation of why the details are what they are that rivals the traditional supposition that they are simply "the God's honest truth."

4 A God you can talk to

The Pope traditionally prays for peace every Easter and the fact that it has never had any effect whatsoever in preventing or ending a war never deters him. What goes through the Pope's mind about being rejected all the time? Does God have it in for him?
—Andy Rooney, *Sincerely, Andy Rooney*

Whatever we may think of Stark's professed agnosticism on this score, surely he is right about the main shortcoming of highly abstract conceptions of God: "Because divine essences are incapable of exchanges, they may present mysteries, but they pose no tactical questions and thus prompt no effort to discover terms of exchange" (p. 16). Who can be loyal to a God who cannot be asked for anything? It doesn't have to be manna from heaven. As the comedian Emo Phillips once said, "When I was a child, I used to pray to God for a bicycle. But then I realized that God doesn't work in that way—so I stole a bike and prayed for forgiveness!" And as Stark observes, "Rewards are always in *limited supply* and some are *entirely unavailable*—at least they are not available here and now through conventional means" (p. 17). A key marketing problem for religions, then, is how to entice the customer to wait.

> Recovery from cancer is rather minor compared with everlasting life. But perhaps the most significant aspect of otherworldly rewards is that the realization of these rewards is postponed (often until after death). Consequently, in pursuit of otherworldly

rewards, humans will accept an *extended exchange relationship* with Gods. That is, humans will make periodic payments over a substantial length of time, often until death. [p. 19]

What can be done to keep people making their payments? Miraculous cures and prayed-for reversals of fortune go a long way, of course, by providing evidence of benefits received in this world by oneself or others, but even in their absence, there are design features that pay for themselves handily. The most interesting is the price-inversion effect described by Stark and Finke (2000).

> The answer can be found in elementary economics. Price is only one factor in any exchange; quality is the other, and combined they yield an estimate of *value*. Herein lies the secret of the strength of higher-tension religious groups: despite being expensive they offer greater value; indeed, they are able to do so *because* they are expensive. [p. 145]

"Tension refers to the degree of distinctiveness, separation, and antagonism between a religious group and the 'outside' world" (p. 143). So, in a spectrum from low to high, large established churches are low-tension, and sects and cults are high-tension. An expensive religion is one that is high in "material, social and psychic costs of belonging." It doesn't just cost time spent on religious duties and money in the collection plate; belonging can incur a loss of social standing and actually exacerbate—not ameliorate—one's anxiety and suffering. But you get what you pay for: unlike the heathen, you get saved for eternity.

> To the extent that one is motivated by religious value, one must prefer a higher-priced supplier. Not only do more expensive religious groups offer more valuable product, but in doing so, they generate levels of commitment needed to maximize individual levels of confidence in the religion—in the truth of the fundamental doctrines, in the efficacy of its practices, and in the certainty of its otherworldly promises. [pp. 146–47]

The more you have invested in your religion, the more you will be motivated to protect that investment. Stark and Finke are not alone in seeing that costliness can sometimes make good economic sense. For instance, the evolutionary economists Samuel Bowles and Herbert Gintis (1998, 2001) have developed formal models of communities that foster *pro-social norms,* "cultural traits governing actions that affect the well-being of others but that cannot be regulated by costlessly enforceable contracts" (2001, p. 345). Their models show that these pro-social effects depend on "low cost access to information about other community members" as well as the tendency to favor interactions with group members, and restrict migration in and out, points that Stark and Finke make as well.[10]

The high entry and exit costs are as crucial to the survival of such arrangements as the membrane surrounding a cell: self-maintenance is costly and is made more efficient by a strict distinction between *me* and *the rest of the world* (in the case of a cell) or between *us* and *them* (in the case of a community). The work by Bowles and Gintis doesn't just provide formal support for some of the propositions defended by Stark and Finke; it shows that the deplorable xenophobia found in "high-tension" religious communities is not a specifically religious feature. Xenophobia, they argue, is the price *any* community or group must pay for a high level of internal trust and harmony, and moreover it is a price we may in the end decide we have to be willing to pay: "Far from being vestigial anachronisms, we think communities may become more rather than less important in the nexus of governance structures in the years to come, since communities may claim some success in addressing governance problems not amenable to market or state solution" (Bowles and Gintis, 2001, p. 364).

Stark and Finke's applications of rational choice theory to many of the trends and disparities observable in American religious denominations are not yet proven, and have spirited detractors, but they are certainly worth further research. And the implications of some of their propositions are provocative indeed. For instance:

Proposition 76. Even where competition is limited, religious firms can generate high levels of participation to the extent that the firms serve as the primary organizational vehicles for social conflict. (Conversely, if religious firms become significantly less important as vehicles for social conflict, they will be correspondingly less able to generate commitment.) [p. 202]

In other words, expect religious "firms" to exploit and exacerbate social conflict whenever possible, since it is a way of generating business. This can be good (Polish Catholic resistance to communism) or bad (the interminable conflict in Ireland). Detractors will say we already knew this about religions, but the claim that this is a *systematic* feature, which follows from other features and interacts with still others in ways that are predictable, is, if true, just the sort of fact we are going to want to understand deeply as we deal with social conflicts in the future. When religious leaders and their critics both inside and outside their religions consider possible reforms and improvements, they are setting themselves up—whether they like it or not—to be memetic engineers, tinkering with the designs they have been bequeathed by tradition in order to adjust the observable effects, and some of the most telling observations in Stark and Finke's book are their biting criticisms of well-intentioned reforms that have backfired. Are they right about the principal reason for the precipitous decline in Catholics seeking a vocation in the church after Vatican II?

Previously, the Catholic Church had taught that priests and the religious [nuns and monks] were in a superior state of holiness. Now, despite their vows, they were just like everyone else [p. 177]. . . . The laity had gained some of the privileges of the priesthood without shouldering the burden of celibacy or a direct accountability to the church hierarchy. For many, the priesthood was no longer a good deal following the renewal efforts of Vatican II. [p. 185]

Or are they wrong? The only way to find out is to do the research. Unpalatability is not a reliable sign of falsehood, and the pious homilies that often guided earlier reformers need to be confirmed, disconfirmed—or else ignored. The stakes are too high for well-meant amateur blundering. As earlier, in my discussions of the work of Boyer, Wilson, and others, I am not declaring a verdict on the soundness or conclusiveness of any of this work, but only introducing what I take to be examples of the work that needs to be taken seriously from now on, and either firmly and fairly refuted or—however begrudgingly—acknowledged for its genuine contributions to our understanding. In the case of Stark's refreshingly candid vision, I myself have deep misgivings, some of which will emerge when we turn to some of the complications that he so resolutely sets aside. Stark and Finke express their fundamental attitude well when they disparage Don Cupitt's *After God: The Future of Religion* (1997), which endorses a brand of religion from which all traces of the supernatural have been removed:

> But why would a religion without God have a future? Cupitt's prescription strikes us as rather like expecting people to continue to buy soccer tickets and gather in the stands to watch players who, for lack of a ball, just stand around. If there are no supernatural beings, then there are no miracles, there is no salvation, prayer is pointless, the Commandments are but ancient wisdom, and death is the end. In which case the rational person would have nothing to do with church. Or, more accurately, a rational person would have nothing to do with a church *like that*. [p. 146]

Strong language, but they must recognize that Cupitt and the others who have turned away from their vision of God the Dealmaker were well aware of its attractions and must have had their reasons (articulated or not) for resisting it so artfully for so long. What can be said in favor of the *God-as-essence* path—or, rather, paths, since there have been many different ways of trying to conceive of God in less anthropomorphic terms? I think the key can be found in some of Stark and Finke's own observations:

> Given the fact that religion is risky goods and that people often can increase their flow of immediate benefits through religious inactivity, it seems unlikely that any amount of pluralism and vigorous marketing can ever achieve anything close to total market penetration. The proportion of Americans who actually belong to a specific church congregation (as opposed to naming a religious preference when asked) has hovered around 65 percent for many decades—showing no tendency to respond even to major economic cycles. [p. 257]

It will be interesting to try to learn more about the 35 percent who are just not cut out for church, as well as the proportion of those churchgoers who are not cut out for high-tension, expensive religions of the sort Stark favors. They exist all over the world; according to Stark and Finke, "There are 'godless' religions, but their followings are restricted to small elites—as in the case of the elite forms of Buddhism, Taoism, and Confucianism" (p. 290n). The attractions of Unitarianism, Episcopalianism, and Reform Judaism are not restricted to the Abrahamic traditions, and if the "elites" find that they just cannot bring themselves to "believe they have experienced long and satisfying exchange relations with" God, why do they persist with (something they call) religion at all?

\\\

Chapter 7 The human proclivity for groupishness is less calculated and prudential than it appears in some economic models, but also more complicated than the evolved herding instinct of some animals. What complicates the picture is human language and culture, and the perspective of memes permits us to comprehend how the phenomena of human allegiance are influenced by a mixture of free-floating and well-tethered rationales. We can make progress by acknowledging that submission to a religion need not be cast as a deliberate economic decision, while also recognizing the analytic and predictive power of the perspective that views religions as designed systems competing in a dynamic marketplace for adherents with different needs and tastes.

\\\

Chapter 8 The stewardship of religious ideas creates a powerful phenomenon: belief in belief, which radically transforms the content of the underlying beliefs, making rational investigation of them difficult if not impossible.

||| \\\ |||

Belief in Belief

1 You better believe it

*I think God honors the fact that I want to believe in Him, whether I feel
sure or not.*
  —Anonymous informant quoted by Alan Wolfe, in *The Transformation of
    American Religion*

*The proof that the Devil exists, acts and succeeds is precisely that we no
longer believe in him.* —Denis de Rougement, *The Devil's Share*

At the end of chapter 1, I promised to return to Hume's question in
his *Dialogues Concerning Natural Religion*, the question of whether
we have *good reasons* for believing in God, and in this chapter, I will
keep that promise. The preceding chapters have laid some new
foundations for this inquiry, but also uncovered some problems be-
setting it that need to be addressed before any effective confronta-
tion between theism and atheism can take place.

Once our ancestors became reflective (and hyperreflective) about
their own beliefs, and thus appointed themselves stewards of the
beliefs they thought most important, the phenomenon of *believing
in belief* became a salient social force in its own right, sometimes

eclipsing the lower-order phenomena that were its object. Consider a few cases that are potent today. Because many of us believe in democracy and recognize that the security of democracy in the future depends critically on *maintaining the belief* in democracy, we are eager to quote (and quote and quote) Winston Churchill's famous line: "Democracy is the worst form of government except all the other forms that have been tried." As stewards of democracy, we are often conflicted—eager to point to flaws that ought to be repaired, yet just as eager to reassure people that the flaws are not that bad, that democracy can police itself, so their faith in it is not misplaced.

The same point can be made about science. Since the belief in the integrity of scientific procedures is almost as important as the actual integrity, there is always a tension between a whistle-blower and the authorities, even when they know that they have mistakenly conferred scientific respectability on a fraudulently obtained result. Should they quietly reject the offending work and discreetly dismiss the perpetrator, or make a big stink?[1]

And certainly some of the intense public fascination with celebrity trials is to be explained by the fact that belief in the rule of law is considered a vital ingredient in our society; so, if famous people are seen to be above the law, this jeopardizes the general trust in the rule of law. Hence we are interested not just in the trial, but in the public reactions to the trial, and the reactions to those reactions, creating a spiraling inflation of media coverage. We who live in democracies have become somewhat obsessed with gauging public opinion on all manner of topics, and for good reason: in a democracy it really matters what the people believe. If the public cannot be mobilized into extended periods of outrage by reports of corruption, or the torturing of prisoners by our agents, for instance, our democratic checks and balances are in jeopardy. In his hopeful book, *Development as Freedom* (1999), and elsewhere (see especially Sen, 2003), the Nobel laureate economist Amartya Sen makes the important point that you don't have to win an election to achieve your political aims. Even in shaky democracies, what the leaders believe about the beliefs that prevail in their countries influences what they take their realistic options to be, so belief maintenance is an important political goal in its own right.

Even more important than political beliefs, in the eyes of many, are what we might call metaphysical beliefs. Nihilism—the belief in nothing—has been seen by many to be a deeply dangerous virus, for obvious reasons. When Friedrich Nietzsche hit upon his idea of the Eternal Recurrence—he thought he had proved that we relive our lives infinitely many times—his first inclination (according to some stories) was to kill himself without revealing the proof, in order to spare others from this life-destroying belief.[2] Belief in the *belief that something matters* is understandably strong and widespread. Belief in free will is another vigorously protected vision, for the same reasons, and those whose investigations seem to others to jeopardize it are sometimes deliberately misrepresented in order to discredit what is seen as a dangerous trend (Dennett, 2003c). The physicist Paul Davies (2004) has recently defended the view that belief in free will is so important that it may be "a fiction worth maintaining." It is interesting that he doesn't seem to think that his own discovery of the awful truth (what he takes to be the awful truth) incapacitates him morally, but believes that others, more fragile than he, will need to be protected from it.

Being the unwitting or uncaring bearer of good news or bad news is one thing; being the self-appointed champion of a meme is something quite different. Once people start committing themselves (in public, or just in their "hearts") to particular ideas, a strange dynamic process is brought into being, in which the original commitment gets buried in pearly layers of defensive reaction and meta-reaction. "Personal rules are a *recursive* mechanism; they continually take their own pulse, and if they feel it falter, that very fact will cause further faltering," the psychiatrist George Ainslie observes in his remarkable book, *Breakdown of Will* (2001, p. 88). He describes the dynamic of these processes in terms of competing strategic commitments that can contest for control in an organization—or an individual. Once you start living by a set of explicit rules, the stakes are raised: When you lapse, what should you do? Punish yourself? Forgive yourself? Pretend you didn't notice?

> After a lapse, the long-range interest is in the awkward position of a country that has threatened to go to war in a particular circumstance that has then occurred. The country wants to avoid war without destroying the credibility of its threat, and may therefore look for ways to be seen as not having detected the circumstance. Your long-range interest will suffer if you catch yourself ignoring a lapse, but perhaps not if you can arrange to ignore it without catching yourself. This arrangement, too, must go undetected, which means that a successful process of ignoring must be among the many mental expedients that arise by trial and error—the ones you keep simply because they make you feel better without your realizing why. [p. 150]

This idea that there are myths we live by, myths that must not be disturbed at any cost, is always in conflict with our ideal of truth-seeking and truth-telling, sometimes with lamentable results. For example, racism is at long last widely recognized as a great social evil, so many reflective people have come to endorse the second-order belief that *belief in the equality of all people regardless of their race* is to be vigorously fostered. How vigorously? Here people of goodwill differ sharply. Some believe that belief in racial differences is so pernicious that *even when it is true* it is to be squelched. This has led to some truly unfortunate excesses. For instance, there are clear clinical data about how people of different ethnicity are differently susceptible to disease, or respond differently to various drugs, but such data are considered off limits by some researchers, and some funders of research. This has the perverse effect that strongly indicated avenues of research are deliberately avoided, much to the detriment of the health of the ethnic groups involved.[3]

Ainslie uncovers strategic belief-maintenance in a wide variety of cherished human practices:

> Activities that are spoiled by counting them, or counting on them, have to be undertaken through indirection if they are to stay valuable. For instance, romance undertaken for sex or even "to be loved" is thought of as crass, as are some of the most lucrative professions if undertaken for money, or performance art if done for effect. Too great an awareness of the motivational contingencies for sex, affection, money, or applause spoils the effort, and not only because it undeceives the other people involved. Beliefs about the intrinsic worth of these activities are valued beyond whatever accuracy these beliefs might have, because they promote the needed indirection. [In press]

Though not at all restricted to religion, belief in belief is nowhere else a more fecund engine of elaboration. Ainslie surmises that it explains some of the otherwise baffling *epistemic* taboos found in religions:

> From priesthood to fortune-telling, contact with the intuitive seems to need some kind of divination. This is all the more true for approaches that cultivate a sense of empathy with a god. Several religions forbid the attempt to make their deity more tangible by drawing pictures of him, and Orthodox Judaism forbids even naming him. The experience of God's presence is supposed to come through some kind of invitation that he may or may not accept, not through invocation. [2001, p. 192]

What do people do when they discover that they no longer believe in God? Some of them don't do anything; they don't stop going to church, and they don't even tell their loved ones. They just quietly get on with their lives, living as morally (or immorally) as they did before. Others, such as Don Cupitt, author of *After God: The Future of Religion*, feel the need to cast about for a religious creed that they can endorse with a straight face. They have a firm belief that *belief in God* is something to preserve, so when they find the traditional concepts of God frankly incredible they don't give up. They seek a substitute. And the search, once again, need not be all that conscious and deliberate. Without ever being frankly aware that a cherished ideal is endangered in some way, people may be strongly moved by a nameless dread, the sinking sense of a loss of conviction, a threat intuited but not articulated that needs to be countered vigorously. This puts them in a state of mind that makes them particularly receptive to novel emphases that somehow seem right or fitting. Like sausage-making and the crafting of legislation in a democracy, creed revision is a process that is upsetting to watch too closely, so it is no wonder that the fog of mystery descends so gracefully over it.

Much has been written over the centuries about the historic processes by which polytheisms turned into monotheism—belief in gods being replaced by belief in God. What is less often stressed is how this belief in God joined forces with the *belief in belief* in God to motivate the migration of the concept of God in the Abrahamic religions (Judaism, Christianity, and Islam) away from concrete anthropomorphism to ever more abstract and depersonalized concepts. What is remarkable about this can be illuminated by contrast with other conceptual shifts that have occurred during the same period. Fundamental concepts can certainly change over time. Our concept of matter has changed quite radically from the days of the ancient Greek atomists. Our scientific conceptions of time and space today, thanks to clocks and telescopes and Einstein and others, are different from theirs as well. Some historians and philosophers have argued that these shifts are not as gradual as they may at first appear but, rather, are abrupt saltations, so drastic that the before and after concepts are "incommensurable" in some way.[4]

Are any of these conceptual revisions actually so revolutionary as to render communication across the ages impossible, as some have argued? The case is hard to make, since we can apparently chart the changes accurately and in detail, understanding them all as we go.

In particular, there seems no reason to believe that our *everyday* conceptions of space and time would be even *somewhat* alien to Alexander the Great, say, or Aristophanes. We would have little difficulty conversing with either of them about today, tomorrow, and last year, or the thousands of yards or paces between Athens and Baghdad. But if we tried to converse with the ancients about God, we would find a much larger chasm separating us. I can think of no other concept that has undergone so dramatic a deformation. It is as if *their* concept of milk had turned into *our* concept of health, or as if *their* concept of fire had turned into *our* concept of energy. You can't *literally* drink health or *literally* extinguish energy, and (today, according to many but not all believers) you can't *literally* listen to God or *literally* sit beside Him, but these would be strange claims indeed to the original monotheists. The Old Testament Jehovah, or Yahweh, was quite definitely a super-*man* (a He, not a She) who could take sides in battles, and be both jealous and wrathful. The original New Testament Lord is more forgiving and loving, but still a Father, not a Mother or a genderless Force, and active in the world, needless to say, through His miracle-performing Son. The genderless Person without a body who nevertheless answers prayers in real time (Stark's conscious supernatural *being*) is still far too anthropomorphic for some, who prefer to speak of a Higher Power (Stark's *essence*) whose characteristics are beyond comprehension—aside from the fact that they are, in some incomprehensible way, good, not evil.[5] Does the Higher Power have (creative) intelligence? In what way? Does It (*not* He or She) care about us? About anything? The fog of mystery has descended conveniently over all the anthropomorphic features that have not been abandoned outright.

And a further adaptation has been grafted on: it is impolite to ask about these matters. If you persist, you are likely to get a response along these lines: "God can see you when you're doing something evil in the dark, but He does *not* have eyelids, and never blinks, you silly rude person, and of course He can read your mind even when you are careful not to talk to yourself, but still He prefers you to pray to Him in words, and don't ask me how or why. These are mysteries we finite mortals will never understand." People of all faiths have been taught that any such questioning is somehow insulting or demeaning to their faith, and must be an attempt to ridicule their views. What a fine protective screen this virus provides—permitting it to shed the antibodies of skepticism effortlessly!

But it doesn't always work, and when the skepticism becomes more threatening, stronger measures can be invoked. One of the most effective is also one of the most transparent: the old *diabolical lie*—the term comes from de Rougemont (1944), who speaks of "the putative proclivity of 'The Father of Lies' for appearing as his own opposite." It is, almost literally, a trick with mirrors, and, like many good magic tricks, it's so simple that it's hard to believe it could ever work. (Novice magicians often have to steel themselves to perform tricks the first time in public—it just doesn't seem possible that audiences will fall for these, but they do.) If I were designing a phony religion, I'd surely include a version of this little gem—but I'd have a hard time saying it with a straight face:

> If anybody ever raises questions or objections about our religion that you cannot answer, that person is almost certainly Satan. In fact, the more reasonable the person is, the more eager to engage you in open-minded and congenial discussion, the more sure you can be that you're talking to Satan in disguise! Turn away! Do not listen! It's a trap!

What is particularly cute about this trick is that it is a perfect "wild card," so lacking in content that *any* sect or creed or conspiracy can use it effectively. Communist cells can be warned that any criticism they encounter is almost sure to be the work of FBI infiltrators in disguise, and radical feminist discussion groups can squelch any unanswerable criticism by declaring it to be phallocentric propaganda being unwittingly spread by a brainwashed dupe of the evil patriarchy, and so forth. This all-purpose loyalty-enforcer is para-noia in a pill, sure to keep the critics muted if not silent. Did any-one invent this brilliant adaptation, or is it a wild meme that domesticated itself by attaching itself to whatever memes were competing for hosts in its neighborhood? Nobody knows, but now it is available for anybody to use—although, if this book has any success, its virulence should diminish as people begin to recognize it for what it is.

(A milder and more constructive response to relentless skepti-cism is the vigorous academic industry of theological discussion and research, very respectfully inquiring into the possible interpre-tations of the various creeds. This earnest intellectual exercise scratches the skeptical itch of those few people who are uncomfort-able with the creeds they were taught as children, and is ignored by everybody else. Most people don't feel the need to examine the de-tails of the religious propositions they profess.)

Mystery is declared to surround the various conceptions of God, but there is nothing mysterious about the process of transforma-tion, which is clear for all to see and has been described (and often decried) by generations of would-be stewards of this important idea. Why don't the stewards just coin new terms for the revised conceptions and let go of the traditional terms along with the dis-carded conceptions? After all, we don't persist in the outmoded medical terminology of *humors* and *apoplexy* or insist on finding something in contemporary physics or chemistry to identify as *phlogiston*. Nobody has proposed that we have discovered the iden-tity of *élan vital* (the secret ingredient that distinguishes living things from mere matter); it's DNA (the vitalists just didn't have the right conception of it, but they knew there had to be something). Why do people insist on calling the Higher Power they believe in "God"? The answer is clear: the believers in the belief in God have appreci-ated that the continuity of *professing* requires continuity of nomen-clature, that *brand loyalty* is a feature so valuable that it would be foolish to tamper with it. So, whatever other reforms you may want to institute, don't try to replace the *word* "God" ("Jehovah," "Theos," "Deus," "the Almighty," "Our Lord," "Allah") when you tinker with your religion.[6] In the beginning was the Word.

I have to say that it has worked pretty well, after a fashion. For a thousand years, roughly, we've entertained a throng of variously deanthropomorphized, intellectualized concepts of God, all more or less peacefully coexisting in the minds of "believers." Since everybody calls his or her version "God," there is something "we can all agree about"—we all believe in God; we're not atheists! But of course it doesn't work *that* well. If Lucy believes that Rock (Hudson) is to die for, and Desi believes that Rock (music) is to die for, they really don't agree on anything, do they? The problem is not new. Back in the eighteenth century, Hume had already decided that "our idea of a deity" had shifted so much that the *gods* of antiquity simply didn't count, being too anthropomorphic:

> To any one, who considers justly of the matter, it will appear, that the gods of all polytheists are not better than the elves and fairies of our ancestors, and merit as little any pious worship or veneration. These pretended religionists are really a kind of superstitious atheists, and acknowledge no being, that corresponds to our idea of a deity. No first principle of mind or thought: No supreme government and administration: No divine contrivance or intention in the fabric of the world. [1777, p. 33]

More recently, and chiding in the opposite direction, Stark and Finke (2000) express dismay at the "atheistic" views of John Shelby Spong, the Episcopal bishop in Newark, whose God is not anthropomorphic enough. In his 1998 book, *Why Christianity Must Change or Die*, Spong dismisses the divinity of Jesus, declares the crucifixion "barbaric," and opines that the God of most traditional Christians is an ogre. Another eminent Episcopal cleric once confided to me that when he found out what some Mormons believed when they said they believed in God, he rather wished they didn't believe in God! Why won't he say this from the pulpit? Because he doesn't want to let down the side. After all, there are lots of evil, "Godless" people out there, and it would never do to upset the fragile fiction that "we are not atheists" (heaven forbid!).

2 God as intentional object

The fool hath said in his heart, There is no God. —Psalms 14:1 (also 53:1)

Belief in belief in God makes people reluctant to acknowledge the obvious: that much of the traditional lore about God is no more worthy of *belief* than the lore about Santa Claus or Wonder Woman. Curiously, it's all right to laugh about it. Consider all the cartoons depicting God as a stern, bearded fellow sitting on a cloud with a pile of lightning bolts at his side, to say nothing of all the jokes, bawdy and clean, about various folks arriving in heaven and having one misadventure or another. This treasury of humor provokes hearty chuckles from all but the most stuffy puritans, but few are comfortable acknowledging just how far we've come from the God of Genesis 2:21, who literally plucks a rib from Adam and closes up the flesh (with his fingers, one imagines) before sculpting Eve on the spot. In *A Devil's Chaplain* Richard Dawkins (2003a), offers some sound advice—but knows in advance it will not be heeded, because people can see the punch line coming:

> ... modern theists might acknowledge that, when it comes to Baal and the Golden Calf, Thor and Wotan, Poseidon and Apollo, Mithras and Ammon Ra, they are actually atheists. We are all atheists about most of the gods that humanity has ever believed in. Some of us just go one god further. [p. 150]

The trouble is that, since this advice won't be heeded, discussions of the existence of God tend to take place in a pious fog of indeterminate boundaries. If theists would be so kind as to make a short list of all the concepts of God they renounce as balderdash before proceeding further, we atheists would know just which topics were still on the table, but, out of a mixture of caution, loyalty, and un-willingness to offend anyone "on their side," theists typically decline to do this.7 Don't put all your eggs in one basket, I guess. This double standard is enabled if not actually licensed by a logical confusion that continues to defy resolution by philosophers who have worked on it: the problem of *intentional objects*.8 In a phrase (which will prove unsatisfactory, as we will soon see), intentional objects are the *things somebody can think about*.

Do I believe in witches? It all depends what you mean. If you mean evil-hearted spell-casting women who fly around supernaturally on broomsticks and wear black pointed hats, the answer is obvious: no, I no more believe in witches than I believe in the Easter Bunny or the Tooth Fairy. If you mean people, both men and women, who practice Wicca, a popular New Age cult these days, the answer is equally obvious: yes, I believe in witches; they are no more supernatural than Girl Scouts or Rotarians. Do I believe these witches cast spells? Yes and no. They sincerely utter imprecations of various sorts, expecting to alter the world in various supernatural ways, but they are mistaken in thinking they succeed, though they may alter their own attitudes and behavior thereby. (If I give you the Evil Eye, you may become seriously unnerved, to the point of serious illness, but if so, that is because you are credulous, not because I have magical powers.)9 So it all depends what you mean. And does it ever!

About forty years ago, in England, I saw a BBC news program in which nursery-school children were interviewed about Queen Elizabeth II. What did they know about her? The answers were charming: the Queen wore her crown while she "hoovered" Buckingham Palace, sat on the throne when she watched telly, and in general behaved like a cross between Mum and the Queen of Hearts. *This* Queen Elizabeth II, the *intentional object* brought into existence (as an abstraction) by the consensus convictions of these children, was much more interesting and entertaining than the real woman. And a more potent political force! Are there, then, two distinct entities, the real woman and the imagined Queen, and if so, are there not millions or billions of distinct entities—the Queen Elizabeth II believed in by teen-agers in Scotland, and the Queen Elizabeth II believed in by the staff at Windsor Castle, and *my* Queen Elizabeth II, and so on? Philosophers have argued vigorously for the better part of a century about how to accommodate such intentional objects into their ontologies—their catalogues of *the things that exist*—with no emerging consensus. Another eminent Briton is Sherlock Holmes, who is often thought about even though he never existed at all. In *one sense or another,* there are both truths and falsehoods about such (mere) intentional objects: It is true that Sherlock Holmes (the intentional object created by Sir Arthur Conan Doyle) lived on Baker Street and smoked, and false that he had a bright-green nose. It is true that Pegasus had wings in addition to four ordinary horse legs, and false that President Truman once owned him and rode him to the White House from Missouri. But of course neither Sherlock Holmes nor Pegasus is or ever was real.

Some people may be under the mistaken impression that Sherlock Holmes actually existed and that Conan Doyle's stories aren't fiction. These people believe in Sherlock Holmes in the strong sense (let us say). Others, known as "Sherlockians," devote their spare time to becoming Sherlock Holmes scholars, and can entertain one another with their encyclopedic knowledge of the Conan Doyle canon, without ever making the mistake of confusing fact with fiction. The most famous society of these scholars is the Baker Street Irregulars, named after the gang of street urchins that Holmes enlisted for various purposes over the years. Members of these societies (for there are many "Sherlockian" societies around the world) delight in knowing which train Holmes took from Paddington on May 12, but know full well that there simply is no fact to be learned about whether he faced forward or backward in the train, since Conan Doyle didn't specify it or anything that would imply it. They know that Holmes is a fictional character, but nevertheless they devote large parts of their lives to studying him, and are eager to explain why their love of Holmes is better justified than some other fan's love of Perry Mason or Batman. They believe in Sherlock Holmes in the weak sense (let us say). They behave very much like the amateur scholars who devote their spare time to trying to figure out who Jack the Ripper was, and an observer who didn't know that the Holmes stories are fiction whereas Jack the Ripper was a real murderer might naturally suppose that the Baker Street Irregulars were investigating a historical person.

It is quite possible for a *mere* intentional object like Sherlock Holmes to obsess people even when they know full well that it isn't real. So it is not surprising that such a thing (if it's right, in the end, to call it a kind of *thing* at all) can dominate people's lives when they believe in it in the strong sense, such as the people who spend fortunes hunting for the Loch Ness Monster or Bigfoot. And whenever a real person, such as Queen Elizabeth II, dominates people's lives, this domination is *usually* accomplished indirectly, by setting up a manifold of beliefs, giving people an intentional object that is featured in their thinking and the decisions they make. I can't hate my rival or love my neighbor without having a pretty clear and largely accurate set of beliefs that serve to pick this person out of the crowd so I can recognize, track, and interact effectively with him or her.

In most circumstances, the things we believe in are perfectly real, and the things that are real we believe in, so we can usually ignore the logical distinction between an intentional object (the object of belief) and the thing in the world that inspired/caused/ grounds/anchors the belief. Not always. The Morning Star turns out to be none other than the Evening Star. "They" are not stars; "they" are one and the same thing—namely, the planet Venus. One planet, two intentional objects? Usually the things that matter to us make themselves securely known to us in a variety of ways that permit us to track them through their trajectories, but other scenarios do occur. I *might* sneak around thwarting your projects, or, alternatively, giving you "good luck," dominating your life one way or another without your ever suspecting that I existed as a *person* or a *thing* or even a *force* in your life, but this is an unlikely possibility. In the main, things that make a difference in a person's life figure in it as intentional objects one way or another, however misidentified or misconstrued. When misconstruals occur, problems arise about how to describe the situation. Suppose you've been surreptitiously doing me good deeds for months. If I "thank my lucky stars" when it is really *you* I should be thanking, it would misrepresent the situation to say that I believe in you and am grateful to you. Maybe I am a fool to say in my heart that it is only my lucky stars that I should thank—saying, in other words, that there is nobody to thank—but that *is* what I believe; there is no intentional object in this case to be identified as you.

Suppose instead that I was convinced that I did have a secret helper but that it wasn't you—it was Cameron Diaz. As I penned my thank-you notes to her, and thought lovingly about her, and marveled at her generosity to me, it would surely be misleading to say that *you* were the object of my gratitude, even though you were in fact the one who did the deeds that I am so grateful for. And then suppose I gradually began to suspect that I had been ignorant and mistaken, and eventually came to the correct realization that you were indeed the proper recipient of my gratitude. Wouldn't it be strange for me to put it this way: "Now I understand: you are Cameron Diaz!" It would indeed be strange; it would be false—unless something else had happened in the interim. Suppose my acquaintances had become so used to my singing the praises of Cameron Diaz and her bountiful works that the term had come, to them and to me, to stand for *whoever it was* who was responsible for my joy. In that case, those syllables would no longer have their original use or meaning. The syllables "Cameron Diaz," purportedly a proper name of a real individual, would have been turned—gradually and imperceptibly—into a sort of wild-card referring expression, the "name" of whoever (or whatever) is responsible for . . . whatever it is I am grateful for. But, then, if the term were truly open-ended in this way, when I thank "my lucky stars" I am thanking exactly the same thing as when I thank "Cameron Diaz"—and you do turn out to be my Cameron Diaz. The Morning Star turns out to be the Evening Star. (How to turn an atheist into a theist by just fooling around with the words—if "God" were just the name of *whatever it is* that produced all creatures great and small, then God might turn out to *be* the process of evolution by natural selection.)

This ambiguity has been exploited ever since the psalmist sang about the fool. The fool doesn't know what he's talking about when he says in his heart there is no God, so he's ignorant in the same way as somebody who thinks that Shakespeare didn't actually write *Hamlet*. (Somebody did; if Shakespeare is *by definition* the author of *Hamlet*, then perhaps Marlowe was Shakespeare, etc.) When people write books about "the history of God" (Armstrong, 1993; Stark, 2001; Debray, 2004, are recent examples), they are actually writing about the history of the *concept* of God, of course, tracing the fashions and controversies about God as intentional object through the centuries. Such a historical survey can be neutral in *two* regards: it can be neutral about which concept of God is *correct* (did Shakespeare write *Hamlet* or did Marlowe write *Hamlet*?), and it can be neutral about whether the whole enterprise concerns fact or fiction (are we the Baker Street Irregulars or are we trying to identify a real murderer?). Rodney Stark opens *One True God: Historical Consequences of Monotheism* with a passage that brandishes this ambiguity:

> All of the great monotheisms propose that their God works through history, and I plan to show that, at least sociologically, they are quite right: that a great deal of history—triumphs as well as disasters—has been made on behalf of One True God. What could be more obvious? [2001, p. 1]

His title suggests that he is not neutral—one *true* God—but the entire book is written "sociologically"—which means that it is not about God, it is about the intentional objects that do all the political and psychological lifting, the God of the Catholics, the God of the

Jews, the God of teen-agers living in Scotland, perhaps. It is indeed obvious that God the intentional object has played a potent role, but that says nothing about whether God exists, and it is disingenuous of Stark to hide behind the ambiguity. The history of disagreement has not all been good clean fun, after all, like the Baker Street Irregulars versus the Perry Mason Fan Club. People have died for their theories. Stark may be neutral, but the comedian Rich Jeni isn't; as he sees it, religious war is pathetic: "You're basically killing each other to see who's got the better imaginary friend." What is Stark's opinion about that? And what is yours? Might it be all right, even obligatory, to fight for a *concept,* whether or not the concept refers to anything real? After all, one might add, hasn't the strife brought us a bounty of great art and literature, in the arms race of competitive glorification?

I find that some people who consider themselves believers actually just believe in the *concept* of God. I myself believe that the concept exists—as Stark says, what could be more obvious? These people believe, moreover, that the concept is worth fighting over. Notice that they *don't* believe in *belief* in God! They are far too sophisticated for that; they are like the Baker Street Irregulars, who don't believe in *belief* in Sherlock Holmes, but just in studying and extolling the lore. They do think that their concept of God is so much better than other concepts of God that they should devote themselves to spreading the Word. But they don't believe in God in the strong sense.

By definition, one would think, theists believe in God. (*Atheism* is the negation of *theism,* after all.) But there is little hope of conducting an effective investigation into the question of whether God exists when there are self-described theists who "think that providing a satisfactory theistic ethics requires giving up the idea that God is some kind of supernatural entity" (Ellis, 2004). If God is not some kind of supernatural entity, then *who knows* whether you or I believe in him (it?)? Beliefs in Sherlock Holmes, Pegasus, witches on broomsticks—these are the easy cases, and they can be quite readily sorted out with a little attention to detail. When it comes to God, on the other hand, there is no straightforward way of cutting through the fog of misunderstanding to arrive at a consensus about the topic under consideration. And there are interesting reasons why people resist having a specific definition of God foisted on them (even for the sake of argument). The mists of incomprehension and failure of communication are not just annoying impediments to rigorous refutation; they are themselves design features of religions worth looking at closely on their own.

3 The division of doxastic labor

Fake it until you make it. —Alcoholics Anonymous

So we have the strange phenomenon, as Kant assures us, of a mind believing with all its strength in the real presence of a set of things of no one of which it can form any notion whatsoever.
—William James, *The Varieties of Religious Experience*

Language gives us many gifts, including the capacity to memorize, transmit, cherish, and in general *protect* formulas that we don't understand. Here is a sentence I firmly believe to be true:

(1)   *Her insan doğar, yaşar, ve ölür.*

I haven't the foggiest idea what (1) means, but I know it's true, because I asked a trusted Turkish colleague to provide me with a true sentence for just this purpose. I would bet a large sum of money on the truth of this sentence—that's how sure I am that it's true. But as I say, I don't know whether (1) is about trees, or people, or history, or chemistry, . . . or God. There is nothing metaphysically peculiar, or difficult, or unseemly, or embarrassing about my state of mind. I just don't know what proposition this sentence expresses, because I'm not "expert" in Turkish. In chapter 7, I noted the methodological problems confronting anthropologists intent on understanding other cultures, and suggested that part of the problem is that individual informants may not view themselves as experts on the doctrines they are asked to elucidate. The problems that arise for such "half-understood ideas" are exacerbated in the case of religious doctrines, but are as often encountered in science as in religion.

Here, one might say, is the ultimate division of labor, the division of *doxastic* labor, made possible by language: we laypeople do the believing—we sign on to the *doxology*—and defer the understanding of those *dogmas* to the experts! Consider the ultimate talismanic formula of science:

$$(2) \qquad\qquad E = mc^2$$

Do you believe that $E = mc^2$? I do. We all know that this is Einstein's great equation, and the heart, somehow, of his theory of relativity, and many of us know what the E and m and c stand for, and could even work out the basic algebraic relationships and detect obvious errors in interpreting it. But only a tiny fraction of those who know that "$E = mc^2$" is a fundamental truth of physics actually understand it in any substantive way. Fortunately, the rest of us don't have to; we have expert physicists around to whom we have gratefully delegated responsibility for understanding the formula. What we are doing, in these instances, is not really *believing the proposition*. For that, you'd have to *understand* the proposition. What we are doing is believing that *whatever proposition is expressed by the formula "$E = mc^2$"* is true.[10]

The difference for me between (1) and (2) is that I know quite a lot—but not enough!—about what (2) is about. In the infinite space of all possible propositions, I can narrow down its meaning to a rather tight cluster of nearly identical variants. A physicist could probably trip me up by getting me to endorse an *almost right* paraphrase that would reveal my ignorance (that's what really tough multiple-choice exams can do, separating the students who really understand the material from those who only sort of understand the material). With (1), however, all I know is that it expresses one of the true propositions—cutting the infinite space of propositions in half, but still leaving infinitely many propositions indistinguishable by me as its best interpretation. (I can guess that it is probably not about how the Red Sox beat the Yankees four straight to win the American League Championship in October 2004, but such whittling away doesn't take us far.)

I drew an example from science to show that this is not an embarrassing foible of religious belief alone. Even scientists rely every day on formulas that they know to be correct but are not themselves expert in interpreting. And they sometimes even foster the separation of understanding and memorization. A vivid instance can be found in Richard Feynman's classic introductory lectures on quantum electrodynamics, *QED: The Strange Theory of Light and Matter* (1985), in which he amusingly cajoles his audience to loosen their grip and not try to understand the method he is teaching:

> So now you know what I'm going to talk about. The next question is, will you *understand* what I'm going to tell you? . . . No, you're not going to be able to understand it. Why, then, am I going to bother you with all this? Why are you going to sit here all this time, when you won't be able to understand what I am going to say? It is my task to convince you *not* to turn away because you don't understand it. You see, my physics students don't understand it either. That is because I don't understand it. Nobody does. . . . It's a problem that physicists have learned to deal with: they've learned to realize that whether they like a theory or they don't like a theory is *not* the essential question. Rather, it is whether or not the theory gives predictions that agree with experiment. It is not a question of whether a theory is philosophically delightful, or easy to understand, or perfectly reasonable from the point of view of common sense. . . . Please don't

turn yourself off because you can't believe Nature is so strange. Just hear me all out, and I hope you'll be as delighted as I am when we're through. [pp. 9–10]

He goes on to describe the methods of calculating probability amplitudes in terms that deliberately discourage understanding— "You will have to brace yourselves for this—not because it is difficult to understand, but because it is absolutely ridiculous: All we do is draw little arrows on a piece of paper—that's all!" (p. 24)—but defends this because the results the methods yield are so impressively accurate: "To give you a feeling for the accuracy of these numbers, it comes out to something like this: If you were to measure the distance from Los Angeles to New York to this accuracy, it would be exact to the thickness of a human hair. That's how delicately quantum electrodynamics has, in the past fifty years, been checked—both theoretically and experimentally" (p. 7).

And *that* is the most important difference between the division of labor in religion and science: in spite of Feynman's uncharacteristically hypermodest denial, the experts *do* understand the methods they use—not everything about them, but enough to explain to one another and to themselves why the amazingly accurate results come out of them. It is only because I am confident that the experts really do understand the formulas that I can honestly and unabashedly cede the responsibility of pinning down the propositions (and hence understanding them) to them. In religion, however, the experts are not exaggerating for effect when they say they don't understand what they are talking about. The fundamental incomprehensibility of God is insisted upon as a central tenet of faith, and the propositions in question are themselves declared to be systematically elusive to everybody. Although we can go along with the experts when they advise us which sentences to say we believe, they also insist that *they themselves* cannot use their expertise to prove—even to one another—that they know what they are talking about. These matters are mysterious *to everybody*, experts and laypeople alike. Why does anybody go along with this? The answer is obvious: *belief in belief.*

Many people believe in God. Many people believe in *belief in God.* What's the difference? People who believe in God are sure that God exists, and they are glad, because they hold God to be the most wonderful of all things. People who moreover believe in belief in God are sure that *belief in God* exists (and who could doubt that?), and they think that this is a good state of affairs, something to be strongly encouraged and fostered wherever possible: If only *belief in God* were more widespread! One *ought* to believe in God. One ought to *strive* to believe in God. One should be uneasy, apologetic, unfulfilled, one should even feel guilty, if one finds that one just doesn't believe in God. It's a failing, but it happens.

It is entirely possible to be an atheist and believe in belief in God. Such a person doesn't believe in God but nevertheless thinks that believing in God would be a wonderful state of mind to be in, if only that could be arranged. People who believe in belief in God try to get others to believe in God and, whenever they find their own belief in God flagging, do whatever they can to restore it.

It is rare but possible for people to believe in something while regretting their belief in it. They don't believe in their own belief! (If I found that I believed in poltergeists or the Loch Ness Monster, I'd be, well, embarrassed. I'd think of this as one of those dirty little secrets about me that I wished were not so, and I'd be glad that nobody else knew! I might take steps to cure myself of this awkward bulge in my otherwise impeccably hardheaded and rational ontology.) People sometimes suddenly awake to the fact that they are racists, or sexists, or have lost their love of democracy. None of us want to discover these things about ourselves. We all have ideals by which we measure the beliefs we discover in ourselves, and belief in God has been one of the most salient ideals for a long time for many people.

In general, if you believe some proposition, you also believe that anybody who disbelieves it is mistaken. And by and large, it's too bad when people are mistaken or ill informed or ignorant. In general, the world would be a better place if people shared more truths and believed fewer falsehoods. That's why we have education and public-information campaigns and newspapers and so forth. There are exceptions—strategic secrets, for instance, cases where I believe something and am grateful that nobody else shares my belief. Some religious beliefs may consist in proprietary secrets, but the general pattern is for people not just to share but to try to persuade others, especially their own children, of their religious beliefs.

4 The lowest common denominator?

God is so great that the greatness precludes existence.
　—Raimundo Panikkar, *The Silence of God*

It is the final proof of God's omnipotence that he need not exist in order to save us.
　—Sermon by the hyperliberal Reverend Mackerel, hero of *The Mackerel Plaza*,
　　by Peter De Vries

The Church Militant and the Church Triumphant has become the Church Social and the Church Bizarre.
　—Robert Benson, personal communication, 1960

Many people believe in God. Many *more* people believe in belief in God! (We can be quite sure that, since just about everybody who believes in God also believes in belief in God, there are actually more people in the world who believe in belief in God than those who believe in God.) The world's literature—including uncounted sermons and homilies—teems with tales of people wracked with doubt and hoping to recover their belief in God. We've just seen that our concept of belief allows that there is a clear empirical difference between these two states of mind, but here is a perplexing question: of all the people who believe in belief in God, what percentage (roughly!) also actually believe in God? Investigating this empirical question turns out to be extremely difficult.

Why? At first it looks as if we could simply give people a questionnaire with a multiple-choice question on it:

I believe in God: _____ *Yes* _____ *No* _____ *I don't know*

Or should the question be:

God exists: _____ *Yes* _____ *No* _____ *I don't know*

Would it make any difference how we framed the questions? (I have begun conducting research on just such questions, and the results are tantalizing but not yet sufficiently confirmed to publish.) The main problem with such a simple approach is obvious. Given the way religious concepts and practices have been designed, the very behaviors that would be clear evidence of belief in God are also behaviors that would be clear evidence of (only) belief in belief in God. If those who have doubts have been enjoined by their church to declare their belief in spite of their doubts, to *say the words* with as much conviction as they can muster, again and again, in hopes of kindling conviction, to join hands and recite the creed, to pray several times a day in public, to do all the things that a believer does, then they will check the "Yes" box with alacrity, even though they really don't believe in God; they fervently believe in belief in God. This fact makes it hard to tell who—if anybody!—actually believes in God in addition to believing in belief in God.

Thanks to the division of labor, it is actually worse than that, as you may already have fathomed. You may find that when you look in your heart you simply do not know whether you *yourself* believe in God. Which God are we talking about? Unless you are an expert, and sure that you understand the formulas that officially express the propositions of your creed, your state of mind must be somewhere in the middle ground between my state of mind with regard to (1) (the sentence in Turkish) and my state of mind with regard to (2) (Einstein's formula). You're not as clueless as I am regarding (1); you have studied and probably even memorized the official formulas, and you believe that these formulas are true (whatever they mean), but you have to admit that you are no *authority* on what they mean. Many Americans find themselves in this position, as Alan Wolfe notes in *The Transformation of American Religion: How We Actually Live Our Faith*, his recent survey of developments in American religion: "These are people who believe, often passionately, in God, even if they cannot tell others all that much about the God in which they believe" (2003, p. 72). If you fall in this category, you must admit, contrary to the way Wolfe puts it, that, although you may well be one of those who believe in belief in God, you aren't really in a good position to judge whether you actually believe (passionately or otherwise) in the God of your particular creed, or in some other God. (And you have almost certainly never taken a tough multiple-choice test to see if you can reliably distinguish the expert's conception of God from the subtle impostors that are almost right.)

Alternatively, you can set yourself up as your own authority: "I know what *I* mean when I utter the creed, and that's good enough for me!" And that's good enough—these days—for a surprising number of organized religions, too. Their leaders have come to realize that the robustness of the institution of religion doesn't depend on uniformity of *belief* at all; it depends on the uniformity of *professing*. This has long been a feature of some strains of Judaism: fake it and *never mind* if you make it (as my student Uriel Meshoulam once vividly put it to me). Recognizing that the very idea of commanding someone to *believe* something is incoherent on its face, an invitation to insincerity or self-deception, many Jewish congregations reject the demand for *orthodoxy*, right *belief*, and settle for *orthopraxy*, right *behavior*. Instead of creating secret pockets of festering guilty skepticism, they make a virtue of candid doubt, respectfully expressed.

As long as the formulas get transmitted down through the ages, the memes will survive and flourish. Much the same attitude has recently been adopted by many evangelical Christian denominations, especially the booming new phenomenon of "mega-churches," which, as Wolfe describes in some detail, go out of their way to give their members plenty of elbow room for personal interpretations of the words they claim to be holy. Wolfe distinguishes sharply between evangelicalism and fundamentalism, which "tends to be more preoccupied with matters of theological substance." His conclusion is intended to be reassuring:

> But those who fear the consequences for the United States of a return to strong religious belief should not be fooled by evangelicalism's rapid growth. On the contrary, evangelicalism's popularity is due as much to its populistic and democratic urges—its determination to find out exactly what believers want and to offer it to them—as it is to certainties of the faith. [2003, p. 36]

Wolfe shows that Stark and Finke's frank marketing approach is not at all foreign to religious leaders themselves. He notes without irony some of the concessions they are willing to make to contemporary secular culture, concessions that go far beyond Web sites and multimillion-dollar television programs, or the introduction of electric guitars, drums, and PowerPoint in their services. For instance, the term "sanctuary" is shunned by one church "because of its strong religious connotations" (p. 28), and more attention is paid to providing plenty of free parking and babysitting than to the proper interpretation of passages of Scripture. Wolfe has conducted many probing interviews with his informants, and they reveal that *revision* of tradition is often hard to distinguish from outright *rejection*. A derisive term has been coined by these memetic engineers to describe the image they are trying hard to shed: "churchianity" (p. 50).

Indeed, Lars and Ann, like many evangelicals throughout the country, say that faith is so important to them that "religion"—

which they associate with discord and disagreement and, there-fore, if often in an unexpected way, with doctrine—cannot be allowed to interfere with its exercise. [p. 73]

There is no denying the results of this marketing expertise. Pastor Chuck Smith's Calvary Chapel has over six hundred churches, some of them with ten thousand worshipers a week (Wolfe, 2003, p. 75). Dr. Creflo Dollar's World Changers Church has twenty-five thousand members, "but only thirty per cent of them were regular tithers" (Sanneh, 2004, p. 48). According to Wolfe, "All of America's religions face the same imperative: Personalize or die. Each does so in different ways" (p. 35). He may be right, but his argument for this sweeping conclusion is sketchy and anecdotal, and though there can be no doubt that the phenomena he describes exist, the question of whether they will be permanent features of religion from now on or a passing fad is a question that cries out for a testable theory, not just a set of observations, however sensitive. Whatever its staying power, and the reasons for it, the example of such *laissez-faire* "noncredal" religion contrasts vividly with the continuing doctrinal emphasis of the Roman Catholic Church.

5 Beliefs designed to be professed

A mountain climber foolishly climbing alone slips off a precipice and finds himself dangling at the end of his safety rope, a thousand feet above a ravine. Unable to climb the rope or swing to a safe resting spot, he calls out in despair: "Hallooo, hallooo! Can anybody help me?" To his astonishment, the clouds part, a beautiful light pours through them, and a mighty voice replies, "Yes, my son, I can help you. Take your knife and cut the rope!" The climber takes out his knife, and then he stops, and thinks and thinks. Then he cries out: "Can anybody else help me?"

According to the old maxim, actions speak louder than words, but this actually doesn't say what it means. Speech acts are actions, too, and a person who *says,* for instance, that infidels deserve death is performing an action with potentially deadly effects, which is about as "loud" as acting can get. What the maxim means, on reflection, is that actions other than speech acts are typically better evidence of what the actor really believes than any words the actor might say. It is easy enough to pay lip service (such a wonderful idiom!), but when the concrete consequences of your actions depend on whether you believe something—whether you believe the gun is loaded, whether you believe the door is unlocked, whether you believe you are unobserved—lip service is a puny datum easily swamped by the nonverbal behavior that expresses—indeed, betrays—your true beliefs.

And here is an interesting fact: the transition from folk religion to organized religion is marked by a shift in beliefs from those with very clear, concrete consequences to those with systematically elusive consequences—paying lip service is just about the only way you *can* act on them. If you really believe that the rain god won't provide rain unless you sacrifice an ox, you sacrifice an ox if you want it to rain. If you really believe that your tribe's god has made you invulnerable to arrows, you readily run headlong into a swarm of deadly arrows to get at your enemy. If you really believe that your God will save you, you cut the rope. If you really believe that your God is watching you and doesn't want you to masturbate, you don't masturbate. (You wouldn't masturbate with your mother watching you! How on earth could you masturbate with God watching you? Do you *really* believe God is watching you? Perhaps not.)

But what could you do to show that you really believe that the wine in the chalice has been transformed into the blood of Christ? You could bet a large sum of money on it and then send the wine to the biology lab to see if there was hemoglobin in it (and recover the genome of Jesus from the DNA in the bargain!)—except that the creed has been cleverly shielded from just such concrete tests. It would be a sacrilege to remove the wine from the ceremony, and, besides, taking the wine out of the holy context would surely untransubstantiate it, turning it back into ordinary wine. There is really only one action you can take to demonstrate this belief: you can *say* that you believe it, over and over, as fervently as the occasion demands.

This topic is broached in a telling way in "Dominus Iesus: On the Unicity and Salvific Universality of Jesus Christ and the Church," a Declaration written by Cardinal Ratzinger (who later was elected Pope Benedict XVI), and ratified by Pope John Paul II at a plenary session on June 16, 2000. Again and again this document specifies what faithful Catholics must *"firmly believe"* (italics in the original), but at several points the Declaration shifts idiom and speaks of what "the Catholic faithful *are required to profess*" (italics in the original). As a professor myself, I find the use of this verb irresistible. What is commonly referred to as "religious belief" or "religious conviction" might less misleadingly be called *religious professing*. Unlike academic professors, religious professors (not just priests, but all the faithful) may not either understand or believe what they are professing. They are *just* professing, because that is the best they can do, and they are *required* to profess. Cardinal Ratzinger cites Paul's letter to the Corinthians: "Preaching the Gospel is not a reason for me to boast; it is a necessity laid on me: woe to me if I do not preach the Gospel!" (1 Corinthians 9:16).

Though lip service is thus required, it is not enough: you must *firmly believe* what you are obliged to say. How is it possible to obey this injunction? Professing is voluntary, but belief is not. Belief— when it is distinguished from believing that some sentence expresses a truth—requires understanding, which is hard to come by, even by the experts in these matters. You can't just make yourself believe something by trying, so what are you to do? Cardinal Ratzinger's Declaration offers some help on this score: "Faith is the acceptance in grace of revealed truth, which 'makes it possible to penetrate the mystery in a way that allows us to understand it coherently' [quoting John Paul II's Encyclical Letter *Fides et Ratio*,

p. 13]." So you should believe *this*. And if you can, believing this should help you believe you *do* understand the mystery (even if it seems to you that you don't), and hence do firmly believe whatever it is you profess you believe. But how do you believe *this*? It takes faith.

Why even try? What if you personally don't happen to share the belief in the belief in the doctrine in question? Here is where the meme's-eye view can provide some explanation. In his original discussion of memes, Dawkins had noted this problem and its traditional solution: "Many children and even some adults believe that they will suffer ghastly torments after death if they do not obey the priestly rules. . . . The idea of hell fire is, quite simply, *self-perpetuating*, because of its own deep psychological impact" (Dawkins, 1976, p. 212). If you have ever received a chain letter that warned of the terrible things that would happen to you if you failed to pass it along, you can appreciate the strategy, even if you didn't fall for it. The assurances of a trusted priest can be much more compelling.

If hellfire is the stick, mystery is the carrot. The propositions to be believed ought to be baffling! As Rappaport has trenchantly put it, "If postulates are to be unquestionable, it is important that they be incomprehensible" (1979, p. 165). Not just *counterintuitive*, in Boyer's technical sense of contradicting only one or two of the default assumptions of a basic category, but downright unintelligible. Prosaic assertions have no bite, and moreover they are too readily checked for accuracy. For a truly awesome and mind-teasing proposition, there is nothing that beats a paradox eagerly avowed. In a later essay, Dawkins drew attention to what we might call the inflation of *credal athleticism*, the boast that my faith is so strong that I can mentally embrace a bigger paradox than you can.

It is easy and non-mysterious to believe that in some symbolic or metaphorical sense the eucharistic wine turns into the blood of

Christ. The Roman Catholic doctrine of transubstantiation, however, claims far more. The "whole substance" of the wine is converted into the blood of Christ; the appearance of wine that remains is "merely accidental", "inhering in no substance". Transubstantiation is colloquially taught as meaning that the wine "literally" turns into the blood of Christ. [Dawkins, 1993, p. 21][11]

There are several reasons why this inflation into incomprehensibility would be an adaptation that would enhance the fitness of a meme. First, as just noted, it tends to evoke wonder and draw attention to itself. It is a veritable peacock's tail of extravagant display, and memetics would predict that something like an arms race of paradoxology should ensue when religions confront waning allegiance. Peacocks' tails are finally limited by the sheer physical inability of the peacocks to carry around still larger ones, and paradoxology must hit the wall, too. People's discomfort with sheer incoherence is strong, so there are always tantalizing elements of sense-making narrative, punctuated with seriously perplexing nuggets of incomprehensibility. The anomalies give the host brains something to gnaw on, like an unresolved musical cadence, and hence something to rehearse, and rehearse again, and baffle themselves deliciously about.[12] Second, as noted in chapter 5, incomprehensibility discourages paraphrase—which can be death to meme identity—by leaving the host with no viable choice but *verbatim* transmission. ("I don't really know what Pope John Paul II *meant*, but I can tell you that what he *said* was: 'Jesus is the Incarnate Word—a single and indivisible person.' ")

Dawkins has noted an extension or refinement of this adaptation: "The meme for blind faith secures its own perpetuation by the simple unconscious expedient of discouraging rational inquiry" (1976, pp. 212–13). At a time when "faith-based initiatives" and other such uses of the term have made "faith" almost synonymous in the minds of many with the term "religion" (as in the phrase "people of all faiths"), it is important to remind ourselves that not all religions have a home for the concept or anything even very close to it. The meme for faith exhibits *frequency-dependent fitness:* it flourishes particularly in the company of rationalistic memes. In a neighborhood with few skeptics, the meme for faith does not attract much attention, and hence tends to go dormant in minds, and hence is seldom reintroduced into the memosphere (Dennett, 1995b, p. 349). Indeed, it is mainly a Christian feature, and as we recently noted, Judaism has actually encouraged vigorous intellectual debate over the meaning, and even the truth, of many of its holy texts. But a similar athleticism is honored in Jewish practice, as explained by a rabbi:

> That most of the Kashrut [kosher] laws are divine ordinances without reason given is 100 per cent to the point. It is very easy not to murder people. Very easy. It is a little bit harder not to steal because one is tempted occasionally. So that is not great proof that I believe in God or am fulfilling His will. But, if He tells me not to have a cup of coffee with milk in it with my mincemeat and peas at lunchtime, that is a test. The only reason I am doing that is because I have been told to so do. It is doing something difficult. [*Guardian*, July 29, 1991, quoted in Dawkins, 1993, p. 22]

Islam, meanwhile, obliges its faithful to stop what they are doing five times a day to pray, no matter how inconvenient or even dangerous that act of loyalty proves to be. This idea that we prove our faith by one extravagant act or another—such as choosing death over recanting an item of doctrine that we don't understand—permits us to draw a strong distinction between religious faith and the sort of faith that I, for one, have in science. My faith in the expertise of physicists like Richard Feynman, for instance, permits me to endorse—and, if it comes to it, bet heavily on the truth of—a proposition that I don't understand. So far, my faith is not unlike religious faith, but I am not in the slightest bit motivated to go to my death rather than recant the formulas of physics. Watch:

E *doesn't* equal mc², it *doesn't*, it *doesn't*! I was lying, so there!

I feel no guilt in making this little joke, unlike people who would find it deeply difficult to utter blasphemous words or recant their creed. But isn't my faith in the truth of the propositions of quantum mechanics that I admit I don't understand a sort of *religious* faith in any case? Let me invent a deeply religious person, Professor Faith,[13] to give a little speech that articulates this charge. Professor Faith wants to teach me a new word, "apophatic":

God is a Something that is Wonderful. He is an appropriate recipient of prayers, and that's about all we can say about Him. My concept of God is *apophatic*! What, you may ask, does that mean? It means I *define* God as ineffable, unknowable, something beyond all human ken. Listen to what Simon Oliver, writing about Denys Turner's recent book, *Faith Seeking* (2003), has to say:

> . . . the God rejected by modern atheism is not the God of orthodox, pre-modern Christianity. God is not any kind of thing whose existence might be rejected in the way that one might reject the existence of Santa Claus. Turner's God—owing much to the medieval mystics—is profoundly apophatic, wholly other and, in the end, unknowable darkness. We begin our journey into that alterity in our realization that our being is a gracious gift. [p. 32]

And here is Raimundo Panikkar, writing about Buddhism:

> The term "apophatic" is usually used in reference to an epistemological apophaticism, positing merely that the ultimate reality is *ineffable*—that human intelligence is incapable of grasping, of embracing it—although this ultimate reality itself may be represented as *intelligible*, even supremely intelligible, *in se*. A gnoseological apophaticism, then, comports an ineffability on the part of the ultimate reality only *quoad nos*. Buddhistic apophaticism, on the other hand, seeks to transport this ineffability to the heart of ultimate reality itself, declaring that this

reality—inasmuch as its *logos* (its expression and communication) no longer pertains to the order of ultimate reality but precisely to the manifestation of that order—is ineffable not merely in our regard, but as such, *quoad se*. Thus Buddhistic apophaticism is an ontic apophaticism. [1989, p. 14]

I claim that these claims really aren't so different from what your scientists say. Physicists have come to realize that matter isn't composed of clusters of hard little spheres (atoms). Matter is much stranger than that, they acknowledge, but still they call it matter, even though they mainly know what matter *isn't*, not what it is. They're still calling them atoms, but they no longer think of them as, well, *atomic*. They've changed their conception of atoms, their conception of matter, quite radically. And if you ask them what they now think matter is, they confess that it's something of a mystery. *Their* concept is apophatic, too! If physicists can move from concreteness to mystery, so can theologians.

I hope Professor Faith has done justice to this theme, which I have often encountered in discussion. I am not at all persuaded by it. There is a big difference between religious faith and scientific faith: what has driven the changes in concepts in physics is not just heightened skepticism from an increasingly worldly and sophisticated clientele, but a tidal wave of exquisitely detailed positive results—the sorts of borne-out predictions that Feynman pointed to in defending his field. And this makes a huge difference because it gives beliefs about the truths of physics a place where the rubber meets the road, where there is more than mere professing that can be done. For instance, you can build something that *depends for its safe operation on the truth of those sentences* and risk your life trying to fly it to the moon. Like the folk religionists' beliefs that they should sacrifice a goat or that they are invulnerable to arrows, these are beliefs that you can act on in ways that speak louder than words. People who give away all their belongings and climb to some mountaintop in anticipation of the imminent End of the World don't just believe in belief in God, but they are the exceptions, not the rule, when it comes to religious convictions.

6 Lessons from Lebanon: the strange cases of the Druze and Kim Philby

There is still more that is systematically curious about the phenomenon that people *call* religious belief but that might better be called religious *professing*. This is a feature that has long captivated me, while further persuading me that Hume's project of natural religion (evaluating arguments for and against the existence of God) is largely wasted effort. My interest in this feature grew out of two experiences, both of which involve events that took place in Lebanon more than forty years ago (though that is a sheer coincidence, so far as I know). I spent some of my earliest days in Beirut, where my father, a historian of Islam, was cultural attaché (and a spy for the OSS). The rhythm of the muezzins calling the faithful to prayer from the nearby minaret was my everyday experience, along with my teddy bear and toy trucks, and the beautifully haunting call never fails to send chills through me when I hear it today. But I left Beirut when I was only five, and didn't return until 1964, when I visited my mother and sister, who were living there then. We spent some time in the mountains outside Beirut in a village that was mostly Druze, with some Christians and Muslims thrown in. I asked some of the non-Druze residents of the town to tell me about the Druze religion, and this is what they said:

> Oh, the Druze are a very sad lot. The first principle of the Druze religion is to lie to outsiders about their beliefs—never tell the truth to an infidel! So you shouldn't take anything a Druze tells you as authoritative. Some of us think, in fact, that the Druze used to have a holy book, their own scripture, but they lost it, and they are so embarrassed by this that they make up all manner of solemn nonsense to keep this from coming out. You will notice

that the women don't participate at all in the Druze ceremonies; that's because they couldn't keep such a secret!

I heard this tale from several people who claimed to know, and I also heard it denied by a few Druze, of course. But if it was true, this would create a dilemma for any anthropologist: the usual method of questioning informants would be a hopeless wild-goose chase, and if he made the ultimate sacrifice and converted to Druze himself so as to gain entrance to the inner sanctum, he would have to admit that we on the outside shouldn't believe his scholarly treatise, *What the Druze Really Believe,* since it was written by a devout Druze (and everybody knows that the Druze lie). As a young philosopher, I was fascinated by this real-life version of the liar paradox (Epimenides the Cretan says that all Cretans are liars; does he speak the truth?), and also by the unmistakable echoes of another famous example in philosophy: Ludwig Wittgenstein's beetle in the box. In *Philosophical Investigations* (1953), Wittgenstein says:

> Suppose everyone had a box with something in it: we call it a "beetle." No one can look into anyone else's box, and everyone says he knows what a beetle is only by looking at *his* beetle.— Here it would be quite possible for everyone to have something different in his box. One might even imagine such a thing constantly changing. —But suppose the word "beetle" had a use in these people's language? —If so it would not be used as the name of a thing. The thing in the box has no place in the language-game at all; not even as a *something*: for the box might even be empty. —No, one can "divide through" by the thing in the box; it cancels out, whatever it is. [Section 293]

Much has been written on Wittgenstein's beetle box, but I don't know if anybody has ever proposed an application to religious belief. In any case, it seems fantastic at first that the Druze might be an actual example of the phenomenon. Am I just inflating a mean-spirited calumny of the Druze by their neighbors to make a dubious philosophical point? Perhaps, but consider what Scott Atran has to say about his attempts, as an anthropologist, to write about the beliefs of the Druze:

> As a graduate student almost three decades ago, I spent some years with the Druze people of the Middle East. I wanted to learn about their religious beliefs, which appeared to weave together ideas from all the great monotheistic faiths in intriguing ways. Learning about Druze religion is a gradual process in the Socratic tradition, involving interpretation of parables in question-and-answer format. Although, as a non-Druze, I could never be formally initiated into the religion, the elders seemed to delight in my trying to understand the world as they conceived it. But every time I reached some level of awareness about a problem, Druze elders reminded me that anything said or learned beyond that point could not be discussed with uninitiated persons, including other Druze. I never did write on Druze religion and wound up with a thesis on the cognitive bases of science. [2002, p. ix]

It seems that we still don't know what the Druze really believe. We may begin to wonder if they themselves know. And we may also begin to wonder if it matters, which brings me to my second lesson from Lebanon.

In 1951, Kim Philby, a senior officer in the British intelligence service (SIS), fell under suspicion of being a double agent, a highly placed traitor working for the Soviet KGB. A secret tribunal was held by SIS, but Philby was found not guilty on the evidence presented. Although SIS had been unable to convict him, they quite reasonably refused to reinstate him to his most sensitive position, and he resigned, and moved to Lebanon, to work as a journalist. In 1963, a Soviet defector to London confirmed Philby's double-agent role, and when the SIS went to Beirut to confront him, he fled to Moscow, where he spent the remainder of his life, working for the KGB.

Or did he? When Philby first showed up in Moscow, he was (apparently) suspected by the KGB of being a British plant—a *triple* agent, if you like. Was he, in fact? For years a story circulated in intelligence circles to this effect. The idea was that when SIS "exonerated" Philby in 1951, they found a brilliant way of dealing with their delicate problem of trust:

> Congratulations, Kim, old chap! We always thought you were loyal to our cause. And for your next assignment, we would like you to *pretend* to resign from SIS—bitter over our failure to reinstate you fully, don't you see—and move to Beirut and take up a position as a journalist in exile. In due course we intend to give you reason to "flee" to Moscow, where you will eventually be appreciated by your comrades because you can spill a lot of relatively innocuous insider information you already know, and we'll provide you with carefully controlled further gifts of intelligence—and disinformation—that the Russians will be glad to accept, even when they have their doubts. Once you're in their good graces, we'd like you to start telling us everything you can about what they're up to, what questions they ask you, and so forth.

Once SIS had given Philby this new assignment, their worries were over. It just *didn't matter* whether he was truly a British patriot pretending to be a disgruntled agent, or truly a loyal Soviet agent pretending to be a loyal British agent (pretending to be a disgruntled agent . . .). He would behave in exactly the same ways in either case; his activities would be interpretable and predictable from either of two mirror-image intentional-stance profiles. In one, he deeply believes that the British cause is worth risking his life for, and in the other, he deeply believes that he has a golden opportunity to be a hero of the Soviet Union by pretending that he deeply believes that the British cause is worth risking his life for, and so on. The Soviets, meanwhile, would no doubt draw the same inference and not bother trying to figure out if Philby was really a double agent or a triple agent or a quadruple agent. Philby, according to this story, had been deftly turned into a sort of human telephone, a mere conduit of information that both sides could exploit for whatever purposes they could dream up, relying on him to be a high-fidelity transmitter of whatever information they gave him, without worrying about where his ultimate loyalties lay.

In 1980, when Philby's standing with his overseers in Moscow was improving (apparently), I was a Visiting Fellow at All Souls College in Oxford, and another Visiting Fellow at the time happened to be Sir Maurice Oldfield, the retired head of MI6, the agency responsible for counterespionage outside Great Britain, and one of the spymasters responsible for Philby's trajectory. (Sir Maurice was the model for Ian Fleming's "M" in the James Bond novels.) One night, after dinner, I asked him whether this story I had heard was true, and he replied quite testily that it was a lot of rubbish. He wished people would just let poor Philby live out his days in Moscow in peace and quiet. I replied that I was pleased to get his answer, but we both had to recognize that it was also what he would have told me had the story been true! Sir Maurice glowered and said nothing.[14]

These two stories illustrate in extreme form the fundamental problem faced by anyone intent on studying religious beliefs. It has been noted by many commentators that typical, canonical religious beliefs cannot be tested for truth. As I suggested earlier, this is as good as a defining characteristic of religious creeds. They have to be "taken on faith" and are not subject to (scientific, historical) confirmation. But, more than that, for this reason and others, religious-belief *expressions* cannot really be taken at face value. The anthropologists Craig Palmer and Lyle Steadman (2004, p. 141) quote the lament of their distinguished predecessor the anthropologist Rodney Needham, who was frustrated in his work with the Penan, in interior Borneo:

I realized that I could not confidently describe their attitude to God, whether this was belief or anything else. . . . In fact, as I had glumly to conclude, I just did not know what was their psychic attitude toward the personage in whom I had assumed they believed. . . . Clearly, it was one thing to report the received ideas to which a people subscribed, but it was quite another matter to say what was their inner state (belief for instance) when they expressed or entertained such ideas. If, however, an ethnographer said that people believed something when he did not actually know what was going on inside them, then surely his account of them must, it occurred to me, be very defective in quite fundamental ways. [Needham, 1972, pp. 1–2]

Palmer and Steadman take this recognition by Needham to signal the need for recasting anthropological theories as accounts of religious behavior, not religious belief: "While religious beliefs are not identifiable, religious behavior is, and this aspect of the human experience can be comprehended. What is needed is an explanation of this observable religious behavior that is restricted to what can be observed" [p. 141]. They go on to say that Needham is virtually alone in realizing the profound implications of this fact about the inscrutability of religious avowal, but they themselves overlook the even more profound implication of it: the natives are in the same boat as Needham! *They* are just as unable to get into the inner minds of their kin and neighbors as Needham is.

When it comes to interpreting religious avowals of others, *everybody is an outsider*. Why? Because religious avowals concern matters that are beyond observation, beyond meaningful test, so the only thing *anybody* can go on is religious behavior, and, more specifically, the behavior of *professing*. A child growing up in a culture is like an anthropologist, after all, surrounded by informants whose professings stand in need of interpretation. The fact that your informants are your father and mother, and speak in your mother tongue, does not give you anything more than a slight circumstantial advantage over the adult anthropologist who has to rely on a string of bilingual interpreters to query the informants.[15] (And think about your own case: weren't you ever baffled or confused about just what you were supposed to believe? You know perfectly well that *you* don't have privileged access to the tenets of the faith that you were raised in. I am just asking you to generalize the point, to recognize that others are in no better position.)

7 Does God exist?

If God did not exist, it would be necessary for us to invent Him.
—Voltaire

At long last I turn to the promised consideration of arguments for the existence of God. And, having reviewed the obstacles—diplomatic, logical, psychological, and tactical—facing anybody who wants to do this constructively, I will give just a brief bird's-eye view of the domain of inquiry, expressing my own verdicts but not the reasoning that has gone into them, and providing references to a few pieces that may not be familiar to many. There is a spectrum of intentional objects to consider, ranging along the anthropomorphism scale from a Guy in the Sky to one Timeless Benign Force or another. And there is a spectrum of arguments, which line up only unevenly with the spectrum of Gods. We can begin with anthropomorphic Gods and the arguments from presumed historical documentation, such as this: according to the Bible, which is the literal truth, God exists, has always existed, and created the universe in seven days a few thousand years ago. The historical arguments are apparently satisfying to those who accept them, but they simply cannot be introduced into a serious investigation, since they are manifestly question-begging. (If this is not obvious to you, ask whether the *Book of Mormon* [1829] or the founding document of Scientology, L. Ron Hubbard's book *Dianetics* [1950], should be taken as irrefutable evidence for the propositions it contains. No text can be conceded the status of "gospel truth" without foreclosing all rational inquiry.)

That leaves us with the traditional arguments discussed at great length by the philosophers and theologians over the centuries, some empirical—such as the Argument from Design—and some purely *a priori* or logical—such as the Ontological Argument and the Cosmological Argument for the necessity of a First Cause.

The logical arguments are regarded by many thinkers, including many philosophers who have looked at them carefully for years, to be intellectual conjuring tricks or puzzles rather than serious scientific proposals. Consider the Ontological Argument, first formulated by Saint Anselm in the eleventh century as a direct response to Psalm 14:1, about what the fool said in his heart. If the fool understands the concept of God, Anselm claims, he should understand that God is (by definition) the *greatest conceivable* being—or, in a famously mind-twisting phrase, the *Being greater than which nothing can be conceived.* But among the perfections such a greatest conceivable Being would have to have is existence, since if God lacked existence it would be possible to conceive of a still greater Being—namely, God-with-all-His-perfections-*plus-existence*! A God that lacked existence would not be the Being greater than which nothing can be conceived, but that is the definition of God, so therefore God *must* exist. Do you find this compelling? Or do you suspect it is some sort of logical "trick with mirrors"? (Could you use the same argument scheme to prove the existence of the most perfect ice-cream sundae conceivable—since if it didn't exist there would be a more perfect conceivable one: namely, one that *did* exist?) If you're suspicious, you're in good company. Ever since Immanuel Kant in the eighteenth century, there has been a widespread—but by no means unanimous—conviction that you can't prove the existence of *anything* (other than an abstraction) by sheer logic. You can prove that *there is* a prime number larger than a trillion, and that *there is* a point at which the lines bisecting the three angles of any triangle all meet, and *there is* a "Gödel sentence" for every Turing machine that is consistent and can represent the truths of arithmetic, but you can't prove that something that has effects in the physical world exists except by methods that are at least partly empirical.[16] There are those who disagree, and continue to champion updated versions of Anselm's Ontological Argument, but the price they pay (willingly, one gathers) for their access to purely logical proof is a remarkably bare and featureless intentional object. Even if a *Being greater than which nothing can be conceived* has to exist, as their arguments urge, it is a long haul from that specification to a Being that is merciful or just or loving—unless you make sure to define it that way from the outset, introducing anthropomorphism by a dodge that will not persuade the skeptics, needless to say. Nor—in my experience—does it reassure the faithful.

The Cosmological Argument, which in its simplest form states that since everything must have a cause the universe must have a cause—namely, God—doesn't stay simple for long. Some deny the premise, since quantum physics teaches us (doesn't it?) that not everything that happens needs to have a cause. Others prefer to accept the premise and then ask: What caused God? The reply that God is self-caused (somehow) then raises the rebuttal: If something can be self-caused, why can't the universe as a whole be the thing that is self-caused? This leads in various arcane directions, into the strange precincts of string theory and probability fluctuations and the like, at one extreme, and into ingenious nitpicking about the meaning of "cause" at the other. Unless you have a taste for mathematics and theoretical physics on the one hand, or the niceties of scholastic logic on the other, you are not apt to find any of this compelling, or even fathomable.

Still, people may want to return to the *a priori* arguments as a safety net of sorts after they see what can be made of the empirical argument, the Argument from Design, including its recent variants invoking the Anthropic Principle. The Argument from Design is surely the most intuitive and popular argument, and has been for centuries. It just stands to reason (doesn't it?) that all the wonders of the living world have to have been arranged by some Intelligent Designer? It couldn't all just be an accident, could it? And even if evolution by natural selection explains the design of living things, doesn't the "fine tuning" of the laws of physics to make all this evolution possible require a Tuner? (The Anthropic Principle Argument.) No, it doesn't stand to reason, and, yes, it could all just be the result of "accidents" exploited by the relentless regularities of nature, and, no, the fine tuning of the laws of physics can be explained without postulating an Intelligent Tuner. I have covered these arguments quite extensively in *Darwin's Dangerous Idea* (especially chapters 1 and 7),[17] so I will repeat not my counterarguments but only my summary of the retreat that Darwin's dangerous idea has propelled during the last century and a half.

We began with a somewhat childish vision of an anthropomorphic, Handicrafter God, and recognized that this idea, taken literally, was well on the road to extinction. When we looked through Darwin's eyes at the actual processes of design of which we and all the wonders of nature are the products to date, we found that Paley was right to see these effects as the result of a lot of design work, but we found a non-miraculous account of it: a massively parallel, and hence prodigiously wasteful, process of mindless, algorithmic design-trying, in which, however, the minimal increments of design have been thriftily husbanded, copied and re-used over billions of years. The wonderful *particularity* or *individuality* of the creation was due, not to Shakespearean inventive genius, but to the incessant contributions of chance, a growing sequence of what Crick (1968) has called "frozen accidents."

That vision of the creative process still apparently left a role for God as Lawgiver, but this gave way in turn to the Newtonian role of Lawfinder, which also evaporated, as we have recently seen, leaving behind no Intelligent Agency in the process at all. What is left is what the process, shuffling through eternity, mindlessly finds (when it finds anything): a timeless Platonic possibility of order. That is, indeed, a thing of beauty, as mathematicians are forever exclaiming, but it is not itself something intelligent but, wonder of wonders, something intelligible. Being abstract and outside of time, it is nothing with an *initiation* or *origin* in need of explanation.[18] What does need its origin explained is the concrete Universe itself, and as Hume's Philo long ago asked: Why not stop at the material world? *It*, we have seen, does perform a version of the ultimate bootstrapping trick; it creates itself *ex nihilo*, or at any rate out of something that is well-nigh indistinguishable from nothing at all. Unlike the puzzlingly mysterious, timeless self-creation of God, this self-creation is a non-miraculous stunt that has left lots of traces. And being not just concrete but the product of an exquisitely particular historical process, it is a creation of utter uniqueness—encompassing and dwarfing all the novels and paintings and symphonies of all the artists— occupying a position in the hyperspace of possibilities that differs from all others.

Benedict Spinoza, in the seventeenth century, *identified* God and Nature, arguing that scientific research was the true path of theology. For this heresy he was persecuted. There is a troubling (or to some, enticing) Janus-faced quality to Spinoza's heretical vision of *Deus sive Natura* (God, or Nature): in proposing his scientific simplification, was he personifying Nature or depersonalizing God? Darwin's more generative vision provides the structure in which we can see the intelligence of Mother Nature (or is it merely apparent intelligence?) as a non-miraculous and non-mysterious—and hence all the more wonderful—feature of this self-creating thing. [Dennett, 1995b, pp. 184–85]

Should Spinoza be counted as an atheist or a pantheist? He saw the glory of nature and then saw a way of eliminating the middleman! As I said at the end of my earlier book:

The Tree of Life is neither perfect nor infinite in space or time, but it is actual, and if it is not Anselm's "Being greater than which nothing can be conceived" it is surely a being that is greater than anything any of us will ever conceive of in detail worthy of its detail. Is something sacred? Yes, say I with Nietzsche. I could not pray to it, but I can stand in affirmation of its magnificence. The world is sacred. [1995b, p. 520]

Does that make me an atheist? Certainly, in the obvious sense. If what you hold sacred is not any kind of Person you could pray to, or consider to be an appropriate recipient of gratitude (or anger, when a loved one is senselessly killed), you're an atheist in my book. If, for reasons of loyalty to tradition, diplomacy, or self-protective camouflage (very important today, especially for politicians), you want to deny what you are, that's your business, but don't kid yourself. Maybe in the future, if more of us brights will just come forward and calmly announce that of course we no longer believe in any of those Gods, it will be possible to elect an atheist to some office higher than senator. We now have Jewish and female senators and homosexual members of Congress, so the future looks bright.

So much for the belief in God. What about belief in belief in God? We still haven't inquired about all the grounds for this belief in belief. Isn't it true? That is, isn't it true that, *whether or not God exists,* religious belief is at least as important as the belief in democracy, in the rule of law, in free will? The very widespread (but far from universal) opinion is that religion is the bulwark of morality and meaning. Without religion we would fall into anarchy and chaos, in a world in which "anything goes."

The last five chapters have exposed a variety of familiar tricks that have been rediscovered over and over and that tend to have the effect of protecting religious practices from extinction or erosion beyond recognition. If the grim side of this is the design of kleptocracies and other manifestly evil organizations that can prey on innocent people, the happy side is the design of humane and useful institutions that do not just deserve the loyalty of people but can effectively secure it. We *still* have not seriously addressed the question of whether religions—some religions, one religion, any religion—are social phenomena that do more good than harm. Now that we can see through some of this protective gauze, we are in position to address that question.

\\\

Chapter 8 The belief that belief in God is so important that it must not be subjected to the risks of disconfirmation or serious criticism has led the devout to "save" their beliefs by making them incomprehensible even to themselves. The result is that even the professors don't really know what they are professing. This makes the goal of either proving or disproving God's existence a quixotic quest—but also for that very reason not very important.

\\\

Chapter 9 The important question is whether religions deserve the continued protection of their adherents. Many people love their religions more than anything else in life. Do their religions deserve this adoration?

PART III

||||\\\||||

RELIGION TODAY

||||\\\\||||

CHAPTER NINE

Toward a Buyer's Guide to Religions

I For the love of God

There is a state of mind, known to religious men, but to no others, in which the will to assert ourselves and hold our own has been displaced by a willingness to close our mouths and be as nothing in the floods and waterspouts of God. In this state of mind, what we most dreaded has become the habitation of our safety, and the hour of our moral death has turned into our spiritual birthday. The time of tension in our soul is over, and that of happy relaxation, of calm, deep breathing, of an eternal present, with no discordant future to be anxious about, has arrived.

—William James, *The Varieties of Religious Experience*

Most people believe in the belief in God, even those who can't manage to believe in God (all the time). Why do they believe this? An obvious answer is that they want to be good. That is, they want to lead good and meaningful lives and they want this for others as well, and they can see no better way to do this than to put themselves in the service of God. This answer may be right, and *they* may be right, but before we can consider this answer with the care it deserves, we need to address a challenge. Some people—and you may be one of them—find this whole setting of the issue objectionable. I will let Professor Faith try to give a fair expression of this point of view:

> You insist on treating the question of religion as if it were like whether or not to switch jobs, or buy a car, or have an operation—a matter that ought to be settled by calmly and objectively considering the pros and cons, and then drawing a conclusion about the best course, "all things considered." That's not how we see it at all. It isn't that belief in the belief in God is our settled conviction, a matter of the best overall life policy we have been able to discover. It goes way beyond that! In the previous chapter you talked about "fake it until you make it," but you never got around to describing the wonderful state of those who *do* "make it," whose honest attempts to imbue themselves with the spirit of God succeed in a burst of glory. Those of us who know the experience know that it is unlike any other experience, a joy warmer than the joy of motherhood, deeper than the joy of victory in sports, more ecstatic than the joys of playing or singing great music. When we *see the light,* it isn't just an "Aha!" experience, like figuring out a puzzle or suddenly seeing the hidden figure in a drawing, or getting a joke, or being persuaded by an argument. It isn't arriving at a *belief* at all. We *know,* then, that God is the greatest thing that could ever enter our lives. It isn't like accepting a conclusion; it's like falling in love.

Yes, I hear you. I deliberately gave this chapter a provocative title to energize this concern and put this objection in the limelight. I recognize the state you're describing, and I would offer a friendly amendment: it isn't just *like* falling in love; it is a *kind* of falling in love. The discomfort or even outrage you feel when confronted by my calm invitation to consider the pros and cons of your religion is the same reaction one feels when asked for a candid evaluation of one's true love: "I don't just *like* my darling because, after due consideration, I believe all her wonderful qualities far outweigh her few faults. I *know* that she is the one for me, and I will always *love* her with all my heart and soul." New England farmers are reputedly as tightfisted with their emotions as they are with their wallets and their words. Here is an old Maine joke:

"How's your wife, Jeb?"

"Compared to what?"

It would appear that Jeb is no longer in love with his wife. And if you are so much as willing to *think about* comparing your religion with others, or with having no religion at all, you must not be in love with your religion. This is a very personal love (not like the love of jazz, or baseball, or mountain scenery), but no single person—not the priest or the rabbi or the imam—or even any group of people—the congregation of the faithful, say—is the beloved. One's undying loyalty is not loyalty to them, singly or together, but to the *system of ideas* that unite them. Of course, people sometimes do fall in love—romantic love—with their priest or with a fellow parishioner, and this can be hard for them to distinguish from love of their religion, but I'm not suggesting that this is the nature of the love most God-loving people experience. I am suggesting, however, that their unquestioning loyalty, their unwillingness even to consider the virtues versus the vices, *is* a type of love, and more like romantic love than brotherly love or intellectual love.

It is surely no accident that the language of romantic love and the language of religious devotion are all but indistinguishable, and it is similarly no accident that almost all religions (with a few austere exceptions, such as the Puritans and the Shakers and the Taliban) have given their lovers a cornucopia of beauty to ravish their senses: soaring architecture, with decoration applied to every surface, music, candles, and incense. The inventory of the world's great works of art is crowned by religious masterpieces. Thanks to Islam, we have the Alhambra, and the exquisite mosques of Isfahan and

Istanbul. Thanks to Christianity, we have the Hagia Sofia and the cathedrals of Europe. You don't have to be a believer to be entranced by Buddhist, Hindu, and Shinto temples of surreal intricacy and sublime proportion. Bach's *Saint Matthew Passion* and Handel's *Messiah* and those miniature marvels the Christmas carols are among the most rapturous love songs ever composed, and the stories they set to music are themselves compositions of extraordinary emotional power. The film director George Stevens may not have been exaggerating when he called his 1965 movie on the life of Jesus *The Greatest Story Ever Told*. The competition is fierce, what with the *Odyssey*, the *Iliad*, Robin Hood, *Romeo and Juliet*, *Oliver Twist*, *Treasure Island*, *Huckleberry Finn*, *The Diary of Anne Frank*, and all the other great narratives of the world's literature, but for joy, danger, pathos, triumph, tragedy, heroes, and villains (but no comic relief), it is hard to beat. And of course the story has a moral. We *love* stories, and Elie Wiesel uses a story to explain this:

When the founder of Hasidic Judaism, the great Rabbi Israel Shem Tov, saw misfortune threatening the Jews, it was his custom to go into a certain part of the forest to meditate. There he would light a fire, say a special prayer, and the miracle would be accomplished and the misfortune averted. Later, when his disciple, the celebrated Maggid of Mezeritch, had occasion, for the same reason, to intercede with heaven, he would go to the same place in the forest and say: "Master of the Universe, listen! I do not know how to light the fire, but I am still able to say the prayer," and again the miracle would be accomplished. Still later, Rabbi Moshe-leib of Sasov, in order to save his people once more, would go into the forest and say, "I do not know how to light the fire. I do not know the prayer, but I know the place, and this must be sufficient." It was sufficient, and the miracle was accomplished. Then it fell to Rabbi Israel of Rizhin to overcome misfortune. Sitting in his armchair, his head in his hands, he spoke to

God: "I am unable to light the fire, and I do not know the prayer, and I cannot even find the place in the forest. All I can do is to tell the story, and this must be sufficient." And it was sufficient. For God made man because He loves stories. [1966, preface (*not* Wiesel, 1972, as many Web sites have it)]

We have been given a lot to love, and not just spectacularly beautiful art and stories and ceremonies. The daily actions of religious people have accomplished uncounted good deeds throughout history, alleviating suffering, feeding the hungry, caring for the sick. Religions have brought the comfort of belonging and companionship to many who would otherwise have passed through this life all alone, without glory or adventure. They have not just provided first aid, in effect, for people in difficulties; they have provided the means for changing the world in ways that remove those difficulties. As Alan Wolfe says, "Religion can lead people out of cycles of poverty and dependency just as it led Moses out of Egypt" (2003, p. 139). There is much for religion lovers to be proud of in their traditions, and much for all of us to be grateful for.

The fact that so many people love their religions as much as, or more than, anything else in their lives is a weighty fact indeed. I am inclined to think that *nothing could matter more* than what people love. At any rate, I can think of no value that I would place higher. I would not want to live in a world without love. Would a world with peace, but without love, be a better world? Not if the peace was achieved by drugging the love (and hate) out of us, or by suppression. Would a world with justice and freedom, but without love, be a better world? Not if it was achieved by somehow turning us all into loveless law-abiders with none of the yearnings or envies or hatreds that are the wellsprings of injustice and subjugation. It is hard to consider such hypotheticals, and I doubt if we should trust our first intuitions about them, but, for what it is worth, I surmise that we almost all want a world in which love, justice, freedom, and peace are all present, as much as possible, but if we had to give up one of these, it wouldn't—and shouldn't—be love. But, sad to say, even if it is true that nothing could matter more than love, it wouldn't follow from this that we don't have reason to question the things that we, and others, love. Love is blind, as they say, and because love is blind, it often leads to tragedy: to conflicts in which one love is pitted against another love, and something has to give, with suffering guaranteed in any resolution.

Suppose I love music more than life itself. Other things being equal, then, I should be free to live my life in pursuit of the exaltation of music, the thing I love most, with all my heart and soul. But that still doesn't give me the right to force my children to practice their instruments night and day, or the right to impose musical education on everybody in the country of which I am the dictator, or to threaten the lives of those who have no love of music. If my love of music is so great that I am simply unable to consider its implications objectively, then this is an unfortunate disability, and others may with good reason assert the right to act as my surrogate, conscientiously deciding what is best for all, since my love has driven me mad, and I cannot rationally participate in the assessment of my own behavior and its consequences. There may well be nothing more wonderful than love, but love is not enough. A world in which baseball fans' love of their teams led them so to hate the other teams and their fans that murderous war accompanied the playoffs would be a world in which a particular love, pure and blameless in itself, led to immoral and intolerable consequences.

So, although I understand and sympathize with those who take offense at my invitation to consider the pros and cons of religion, I insist that they have no right to indulge themselves by declaring their love and then hiding behind the veil of righteous indignation or hurt feelings. Love is not enough. Have you ever had to face the heart-wrenching problem of a dear friend who has fallen head over heels in love with somebody who is just not worthy of her love? If you suggest this to her, you risk losing a friend and getting slapped in the face for your trouble, for people in love often make it a point of honor to respond irrationally and violently to any perceived slight of their beloved. It's part of the whole point of being in love, after all. When they say that love is blind, they say it without regret. It is commonly understood that love *should be* blind; the whole idea of assessment should be off limits when it comes to true love. But why? Common wisdom doesn't answer, and hardheaded economists have long dismissed the idea as romantic nonsense, but the evolutionary economist Robert Frank has pointed out that there is in fact an excellent (free-floating) rationale for the phenomenon of romantic love in the unruly marketplace of human mate-finding:

> Because search is costly, it is rational to settle on a partner before having examined all potential candidates. Once a partner is chosen, however, the relevant circumstances will often change.... The resulting uncertainty makes it imprudent to undertake joint investments that would otherwise be strongly in each party's interest. In order to facilitate these investments, each party wants to make a binding commitment to remain in the relationship.... Objective personal characteristics may continue to play a role in determining which people are initially most attracted to one another, as much evidence suggests. But the poets are surely correct that the bond we call love does not consist of rational deliberations about these characteristics. It is instead an intrinsic bond, one in which the person is valued for his or her own sake. And precisely therein lies its value as a solution to the commitment problem. [1988, pp. 195–96]

As Steven Pinker says, "Murmuring that your lover's looks, earning power, and IQ meet your minimal standards would probably kill the romantic mood, even though the statement is statistically true. The way to a person's heart is to declare the opposite—that you're in love because you can't help it" (1997, p. 418). This demonstrated (or at least passionately professed) helplessness is as close as you can muster to a guarantee that you are not still shopping around. Like all communicative signals, however, if it can be cheaply faked, your commitment signal will not be effective, and the result, as so often in the world of animal signaling, is the inflationary spiral of costly signaling (Zahavi, 1987). It is not just lovestruck young men who shower their beloveds with presents they can barely afford; the bowerbirds' bowers are costly investments, and so are the "nuptial gifts" of food and other goods provided by male moths, beetles, crickets, and many other creatures.

Has our evolved capacity for romantic love been exploited by religious memes? It would surely be a Good Trick. It would get people to think that it was actually honorable to *take offense,* to attack all skeptics with fury, to lash out wildly and without concern for their own safety—let alone the safety of the person they are attacking. Their beloved deserves nothing less than this, they think: a total commitment to eradicating the blasphemer. Of such stuff are fatwas made, but this meme is not at all restricted to Islam. There are plenty of misguided Christians, for instance, who will contemplate with relish the prospect of demonstrating the depths of their commitment by raining abuse on me for daring to question the love they have for their Jesus. Before they act on their self-indulgent fantasies, I hope they will pause to consider that any such action would actually bring dishonor to their faith.

Some of the saddest spectacles of the last century have been the way zealots of all faiths and ethnicities have defiled their own shrines and holy places, and brought shame and dishonor to their causes, by their acts of fanatical loyalty. Kosovo may have been a holy place to Serbs since the battle of 1389, but it is hard to see how Serbs can continue to cherish its memory after recent history. By destroying the "idolatrous" Buddhist monuments in Afghanistan, the Taliban dishonored themselves and their tradition in ways that will take centuries of good works to expiate. The killing of hundreds of Muslims in reprisal for the killing of dozens of Hindus in the Akshardham temple in Gujarat besmirches the reputations of both religions, whose fanatical devotees should be reminded that the rest of the world is not just unmoved by, but sick and tired of, their respective demonstrations of their devotion. What would *really* impress us infidels would be an announcement, unilateral or joint, that the contested site was henceforth to be considered the Hall of Shame, no longer holy but, rather, a reminder to all of the evils of zealotry.

Since September 11, 2001, I have often thought that perhaps it was fortunate for the world that the attackers targeted the World Trade Center instead of the Statue of Liberty, for if they had destroyed our sacred symbol of democracy I fear that we Americans would have been unable to keep ourselves from indulging in paroxysms of revenge of a sort the world has never seen before. If that had happened, it would have befouled the meaning of the Statue of Liberty beyond any hope of subsequent redemption—if there were any people left to care. I have learned from my students that this upsetting thought of mine is subject to several unfortunate misconstruals, so let me expand on it to ward them off. The killing of thousands of innocents in the World Trade Center was a heinous crime, much more evil than the destruction of the Statue of Liberty would have been. And, yes, the World Trade Center was a much more appropriate symbol of Al Qaeda's wrath than the Statue of Liberty would have been, but for that very reason it didn't mean as much, as a symbol, *to us.* It was Mammon and Plutocrats and Globalization, not Lady Liberty. I do suspect that the fury with which many Americans would have responded to the unspeakable defilement of our cherished national symbol, the purest image of our aspirations as a democracy, would have made a sane and measured response extraordinarily difficult. This is the great danger of symbols—they can become *too* "sacred." An important task for religious people of all faiths in the twenty-first century will be spreading the conviction that there are no acts more *dis*honorable than harming "infidels" of one stripe or another for "disrespecting" a flag, a cross, a holy text.

By asking for an accounting of the pros and cons of religion, I risk getting poked in the nose or worse, and yet I persist. Why? Because I believe that it is very important to break *this* spell and get us all to look carefully at the question with which I began this section: *are people right* that the best way to live a good life is through religion? William James confronted the same problem squarely when he gave the Gifford Lectures that became his great book, *The Varieties of Religious Experience*, and I will echo his plea for forbearance:

> I am no lover of disorder and doubt as such. Rather do I fear to lose truth by this pretension to possess it already wholly. That we can gain more and more of it by moving always in the right direction, I believe as much as any one, and I hope to bring you all to my way of thinking before the termination of these lectures. Till then, do not, I pray you, harden your minds irrevocably against the empiricism which I profess. [1902, p. 334]

2 The academic smoke screen

The word God *refers to a "depth" and "wholeness" unlike anything that* we humans know or can know. *Certainly it is beyond our ability to discriminate and label.*

—James B. Ashbrook and Carol Rausch Albright, *The Humanizing Brain*

A mystery is a mystery. If, on the other hand, we consider that it is important to study how people communicate about the idea of something being a mystery, there is no a priori reason why this should be beyond the reach of scientific method. —Ilkka Pyysièainen, *How Religion Works*

To oppose the torrent of scholastic religion by such feeble maxims as these, that it is impossible for the same thing to be and not to be, that the whole is greater than a part, that two and three make five; is pretending to stop the ocean with a bullrush. Will you set up profane reason against sacred mystery? No punishment is great enough for your impiety. And the same fires, which were kindled for heretics, will serve also for the destruction of philosophers.

—David Hume, *The Natural History of Religion*

James was trying to forestall dismissal by the devout, but they are not the only ones who resort to protectionism. A subtler, less forthright, but equally frustrating barrier to straightforward inquiry into the nature of religion has been erected and maintained by the scholarly friends of religion, many of whom are atheistic or agnostic connoisseurs, not champions of any creed. They do want to study religion, but only *their way*, not the way I am proposing, which by their lights is "scientistic," "reductionistic," and, of course, philistine. I alluded to this opposition in chapter 2, when I discussed the legendary gap that many want to see between the natural sciences and the interpretive sciences, *Naturwissenschaften* and *Geisteswissenschaften*. Anyone who tries to bring an evolutionary perspective to bear on any item of human culture, not just religion, can expect rebuffs ranging from howls of outrage to haughty dismissal from the literary, historical, and cultural experts in the humanities and social sciences.

When the cultural phenomenon is religion, the most popular move is pre-emptive disqualification, and it has been well known since the eighteenth century, when it was used to discredit the earliest atheists and deists (such as David Hume and Baron d'Holbach, and some great American heroes, Benjamin Franklin and Thomas Paine). Here is an early-twentieth-century version from Emil Durkheim: "He who does not bring to the study of religion a sort of religious sentiment cannot speak about it! He is like a blind man trying to talk about color" (1915, p. xvii). And here, half a century later, is an oft-quoted version from the great religious scholar Mircea Eliade:

A religious phenomenon will only be recognized as such if it is grasped at its own level, that is to say, if it is studied *as* something religious. To try to grasp the essence of such a phenomenon by means of physiology, psychology, sociology, economics, linguistics, art or any other study is false; it misses the one unique and irreducible element in it—the element of the sacred. [1963, p. iii]

You can find similar claims of pre-emptive disqualification protecting other topics. Only women are qualified to do research on women (according to some radical feminists), because only they can overcome the *phallocentrism* that renders males obtuse and biased in ways they can never acknowledge and counteract. Some multiculturalists insist that Europeans (including Americans) can never really cancel out their disabling Eurocentrism and understand the subjectivity of Third World people. It takes one to know one, according to this theme in all its variations. Well, then, should we all just hunker down in our isolationist enclaves and wait for death to overtake us, since we can never understand one another? And then there is the brand of defeatism in my own home discipline, philosophy of mind, the *mysterian* doctrine that insists that the human brain is simply not up to the task of understanding the human brain, that consciousness is not a puzzle but an insoluble mystery (so stop trying to explain it). What is transparent in all these claims is that they are not so much defeatist as protectionist: don't even try, because we're afraid you might succeed! "You'll never understand Indian street magic if you're not an Indian born into the caste of magicians. It is impossible." But of course it is possible (Siegel, 1991). "You'll never understand music unless you are born with a great ear for music—and perfect pitch." Nonsense. In fact, people who have difficulty training themselves as musicians sometimes grind out insights into the nature of the music and how to perform it that were unavailable to those who glide effortlessly to musical mastery. Similarly, Temple Grandin (1996), who is autistic and hence has a tin ear for the intentional stance and folk psychology, has come up with striking observations about how people present themselves and interact, insights that had escaped the rest of us *normal* folk.

We would never let business tycoons get away with saying that since we weren't plutocrats ourselves we couldn't hope to understand the world of high finance and were hence disqualified from investigating their deals. Generals can't escape civilian oversight by claiming that only those in uniform can appreciate what they are doing, and doctors have had to open up their methods and practices to the scrutiny of experts who are not themselves M.D.'s. It would be dereliction of duty for us to let pedophiles insist that only those who appreciate a commitment to pedophilia can really understand them at all. So what we may say to those who insist that only those who believe, only those with a deep appreciation of the sacred, are to be entrusted with the investigation of religious phenomena, is that they are simply wrong, about both facts and principles. They are mistaken about the imaginative and investigative powers of those they would exclude, and they are wrong to suppose that it might be justifiable on any grounds to limit the investigation of religion to those who are religious. If we say this politely, firmly, and often, they may eventually stop playing this card and let us get on with our investigations, hampered though we may be by our lack of faith. We'll just have to work harder.

A related smoke screen is the more general declaration that the methods of the natural sciences cannot possibly make progress on human culture, which requires "semiotics" or "hermeneutics," not experiments. A favorite exponent of this position is the anthropologist Clifford Geertz, who has put it this way:

> Believing, with Max Weber, that man is an animal suspended in webs of significance he himself has spun, I take culture to be those webs, and the analysis of it to be, *therefore* [emphasis added], not an experimental science in search of law but an interpretative one in search of meaning. [1973, p. 5]

"Therefore"? Back in 1973, it might have passed muster, but this argument is way out of date. That we human beings spin webs of significance is not in doubt, but those webs *can* be analyzed by methods that critically involve experiments and the disciplined methods of the natural sciences. Interpretation in the natural sciences is not opposed to experiment, and science isn't all subsumption under some covering law. All of cognitive science and all of evolutionary biology, for instance, is interpretive in ways that closely parallel some of the interpretive strategies of the humanities and anthropology (Dennett, 1983, 1995b).

In fact, one of the few serious differences between the natural sciences and the humanities is that all too many thinkers in the humanities have decided that the postmodernists are right: it's all just stories, and all truth is relative. A cultural anthropologist who will go unnamed recently announced to his students that one of the great things about his field is that, given the same set of data, no two anthropologists would arrive at the same interpretation. End of story. Scientists often have just such disagreements about how to interpret a shared pool of unchallenged data, but for them it is the beginning of a task of resolution: which of them is wrong? Experiments and further statistical analyses and the like are then designed to *answer* the question—by discovering the *truth* (not the capital "T" Truth about everything, but just the ho-hum truth about this particular little factual disagreement). It is this subsequent process (which may take years) that has been declared impossible or unnecessary by these ideologues, who scoff at the very idea that there are objective truths about such matters to be discovered. They couldn't claim to *prove* that there is no such thing as objective truth, of course, for that would be to contradict themselves blatantly, and they have at least *that* much respect for logic. So they content themselves with clucking at the presumption and naïveté of anyone who still believes in truth. It is hard to convey how boring this relentless barrage of defensive sneering is, so it is not surprising that some investigators have stopped trying to rebut it, and settle for poking fun at it instead:

> For example, right now I am typing on my keyboard with the intention of creating a coherent story about the logic of postmodernism. Were someone to study me, they might look beyond that surface level intention I just offered and infer instead that what I really am doing is inventing a story from my personal experi-

ences for the purposes of advancing my academic career. To accomplish this, they might argue, I am constructing a discourse that sets me apart from other people and thus increases my value as a writer. (The more I confuse you, the smarter I appear!) Why do I do this? Because I am a self-interested white heterosexual privileged Protestant male who uses knowledge for power (a strategy not of savvy but of manipulation and exploitation). For postmodernists, that which gets presented as truth (e.g., this book) is an invention, just a take on reality, that masks what I am really doing—tricking everyone in order to acquire and maintain power. [Slone, 2004, pp. 39–40]

The pioneers whose scientific work on religion I have been introducing—Atran, Boyer, Diamond, Dunbar, Lawson, McCauley, McClenon, Sperber, Wilson, and the rest—all have to deal with this. It can be amusing, in the end, to see how they all brace themselves against this onslaught and, following in William James's footsteps, beg for an open-minded audience. So much pleading! The irony is that these intrepid interlopers have been far more conscientious in their attempts to get a sympathetic, informed view of religion than the self-appointed defenders of religion have been in trying to understand the point of view and methods of those they are resisting. When the humanist defenders have studied evolutionary biology and cognitive neuroscience (and statistics and the rest) with the same energy and imagination that the scientists have devoted to studying the histories, rites, and creeds of the various religions, they will become worthy critics of the work they fear.

When the Zurich classicist Walter Burkert dared to expose his fellow humanists to biological thinking about the origins of religion in his Gifford Lectures of 1989, he became really the first humanist to attempt to cross the chasm going in the other direction. Burkert is a distinguished historian of ancient religion, widely read in anthropology, linguistics, and sociology, and he has begun educating himself in the evolutionary biology that he sees clearly must ground his own efforts at theorizing. One of the delights of reading his book *Creation of the Sacred: Tracks of Biology in Early Religions* (1996) is seeing how valuable his treasure trove of historical insights turns out to be when placed in the context of biological questions. And one of the causes for dismay is seeing how gingerly he thinks he must tiptoe around the hair-trigger sensitivities of his fellow humanists when he introduces these dread biological notions into their world (Dennett, 1997, 1998b).

Scientists have much to learn from the historians and the cultural anthropologists. The infrastructure for constructive collaborations already exists in the form of interdisciplinary journals, such as *Journal for the Scientific Study of Religion* and *Method & Theory in the Study of Religion*, and *Journal of Cognition and Culture*, as well as professional societies and Web sites. One of my goals in this book is to make it easier for subsequent researchers to enter these forbidden zones and find friendly natives with whom to collaborate, without having to hack their way through a jungle of hostile defenders. They will discover that the anthropologists and historians have already thought of most of their "new" ideas and have plenty to say about what the problems with them are, so I recommend that they behave modestly, ask lots of questions, and just ignore the often breathtakingly rude and condescending put-downs they inspire in those who dread their approach.[1]

3 Why does it matter what you believe?

To-day we have to change our attitude from that of description to that of appreciation; we have to ask whether the fruits in question can help us to judge the absolute value of what religion adds to human life.

—William James, *The Varieties of Religious Experience*

It isn't just that I don't believe in God and naturally, hope there is no God! I don't want there to be a God; I don't want the universe to be like that. —Thomas Nagel, *The Last Word*

We have one last deflector to set aside before we can safely address the main question. Why believe in belief in God? Many people would answer: Simply because God exists! They believe in trees and mountains, tables and chairs, people and places, wind and water—and God. This would indeed explain their belief in God, but not the fact that they take believing in God to be so important. In particular, why do people care so much what *other* people believe about God? I believe that the center of the Earth consists mainly of molten iron and nickel. Relative to other things I believe, this is a pretty big and exciting fact. Just imagine: there's a ball of molten iron and nickel nearby; it's about the size of the moon and a lot closer; in fact, it's between me and Australia! Lots of people don't know this, and too bad for them—since it's quite a delightful fact. But it really doesn't *bother* me that they don't share my belief, or my delight. Why should it matter so much whether others share your belief in God?

Does God care? I can see that Jehovah might be really peeved if He found lots of people oblivious to His power and greatness. Part of what makes Jehovah such a fascinating participant in stories of the Old Testament is His kinglike jealousy and pride, and His great appetite for praise and sacrifices. But we have moved beyond this God (haven't we?). The Creative Intelligence that is supposed by many to have done all the design work we evolutionists attribute to natural selection is not the sort of Being that could be jealous, is it? I know professors who can get mighty annoyed if you pretend you haven't heard of their published work, but it is hard to see why the Creative Intelligence that invented DNA and the metabolic cycle and mangrove trees and sperm whales would care whether any of Its creatures recognized Its authorship. The second law of thermodynamics can't care whether anybody believes in it, and I would think that the Ground of All Being must be a similarly unmoved mover.

An anthropologist once told me about an African tribe (I can't remember their name) whose dealings with their neighbors proceeded at a stately pace. An emissary sent by foot to the settlement of a neighboring tribe would rest for a day after his arrival before conducting any official business, since he had to wait for his soul to catch up. Souls in that culture are slow walkers, apparently. We can see a similar time lag in the migration many believers have made from a highly anthropomorphic God to a more abstract and hard-to-imagine God. They still use anthropomorphic language when speaking of a God who (*sic*) is not a supernatural *being* at all but just an *essence* (to use Stark's useful if philosophically misbegotten terminology). It is obvious enough why they do this: it permits them to carry over all the connotations required to make any sense of a *personal* love of God. One can feel, I suppose, a certain *affection* or *gratitude* for a law of nature—"Good ol' gravity, she just never lets you down!"—but the proper object of adoration really has to be some sort of *person*, however inconceivably unlike us talking, featherless bipeds. Only a person could be literally disappointed in you if you misbehaved, or could answer your prayers, or forgive you, so the "theological incorrectness" that persists in imagining God to be a Wise Old Guy in the Sky is not only tolerated by the experts, but subtly encouraged.

William James opined at the turn of the twentieth century, "Today a deity who should require bleeding sacrifices to placate him would be too sanguinary to be taken seriously" (1902, p. 328), but a century later, few would agree publicly with Thomas Nagel when he candidly says that he would not want such a God to exist. (I doubt if Nagel finds Spinoza's *Deus sive Natura*—God, or Nature—repugnant, and he may well be as indifferent as I am to the Ground of All Being, whatever that is.) If pressed, many people insist that the anthropomorphic language used to describe God is metaphorical, not literal. One might suppose, then, that the curious adjective "God-fearing" would have faded into disuse over the years, a fossil trace of a rather embarrassingly juvenile period in our religious past, but far from it. People want a God who can be loved and feared the way you love or fear another person. "Religion, in short, is a monumental chapter in the history of human egotism. The gods believed in—whether by crude savages or by men disciplined intellectually—agree with each other in recognizing personal calls," James observed. "Today, quite as much as at any previous age, the religious individual tells you that the divine meets him on the basis of his personal concerns" (1902, p. 491).

For many believers, of course, this is all just obvious. God is—of course—a person who talks to them directly—if not on a daily basis, then at least in a once-in-a-lifetime revelation. But as James pointed out, the believers themselves shouldn't put too much stock in such experiences:

> The super-normal incidents, such as voices and visions and over-powering impressions of the meaning of suddenly presented scripture texts, the melting emotions and tumultuous affections connected with the crisis of change, may all come by way of nature, or worse still, be counterfeited by Satan. [1902, p. 238]

So, however convinced some people may be by their powerful personal experiences, such revelations don't travel well. They can't be used as contributions to the communal discussion that we are now conducting. Philosophers and theologians have often debated the question of whether acts are good because God loves them or God loves them because they are good, and although these inquiries may make some sense within a theological tradition, in any ecumenical setting where we aspire for "universal" consensus we have to choose the latter presumption. Moreover, the evidence of history makes it clear that, as time has passed, people's moral sense about what is permissible and what is heinous has shifted, and along with it their convictions about what God loves and hates. Those who see either blasphemy or adultery as a crime deserving the death penalty are today a dwindling minority, thank heaven. Still, the reason people care so much what other people believe about God is a fine reason, so far as it goes: they want the world to be a better place. They think that getting others to share their beliefs about God is the best way to achieve that end, and this is far from obvious.

I, too, want the world to be a better place. This is my reason for wanting people to understand and accept evolutionary theory: I believe that their salvation may depend on it! How so? By opening their eyes to the dangers of pandemics, degradation of the environment, and loss of biodiversity, and by informing them about some of the foibles of human nature. So isn't my belief that belief in evolution is the path to salvation a religion? No; there is a major difference. We who love evolution do not honor those whose love of evolution prevents them from thinking clearly and rationally about it! On the contrary, we are particularly critical of those whose misunderstandings and romantic misstatements of these great ideas mislead themselves and others. In our view, there is no safe haven for mystery or incomprehensibility. Yes, there is humility, and awe, and sheer delight, at the glory of the evolutionary landscape, but it is not accompanied by, or in the service of, a willing (let alone thrilling) abandonment of reason. So I feel a moral imperative to spread the word of evolution, but evolution is not my religion. I don't have a religion.

So, now, with apologies to those whose equanimity is disturbed by my asking such a fundamental question: What *are* the pros and cons of religion? Is it *worthy* of the intense loyalty it has inspired in most of the people of the world? William James led the way in this inquiry as well, and I will use his words to frame the issues for us, because they are wonderful in themselves but also because they reveal some of the progress we have made in the last century, clarifying and sharpening our thinking in a number of regards. Long before anybody talked of memes or memetics, James noted that religions had indeed evolved, in spite of all their claims to "eternal" and "immutable" principles, and he noted that this evolution had always been responsive to human value judgments:

What I then propose to do is, briefly stated, to test saintliness by common sense, to use human standards to help us decide how far the religious life commends itself as an ideal kind of human activity. . . . It is but the elimination of the humanly unfit, and the survival of the humanly fittest, applied to religious beliefs; and if we look at history candidly and without prejudice, we have to admit that no religion has ever in the long run established or proved itself in any other way. [1902, p. 331]

When James speaks of what is "humanly unfit" he means something like "unfit for human use" rather than "biologically" or "genetically" unfit, and this choice of words blurs his vision. In spite of his desire to look at history without prejudice, his phrase biases his judgment in the direction of optimism: the memes that have resisted extinction over the centuries are only those memes that actually somehow enhance humanity. What do they enhance exactly: human genetic fitness? human happiness? human well-being? James gives us a very Victorian version of Darwinism: what survives *must* be good, because evolution is always a matter of *progress toward the better*. Does evolution foster the good? It all depends, as we have seen, on how we ask and answer the *cui bono?* question.

But now, for the first time in the book, we are stepping aside from explanation and description and turning to appreciation, as James said, asking what *ought* to be, not just what *is* (and how it got that way):

If the *fruits for life* of the state of conversion are good, we ought to idealize and venerate it, even though it be a piece of natural psychology; if not, we ought to make short work of it, no matter what supernatural being may have infused it. [1902, p. 237]

Does religion *make us better*? James distinguished two main ways in which this might be true. It might make people *more effective* in their daily lives, healthier, both physically and mentally, more steadfast and composed, more strong-willed against temptation, less tormented by despair, better able to bear their misfortunes without giving up. He calls this the "mind-cure movement." Or it might make people *morally* better. The ways in which religion purports to accomplish this he calls "saintliness." Or it could accomplish both ends, in varying degrees under different circumstances. There is a lot to be said regarding both of them, and the rest of this chapter will be devoted to the first claim, leaving the hugely important question of the role of religion in morality to another chapter.

4 What can your religion do for you?

Religion in the shape of mind-cure gives to some of us serenity, moral poise, and happiness, and prevents certain forms of disease as well as science does, or even better in a certain class of persons.
  —William James, *The Varieties of Religious Experience*

No one dares suggest that neon signs blinking the message that "Jesus Saves" may be false advertising. —R. Laurence Moore, *Selling God*

Pray: To ask that the laws of the universe be annulled in behalf of a single petitioner confessedly unworthy. —Ambrose Bierce, *The Devil's Dictionary*

In a dangerous world there will always be more people around whose prayers for their own safety have been answered than those whose prayers have not.
  —Nicholas Humphrey's Law of the Efficacy of Prayer (2004)[2]

James speculated that there may be two entirely different types of people, the healthy-minded and the sick-minded, who need different things from religion, and noted that churches face "an everlasting inner struggle of the acute religion of the few against the chronic religion of the many" (p. 114). You can't please everybody all the time, so every religion must make its compromises. His informal surveys and inquiries were the forerunners of the intensive and sometimes quite sophisticated market research undertaken by religious leaders in recent years, as well as the more academic investigations conducted by psychologists and other social scientists trying to assess the claims made on behalf of religion. Religious revival movements flourished in James's day, but so did secular promoters of all manner of fantastical products and regimens. The self-help "infomercials" on television today are the descendants of a long line of earlier hucksters who plied their wares in tent shows and rented theaters.

> One hears of the "Gospel of Relaxation," of the "Don't Worry Movement," of people who repeat to themselves, "Youth, health, vigor!" when dressing in the morning, as their motto for the day. [p. 95]

James asked if the religions provided bracing as good as or better than that of their secular counterparts, and observed that, whatever they may protest about their aloofness from science, in fact religions do rely on "experiment and verification" at every turn: "Live as if I were true, [religion] says, and every day will practically prove you right" (p. 119). In other words: you'll see the results for yourself; try it, you'll like it. "Here, in the very heyday of science's authority, it carries on an aggressive warfare against the scientific philosophy, and succeeds by using science's own peculiar methods and weapons" (p. 120).

The best salespeople are satisfied customers, and even if that is not the point of being a member of a church, there is nothing wrong with paying close attention to any factors that may improve the health, both spiritual and physical, of those who are loyal and active members. If I were to try to design a secular organization for furthering world peace, for instance, I would certainly keep my eyes open for any features that would have the incidental benefit of boosting members' health or prosperity, since I recognize that the organization would always be competing against all the other ways people can spend their time and energy. Even if I expect and encourage sacrifices from those who join, I should weigh the sacrifices carefully, and eliminate any gratuitous shortcomings—and replace them with benefits, if possible—so as to give greater leverage to the essential sacrifices.[3]

So is religion good for your health? There is growing evidence that many religions have succeeded remarkably well on this score, improving both the health and the morale of their members, quite independently of the good works they may have accomplished to benefit others. For instance, eating disorders such as anorexia nervosa and bulimia are much less common among women in Muslim countries, in which the physical attractiveness of women plays a muted role relative to that in Westernized countries (Abed, 1998). A current surge of interest is bringing to bear all the statistical tools of epidemiology and public health on such questions as whether regular churchgoers live longer, are less likely to have heart attacks, and so forth, and in most of the surveys the results are positive, often strongly so. (For an extensive overview, now rapidly becoming out of date, see Koenig et al., 2000.) The early results are impressive enough to have provoked knee-jerk skeptical dismissals from some atheists who haven't stopped to consider how independent these questions are from whether or not any religious beliefs are *true*. We already know from studies involving many different kinds of performance that if you randomly tell half a group of subjects that they are "above average" on the task in question, they will do better, so the power of false belief to improve human capacities is already established. There are studies that demonstrate, according to some (e.g., Taylor and Brown, 1988), that positive illusions improve mental health, but there are critics who say the case is not yet secure (Colvin and Block, 1994).

It might well be that believing in God (and engaging in all the practices that go with that belief) improves your state of mind and thereby improves your health by, say, 10 percent. We should do the research to find out for sure, bearing in mind that it also may be true that believing that Earth is being invaded by space aliens who plan to take us to their planet and teach us all how to fly (and engaging in all the behaviors that are appropriate to *that* belief) improves your state of mind and health by 20 percent! We won't know until we run the experiments, but since the world's literature is overflowing with stories of people who have benefited greatly from being deceived by well-wishing acquaintances, we shouldn't be surprised to find positive effects for well-chosen falsehoods, and if such concoctions were more effective than any known religious creed, we would have to confront the ethical question of whether *any amount* of health benefits could justify such deliberate misrepresentation.

The results so far are strong but in need of further investigation.[4] Since the benign effects that religions do seem to be having would probably diminish if skepticism took hold, regardless of whether it was justified, caution is called for. Many effects studied by psychologists depend on *naïve subjects* who are relatively uninformed about the mechanisms and conditions of the phenomena. The effects are diminished or entirely obliterated when subjects are given more information. We should be alert to the possibility that the good effects, if they hold up to further scrutiny, *might* be jeopardized by anything that throws too strong a light of public scrutiny on them. On the other hand, the effects may be robust under a barrage of skeptical attention. We will just have to see. And, of course, if the results tend to evaporate as we study them more intensively, we can anticipate that those who are sure that the effects are real will protest that the "climate of skepticism" is inimical to the effects, making perfectly real phenomena vanish under the harsh light of science. And they may be right. And they may be wrong. This, too, is indirectly testable.

Here, more than in any other area of conflict between science and religion, those who are dubious about, or fearful of, the authority of science will have to search their souls. Do they acknowledge the power of science, properly conducted, to settle such controversial factual questions, or do they reserve judgment, wait-ing to see what the verdict will be? If it turns out that, in spite of the anecdotal evidence, and mountains of testimonials, religion is no better than alternative sources of well-being, will they be willing to accept that result and *drop the advertising*? Some major pharmaceu-tical companies are currently under fire for trying to suppress the publication of studies they funded that fail to show the effective-ness of their products. In the future, it now seems clear, these com-panies will be obliged to consent in advance to the publication of all the research they fund, however it comes out. That is the ethos of science: the price you pay for the authoritative confirmation of your favorite hypothesis is risking an authoritative refutation of it. Those who want to make claims about the health benefits of religion will have to live by the same rules: prove it or drop it. And, if you set out to prove it and fail, you are obliged to tell us.

The potential benefits of joining the scientific community on these issues are enormous: getting the authority of science in support of what you say you believe with all your heart and soul. Not for nothing have the new religions of the last century or two been given names like Christian Science and Scientology. Even the Roman Catholic Church, with its unfortunate legacy of persecu-tion of its own scientists, has recently been eager to seek scientific confirmation—and accept the risk of disconfirmation—of its tradi-tional claims about the Shroud of Turin, for example.[5]

One strand in the current wave of research on religion raises a much more fundamental issue, in undeniable terms. Studies are now under way on the efficacy of intercessory prayer, "praying with the real hope and real intent that God would step in and act for the good of some specific other person(s) or other entity" (Longman, 2000). These are quite unlike the studies mentioned above in their import. As we have just noted, scientists have plenty of resources already well in hand that could explain general health benefits to those who pray and practice and tithe; no supernatural forces

would need to be invoked to account for such ambient health benefits. But if a properly conducted, double-blind, rigorously controlled
test with a sufficiently large population of subjects were to demonstrate that people who are prayed for are significantly more likely to
get well than people who get the same medical treatments but are
not prayed for, this would be all but impossible for science to account for without a major revolution.

Many atheists and other skeptics are so confident that no such
effects could possibly exist that they are eager to see these tests
performed. Those, in contrast, who believe in the efficacy of intercessory prayer have a tough call here. The stakes are high, since, if
the studies are performed properly *and show no positive effect*, the
religions that practice intercessory prayer would be obliged by
the principles of truth in advertising to renounce all claims to its
efficacy—just like the pharmaceutical companies. On the other
hand, a positive result would stop science in its tracks. After five
hundred years of steady retreat in the face of advancing science, religion could demonstrate, in terms that the scientists would have to
respect, that its claims to truth were not all vacuous.

In October 2001, the *New York Times* reported a remarkable Columbia University study that purportedly showed that infertile
women who were prayed for became pregnant twice as often as
those who were not prayed for. Published in a major scientific journal, the *Journal of Reproductive Medicine*, the results were worth the
headlines, since Columbia University is not a Bible Belt college that
would be instantly under suspicion in many quarters. Its medical
school, a bastion of the medical establishment, supported the results in a news release that described the safeguards that had been
taken to ensure that this was a properly controlled investigation.
But, to make a long and sordid story short, it has subsequently
turned out that this was a case of scientific fraud. Of the three authors of the study, two have now left their positions at Columbia
University, and the third, Daniel Wirth, who had no connection

with Columbia, has recently pled guilty, in an unrelated case, to conspiracy to commit mail fraud and conspiracy to commit bank fraud—and turns out not to have any medical credentials at all (Flamm, 2004). One study is discredited, and others have been severely criticized, but there are still others under way, including a major study by Dr. Herbert Benson and his colleagues at Harvard Medical School funded by the Templeton Foundation, so there is no verdict yet on the hypothesis that intercessory prayer actually works (see, e.g., Dusek et al., 2002). Even if studies eventually show that it doesn't, there will still be plenty of evidence of less miraculous benefits of being an active member of a church, which is all that many churches have ever maintained. The Reverend Raymond J. Lawrence, Jr., director of pastoral care at New York–Presbyterian Hospital/Columbia University Medical Center, expresses the liberal view:

> There's no way to put God to the test, and that's exactly what you're doing when you design a study to see if God answers your prayers. This whole exercise cheapens religion, and promotes an infantile theology that God is out there ready to miraculously defy the laws of nature in answer to a prayer. [Carey, 2004, p. 32]

Prolonged exposure to the fumes of incense and burning candles may have some detrimental health effect, concluded one recent study (Lung et al. [I'm not kidding], 2003), but there is plenty of other evidence that active participation in religious organizations can improve the morale, and hence the health, of participants. Moreover, the defenders of religion can rightly point to less tangible but more substantial benefits to their adherents, such as having a meaning for their lives provided! People who are suffering, even if their morale is not improved in measurable ways, may well gain some solace from nothing more than the knowledge that they are being acknowledged, noticed, thought about. It would be a mistake to suppose that these "spiritual" blessings have no place in the inventory of reasons that we skeptics are trying to assay, just as it would be a mistake to suppose that the nonexistence of an intercessory-prayer effect would show that prayer is a useless practice. There are subtler benefits to be evaluated—but they do need to be identified.

\\\

Chapter 9 Before we can ask the question of whether religion is, all things considered, a good thing, we must first work through several protective barriers, such as the love barrier, the academic-territoriality barrier, and the loyalty-to-God barrier. Then we can calmly consider the pros and cons of religious allegiance, looking first at the question, Is religion good for people? And the evidence to date on that question is mixed. It does seem to provide some health benefits, for instance, but it is too early to say whether there are other, better ways of delivering these benefits, and too early to say if the side effects outweigh the benefits.

\\\

Chapter 10 The more important question, finally, is whether religion is the foundation of morality. Do we get the content of morality from religion, or is it an irreplaceable infrastructure for organizing moral action, or does it provide moral or spiritual strength? Many think the answers are obvious, and positive, but these are questions that need to be re-examined in the light of what we have learned.

CHAPTER TEN

Morality and Religion

1 Does religion make us moral?

Then Jesus beholding him loved him, and said unto him, One thing thou lackest: go thy way, sell whatsoever thou hast, and give to the poor, and thou shalt have treasure in heaven: and come, take up the cross, and follow me. —Mark 10:21

The Lord trieth the righteous: but the wicked and him that loveth violence his soul hateth. Upon the wicked he shall rain snares, fire and brimstone, and an horrible tempest, this shall be the portion of their cup.
—Psalms 11:5–6

Believing as I do that man in the distant future will be a far more perfect creature than he now is, it is an intolerable thought that he and all other sentient beings are doomed to complete annihilation after such long-continued slow progress. To those who fully admit the immortality of the human soul, the destruction of our world will not appear so dreadful.
—Charles Darwin, *Life and Letters*

Non-Muslims love their life too much, they can't fight, and they are cowards. They don't understand that there will be life after death. You cannot live forever, you will die. Life after death is forever. If life after death were an ocean, the life you live is only a drop in the ocean. So it's very important that you live your life for Allah, so you are rewarded after death.

—A young mujaheed from Pakistan, quoted by Jessica Stern, *Terror in the Name of God*

Good people will do good things, and bad people will do bad things. But for good people to do bad things—that takes religion.

—Steven Weinberg, 1999

Religion plays its most important role in supporting morality, many think, by giving people an unbeatable reason to do good: the promise of an infinite reward in heaven, and (depending on tastes) the threat of an infinite punishment in hell if they don't. Without the divine carrot and stick, goes this reasoning, people would loll about aimlessly or indulge their basest desires, break their promises, cheat on their spouses, neglect their duties, and so on. There are two well-known problems with this reasoning: (1) it doesn't seem to be true, which is good news, since (2) it is such a demeaning view of human nature.

I have uncovered no evidence to support the claim that people, religious or not, who *don't* believe in reward in heaven and/or punishment in hell are more likely to kill, rape, rob, or break their promises than people who do.[1] The prison population in the United States shows Catholics, Protestants, Jews, Muslims, and others—including those with no religious affiliation—represented about as they are in the general population. Brights and others with no religious affiliation exhibit the same range of moral excellence and turpitude as born-again Christians, but, more to the point, so do members of religions that de-emphasize or actively deny any relationship between moral behavior "on earth" and eventual postmortem reward and punishment. And when it comes to "family values," the available evidence to date supports the hypothesis that brights have the lowest divorce rate in the United States, and born-again Christians the highest (Barna, 1999). Needless to say, these results strike so hard at the standard claims of greater moral virtue among the religious that there has been a considerable surge of further research initiated by religious organizations attempting to refute them. At this time, nothing very surprising has emerged, and nothing approaching a settled consensus among researchers has been achieved, but one thing we can be sure of is that *if* there is a significant positive relationship between moral behavior and religious affiliation, practice, or belief, it will soon be discovered, since so many religious organizations are eager to confirm their traditional beliefs about this scientifically. (They are quite impressed with the truth-finding power of science when it supports what they already believe.) Every month that passes without such a demonstration underlines the suspicion that it just isn't so.

It is clear enough why believers might want to come up with evidence that belief in heaven and hell has benign effects. Everybody already knows the evidence for the countervailing hypothesis that the belief in a reward in heaven can sometimes motivate acts of monstrous evil. Nevertheless, there are many in the religious community who would not welcome the demonstration that a belief in God's reward in heaven or punishment in hell makes a significant difference, since they view this as an infantile concept of God in the first place, pandering to immaturity instead of encouraging genuine moral commitment. As Mitchell Silver notes, the God who rewards goodness in heaven bears a striking resemblance to the hero of the popular song "Santa Claus Is Coming to Town."

Like Santa, God "knows if you are sleeping, he knows if you're awake, he knows if you've been bad or good" ... The lyrics continue "so be good for goodness' sake." Catchy but a logical solecism. In logic the song should have continued "so be good for the sake of the electronic equipment, dolls, sports gear and other gifts you hope to get but will get only if the omniscient and just

Santa judges you worthy of receiving." If you were good for goodness' sake, the all-seeing Santa would be irrelevant as a motivator
of your virtue. [In press]

Moral philosophers who have agreed about little else, from the
days of Hume and Kant through Nietzsche to the present, have regarded this pie-in-the-sky vision of morality as something of a trap,
a *reductio ad absurdum* into which only the most unwary moralist
would fall. Many religious thinkers agree: a doctrine that trades in a
person's good intentions for the prudent desires of a rational maxi-
mizer shopping around for eternal bliss may win a few cheap victo-
ries, luring a few selfish and unimaginative souls into behaving
themselves for a while, but at the cost of debasing their larger campaign for goodness. We see an echo of this familiar recognition in
the derision heaped by many commentators on the Al Qaeda hi-
jackers of 9/11 for their purported goal of luxuriating in heaven
with seventy-two virgins (each) as the reward for their martyrdom.[2]
We may shun this theme as a foundation of our morality *today*
yet still honor it for having played a founding role in the past, as a
ladder that, once climbed, may be discarded. How could this work?
The economist Thomas Schelling has pointed out that "belief in a
deity who will reward goodness and punish evil transforms many
situations from subjective to secured, at least in the believer's
mind" (quoted in Nesse, ed., 2001, p. 16). Consider a situation in
which two parties confront each other with a prospect for cooperat-
ing on something both parties would want, but each is afraid the
other will renege on any bargain struck, and there are no authorities or stronger parties around to enforce it. Promises can be made
and then broken, but sometimes they can be *secured*. A commitment may be secured by being self-enforcing; for instance, you can
burn your bridges behind you so you can't escape even if you
change your mind. Or it may be secured by your greater desire to
preserve your reputation. You may have good reason to fulfill your

side of a contract even if your reason for signing it in the first place has lapsed, simply because your reputation is also at stake, a valuable social commodity indeed. Or—and this is Schelling's point—a promise made "in the eyes of God" may well convince those who believe in that God that a sort of virtual escrow account has been created, protecting both parties and giving each the confidence to move ahead without fear of reneging by the other party.

Consider the current situation in Iraq, where a security force is supposed to provide a temporary scaffolding on which to construct a working society in post-Saddam Iraq. It might actually have worked from the outset if the force had been large enough and well enough trained and deployed to reassure people without having to fire a shot. With insufficient forces, the credibility of the peacekeepers was diminished, however, and a positive feedback cycle of violence was put in motion, destroying confidence in security. How can you break out of such a downward spiral? It is hard to say. The flawed and fragile democracy that has been installed may still overcome its corrupt and violence-ridden beginnings, if the world is lucky, however forlorn it looks today. Failed states have a way of perpetuating themselves, and perpetuating both the misery of their inhabitants and the insecurity of their neighbors. In the distant past, the *very idea* of an overseeing God might often have permitted an otherwise chaotic and ungovernable population to bootstrap itself into a working state, with enough law and order so that credible *promising* could take hold. Only in such a climate of trust can investment and commerce and free passage, and all the other things we take for granted in a working society, flourish. Such a meme would be vulnerable to collapse if its credibility was threatened, just as surely as the occupying forces in Iraq depend on their (problematic) credibility for their own effectiveness. The rationale for incorporating whatever doubt-suppression devices could be found would have been obvious (to the blind forces of cultural selection, and probably to the authorities themselves).

Today, when patterns of mutual trust are quite securely established in modern democratic states independently of any shared religious belief, the bristling defenses of religions against corrosive doubt begin to look vestigial, like fossil traces of an earlier epoch. We no longer need God the Policeman to create a climate in which we can make promises and conduct human affairs on their basis, but He lives on in legal oaths—and in the imaginations of many who are terrified of the prospect of abandoning religion.

But reward in heaven is not the only—and certainly not the best—inspirational theme in religious doctrine. The God who is watching you need not be seen to be either list-making Santa or Orwell's Big Brother, but instead a hero or "role model," as we say today, someone to emulate rather than fear. If God is just, and merciful, and forgiving, and loving, and the most wonderful Being imaginable, then anyone who loves God should want to be just, and merciful, and forgiving, and loving, *for goodness' sake.* Blurring these two very different views of God's motivating role into one is yet another casualty of the gauze curtains of soft-focus veneration through which we traditionally inspect religion.

Still, there may be the best of (free-floating) reasons for not peering too closely at these fine differences between doctrines. Why create dissension where none need exist? Don't rock the boat. It is widely agreed that all religions provide social infrastructures for creating and maintaining moral teamwork. Perhaps their value as organizers and amplifiers of good intentions far outweighs any deficits created by the putative incoherence created by contradictions between (some of) their doctrines. Perhaps it would be foolish perfectionism, and an act of moral ineptitude, to distract ourselves with minor conflicts of dogma when there is so much work to be done making the world a better place.

This is a persuasive claim, but it has the disadvantage of undercutting itself somewhat in public, since it amounts to making the acknowledgment that "good as we are, we aren't perfect, but we have more important things to do than fix our foundations"—a modest admission that jars with the traditional claims of purity that

religions find irresistible. Moreover, any such lapses from abso-
lutism threaten to undermine the chief psychological source of the
very organizational power that is being recognized. Today's religious warriors may be too sophisticated to expect their God to stop
the bullets in midair at their behest, but their belief in the *absolute*
rightness of their cause may well be a crucial ingredient in creating
the calm with which truly effective soldiers go into battle. As William James puts it:

> Whoever not only says, but *feels,* "God's will be done," is mailed
> [armored] against every weakness; and the whole historic array of
> martyrs, missionaries, and religious reformers is there to prove
> the tranquil-mindedness, under naturally agitating or distressing
> circumstances, which self-surrender brings. [1902, p. 285]

This heroic state of mind does not harmonize well with secular
modesty, and though many think it is true that religious fanatics
make the most reliable soldiers, we may well wonder whether, all
things considered, James is right when he goes on to note (quoting
"a clear-headed Austrian officer"), "Far better is it for an army to be
too savage, too cruel, too barbarous, than to possess too much sen-
timentality and human reasonableness" (p. 366). Here is a morally
relevant question well worth careful empirical investigation: can a
secular armed force, motivated in the main by a love of liberty or
democracy, not of God (or Allah), maintain its credibility, and hence
its effectiveness, with a minimum of bloodshed, against an army of
fanatics? Until we know the answer, we risk being blackmailed by
sheer fear into indoctrinating the troops with barbarism. It will take
a combination of courage and wise planning—and maybe a large
helping of luck—even to do the research needed to find out. But the
alternative is even more grim: perpetuating the fatal downward spiral of "righteous" wars, fought by misguided young people sent
into dubious battle by leaders who don't really believe the myths
that sustain those who are risking their lives. As the Grand Inquisi-
tor says in Dostoevski's *The Brothers Karamazov,* "Beyond the grave

they will find nothing but death. But we shall keep the secret, and for their happiness we shall allure them with the reward of heaven and eternity."

There is a further allure for the zealot, and it is probably—who knows?—a more robust motivator than the prospect of heavenly reward: the *license to kill* (to adapt Ian Fleming's all-too-appealing fantasy about the official status of James Bond). Some people, it seems—who knows?—are just bloodthirsty, or thrill-seeking, and as our customs become ever more civilized and opposed to violence, such people are highly motivated to find a cause that can provide them with a "moral" justification for their swashbuckling, whether it is "liberating" laboratory animals (whose subsequent welfare seems not to motivate the activists sufficiently), avenging Ruby Ridge with the Oklahoma City bombing, murdering doctors who perform abortions, sending anthrax to "evil" federal employees, murdering an innocent person under cover of fatwa, achieving martyrdom in jihad, or becoming a "settler" (armed to the teeth) in the West Bank territory. Religion may well not be the root cause of this dangerous yearning; the Hollywood-inspired desire to lead an adventurous and hence "meaningful" life may play a larger role in multiplying the number of young people who decide to frame their lives in such terms. But religions are certainly the most prolific source of the "moral certainties" and "absolutes" that such zealotry depends on. And although people who can see the shades of gray are less apt to be able to find excuses for committing criminal acts themselves, they are also, today, all too likely to see devout religious conviction as a significantly mitigating factor when meting out punishment. (We can hope that this will change swiftly if given sufficient public attention. We used to regard drunks as somewhat diminished in their responsibility for their actions—they were too drunk to know what they were doing, after all—but we now see them, and the bartenders who served them, as fully responsible. We need to spread the word that religious intoxication is no excuse either.)

2 Is religion what gives meaning to your life?

*A puppet of the gods is a tragic figure, a puppet suspended on his chro-
mosomes is merely grotesque.* —Arthur Koestler, *The Sleepwalkers*

*Ohhh, McTavish is dead and his brother don't know it;
   His brother is dead and McTavish don't know it.
They're both of them dead and they're in the same bed,
   And neither one knows that the other is dead!*
   —Lyrics to the "Irish Washerwoman" jig

According to surveys, *most of the people in the world* say that religion
is very important in their lives. (See, e.g., the Web site of the Pew
Research Center, http://people-press.org/.) Many of these people
would say that without their religion their lives would be meaningless. It's tempting just to take them at their word, to declare that in
that case there is really nothing more to be said—and tiptoe away.
Who would want to interfere with whatever it is that gives their
lives meaning? But if we do that, we willfully ignore some serious
questions. Can just *any* religion give lives their meaning, in a way
that we should honor and respect? What about people who fall into
the clutches of cult leaders, or who are duped into giving their life
savings to religious con artists? Do their lives still have meaning
even though their particular "religion" is a fraud?

   In *Marjoe*, the 1972 documentary about the bogus evangelist
Marjoe Gortner mentioned in chapter 6, we see poor people empty-
ing their wallets and purses into the collection plate, their eyes glis-
tening with tears of joy, thrilled to be getting "salvation" from this
charismatic phony. The question that has been troubling me ever
since I saw the film when it first came out is: who is committing
the more reprehensible act—Marjoe Gortner, who lies to these people in order to get their money, or the filmmakers who expose these
lies (with Gortner's enthusiastic complicity), thereby robbing these

good folk of the meaning they *thought* they had found for their lives? Were they not getting their money's worth and then some before the filmmakers came along? Consider their lives (I am imagining these details, which are not in the documentary): Sam is a high-school dropout, pumping gas at the station at the crossroads and hoping someday to buy a motorcycle; he is a Dallas Cowboys fan, and likes to have a few beers while watching the games on TV. Lucille, who never married, is in charge of the night-shift shelf-stockers at the local supermarket and lives in the modest house she has always lived in, caring for her aged mother; they follow the soap operas together. No adventurous opportunities beckon in the futures of Sam or Lucille, or most of the others in the blissful congregation, *but they have now been put in direct contact with Jesus* and are now saved for eternity, beloved members in good standing of the community of the born-again. They have turned over a new leaf, in a most dramatic ceremony, and they face their otherwise uninspiring lives refreshed and uplifted. Their lives now tell a story, and it's a chapter of the Greatest Story Ever Told. Can you imagine anything *else* they could buy with those twenty-dollar bills they deposit in the collection plate that would be remotely as valuable to them?

Certainly, comes the reply. They could donate their money to a religion that was honest, and that actually used their sacrifices to help others who were still needier. Or they could join any secular organization that put their free time, energy, and money to effective use in ameliorating some of the world's ills. Perhaps the main reason that religions do most of the heavy lifting in large parts of America is that people really do want to help others—and secular organizations have failed to compete with religions for the allegiance of ordinary people. That's important, but it's the easy part of the answer, leaving untouched the hard part: what should we do about those we honestly think are being conned? Should we leave them to their comforting illusions or blow the whistle? I have eventually come to the tentative conclusion that Marjoe Gortner and his filmmaking collaborators performed a great public service in spite of the pain and humiliation the film no doubt caused to many basically innocent people, but further details, or just further reflection on the details that are known, might lead me to change my mind.

Dilemmas like this are all too familiar in somewhat different contexts, of course. Should the sweet old lady in the nursing home be told that her son has just been sent to prison? Should the awkward twelve-year-old boy who wasn't cut from the baseball team be told about the arm-twisting by all the parents that persuaded the coach to keep him on the squad? In spite of ferocious differences of opinion about other moral issues, there seems to be something approaching consensus that it is cruel and malicious to interfere with the life-enhancing illusions of others—unless those illusions are themselves the cause of even greater ills. The disagreements come over what these greater ills might be—and this has led to the breakdown of the whole rationale. Keeping secrets from people for their own good can often be wise, but it takes only one person to give away a secret, and since there are disagreements about which cases warrant discretion, the result is an unsavory miasma of hypocrisy, lies, and frantic but fruitless attempts at distraction.

What if Marjoe Gortner were to con a cadre of *sincere* evangelical preachers into doing his dirty work for him? Would their personal innocence change the equation and give genuine meaning to the lives of those whose sacrifices they encouraged and collected? For that matter, aren't *all* evangelical preachers just as false as Marjoe Gortner? Certainly Muslims think so, even though they are generally too discreet to say it. And Catholics think that Jews are just as deluded, and Protestants think that Catholics are wasting their time and energy on a largely false religion, and so forth. *All* Muslims? *All* Catholics? *All* Protestants? *All* Jews? Of course not. There are vocal minorities in every faith who blurt it out, like the Catholic movie star Mel Gibson, who was interviewed by Peter Boyer (2003) in a profile in *The New Yorker*. Boyer asked him if Protestants are denied eternal salvation.

"There is no salvation for those outside the Church," Gibson replied. "I believe it." He explained: "Put it this way. My wife is a saint. She's a much better person than I am. Honestly. She's, like, Episcopalian, Church of England. She prays, she believes in God, she knows Jesus, she believes in that stuff. And it's just not fair if she doesn't make it, she's better than I am. But that is a pronouncement from the chair. I go with it."

Such remarks deeply embarrass two groups of Catholics: those who believe it but think it is best left unsaid, and those who don't believe it at all—no matter what "the chair" may pronounce. And which group of Catholics is larger, or more influential? That is utterly unknown and currently unknowable, a part of the unsavory miasma.

It is equally unknown how many Muslims truly believe that all infidels and especially kafirs (apostates from Islam) deserve death, which is what the Koran (4:89) undeniably says. Johannes Jansen (1997, p. 23) points out that in earlier times Judaism (see Deuteronomy 18:20) and Christianity (see Acts 3:23) also regarded apostasy as a capital offense, but of the Abrahamic faiths, Islam stands alone in its inability to renounce this barbaric doctrine convincingly. The Koran does not explicitly commend killing apostates, but the hadith literature (the narrations of the life of the Prophet) certainly does. Most Muslims, I would *guess,* are sincere in their insistence that the hadith injunction that apostates are to be killed is to be disregarded, but it's disconcerting, to say the least, that fear of being regarded as an apostate is apparently a major motivation in the Islamic world. As Jansen puts it, "There can be no Hare Krishna or Baghwan, no Scientology, Mormonism or Transcendental Meditation in Mecca or Cairo. Within the world of Islam religious renewal has to steer clear of anything that implies or suggests apostasy" (pp. 88–89). So it is not just we outsiders who are left guessing. Even Muslims "on the inside" really don't know what Muslims think about apostasy— they mostly aren't prepared to bet their lives on it, which is the surest sign of belief, as we saw in chapter 8.

Here, then, we see a different face of the epistemological problem we encountered in chapter 8, on belief in belief. There we discovered that it is all but impossible to distinguish those who genuinely believe and those who (merely) believe in belief, since the beliefs in question are conveniently removed from the world of action. Now we see that one reason, free-floating or not, for such systematically masked creeds is to avoid—or at least postpone—the collision between contradictory creeds that would otherwise oblige the devout to behave far more intolerantly than most people *today* want to behave. (It is always worth reminding ourselves that not so very long ago people were banished, tortured, and even executed for heresy and apostasy in the most "civilized" corners of Christian Europe.)

So what *is* the prevailing attitude today among those who call themselves religious but vigorously advocate tolerance? There are three main options, ranging from the disingenuous Machiavellian—

1. As a matter of political strategy, the time is not ripe for candid declarations of religious superiority, so we should temporize and let sleeping dogs lie in hopes that those of other faiths can gently be brought around over the centuries.

—through truly tolerant Eisenhowerian "Our government makes no sense unless it is founded on a deeply held religious belief—and I don't care what it is"—

2. It really doesn't matter which religion you swear allegiance to, as long as you have *some* religion.

—to the even milder Moynihanian benign neglect—

3. Religion is just too dear to too many to think of discarding, even though it really doesn't do any good and is simply an empty historical legacy we can afford to maintain until it quietly extinguishes itself sometime in the distant and unforeseeable future.

It is no use asking people which they choose, since both extremes are so undiplomatic we can predict in advance that most people will go for some version of ecumenical tolerance whether they believe it or not. (It's just like Sir Maurice Oldfield's predictable denunciation of my subversive hypothesis about Kim Philby.)

We've got ourselves caught in a hypocrisy trap, and there is no clear path out. Are we like the families in which the adults go through all the motions of believing in Santa Claus for the sake of the kids, and the kids all pretend still to believe in Santa Claus so as not to spoil the adults' fun? If only our current predicament were as innocuous and even comical as that! In the adult world of religion, people are dying and killing, with the moderates cowed into silence by the intransigence of the radicals in their own faiths, and many afraid to acknowledge what they actually believe for fear of breaking Granny's heart, or offending their neighbors to the point of getting run out of town, or worse.

If *this* is the precious meaning our lives are vouchsafed thanks to our allegiance to one religion or another, it is not such a bargain, in my opinion. Is this the best we can do? Is it not tragic that so many people around the world find themselves enlisted against their will in a conspiracy of silence, either because they secretly believe that most of the world's population is wasting their lives in delusion (but they are too tenderhearted—or devious—to say so), or because they secretly believe that their own tradition is just such a delusion (but they fear for their own safety if they admit it)?

What alternatives are there? There are the moderates who revere the tradition they were raised in, simply because it is *their* tradition, and who are prepared to campaign, tentatively, for the details of their tradition simply because, in the marketplace of ideas, somebody should stick up for each tradition until we can sort out the good from the better and settle for the best we can find, *all things considered*. This is like allegiance to a sports team, and it, too, can give meaning to a life—if not taken too seriously. I am a Red Sox fan simply because I grew up in the Boston area and have happy memories of Ted Williams and Jimmy Piersall and Jackie Jensen and Carl Yastrzemski and Wade Boggs and Luis Tiant and Pudge

Fisk, among others. My allegiance to the Red Sox is enthusiastic, but cheerfully arbitrary and undeluded. The Red Sox aren't my team because they are, in fact, the Best; they are "the Best" (in my eyes) because they are my team. I bask in the glory of their victory in 2004 (which was, of course, the Most Amazing and Inspiring Come-from-Behind Saga Ever), and if the team were ever to disgrace itself, I would be not just deeply chagrined but *personally* ashamed—as if I had something to do with it. And of course I *do* have something to do with it; my tiny personal contribution to the ocean of local enthusiasm and pride actually does buoy the players' spirits (as they always insist).

This is a kind of love, but not the rabid love that leads people to lie, and torture, and kill. Those who feel guilty contemplating "betraying" the tradition they love by acknowledging their disapproval of elements within it should reflect on the fact that the very tradition to which they are so loyal—the "eternal" tradition introduced to them in their youth—is in fact the evolved product of many adjustments firmly but delicately made by earlier lovers of the same tradition.

3 What can we say about sacred values?

We are here on Earth to do good to others. What the others are here for, I don't know. —W. H. Auden

For many years now, you and I have been shushed like children and told there are no simple answers to the complex problems that are beyond our comprehension. Well, the truth is there are simple answers. They are just not easy ones.

—Ronald Reagan, inaugural address as governor of California, January 1977

If our tribalism is ever to give way to an extended moral identity, our religious beliefs can no longer be sheltered from the tides of genuine inquiry

and genuine criticism. It is time we realized that to presume knowledge where one has only pious hope is a species of evil. Wherever conviction grows in inverse proportion to its justification, we have lost the very basis of human cooperation. —Sam Harris, *The End of Faith*

In order to adopt such a moderate position, however, you have to loosen your grip on the absolutes that are apparently one of the main attractions of many religious creeds. It isn't easy being moral, and it seems to be getting harder and harder these days. It used to be that most of the world's ills—disease, famine, war—were quite beyond the capacities of everyday people to ameliorate. There was nothing they could do about it, and since "'ought' implies 'can,'" people could ignore the catastrophes on the other side of the globe—if they even knew about them—with a clear conscience, since they were powerless to avert them in any way. Living by a few simple, locally applicable maxims could more or less guarantee that one lived about as good a life as was possible at the time. No longer.

Thanks to technology, what almost anybody *can* do has been multiplied a thousandfold, and our moral understanding about what we *ought* to do hasn't kept pace (Dennett, 1986, 1988). You *can* have a test-tube baby or take a morning-after pill to keep from having a baby; you *can* satisfy your sexual urges in the privacy of your room by downloading Internet pornography, and you *can* copy your favorite music for free instead of buying it; you *can* keep your money in secret offshore bank accounts and purchase stock in cigarette companies that are exploiting impoverished Third World countries; and you *can* lay minefields, smuggle nuclear weapons in suitcases, make nerve gas, and drop "smart bombs" with pinpoint accuracy. Also, you *can* arrange to have a hundred dollars a month automatically sent from your bank account to provide education for ten girls in an Islamic country who otherwise would not learn to read and write, or to benefit a hundred malnourished people, or provide medical care for AIDS sufferers in Africa. You *can* use the

Internet to organize citizen monitoring of environmental hazards, or to check the honesty and performance of government officials—or to spy on your neighbors. Now, what ought we to do?

In the face of these truly imponderable questions, it is entirely reasonable to look for a short set of simple answers. H. L. Mencken cynically said, "For every complex problem, there is a simple answer . . . and it is wrong." But maybe he was wrong! Maybe one Golden Rule or Ten Commandments or some other short list of absolutely nonnegotiable Dos and Don'ts resolves all the predicaments just fine, once you figure out how to apply them. Nobody would deny, however, that it is far from obvious how any of the favored rules or principles can be interpreted to fit all our quandaries. As Scott Atran points out, the commandment "Thou shalt not kill" is cited by religious opponents of the death penalty, and by religious proponents as well (2002, p. 253). The principle of the Sanctity of Human Life sounds bracingly clear and absolute: every human life is equally sacred, equally inviolable; as with the king in chess, no price can be placed on it—aside from "infinity," since to lose it is to lose everything. But in fact we all know that life isn't, and can't be, like chess. There are multitudes of interfering "games" going on at once. What are we to do when more than one human life is at stake? If each life is *infinitely* valuable and none more valuable than another, how are we to dole out the *few* transplantable kidneys that are available, for instance? Modern technology only exacerbates the issues, which are ancient. Solomon faced tough choices with notable wisdom, and every mother who has ever had less than enough food for her own children (let alone her neighbor's children) has had to confront the impracticality of applying the principle of the Sanctity of Human Life.

Surely just about everybody has faced a moral dilemma and secretly wished, "If only somebody—somebody I trusted—could just *tell* me what to do!" Wouldn't this be morally inauthentic? Aren't we responsible for making our *own* moral decisions? Yes, but the virtues of "do it yourself" moral reasoning have their limits, and if you decide, after conscientious consideration, that your moral decision is to delegate further moral decisions in your life to a trusted expert, then you *have* made your own moral decision. You have decided to take advantage of the division of labor that civilization makes possible and get the help of expert specialists.

We applaud the wisdom of this course in all other important areas of decision-making (don't try to be your own doctor; the lawyer who represents himself has a fool for a client, and so forth). Even in the case of political decisions, like which way to vote, the policy of delegation can be defended. When my wife and I go to Town Meeting, I know that she has studied the issues that confront our town so much more assiduously than I have that I routinely follow her lead, voting the way she tells me to vote, even if I'm not sure just why, because I have plenty of evidence for my conviction that if we did take the time and energy to thrash it all out she'd persuade me that, all things considered, her opinion was correct. Is this a dereliction of my duties as a citizen? I don't think so, but it does depend on my having good grounds for trusting her judgment. Love is not enough. That's why those who have an *unquestioning* faith in the correctness of the moral teachings of their religion are a problem: if they themselves haven't conscientiously considered, on their own, whether their pastors or priests or rabbis or imams are worthy of this delegated authority over their own lives, then they are in fact taking a personally *immoral* stand.

This is perhaps the most shocking implication of my inquiry, and I do not shrink from it, even though it may offend many who think of themselves as deeply moral. It is commonly supposed that it is entirely exemplary to adopt the moral teachings of one's own religion *without question,* because—to put it simply—it is the word of God (as interpreted, always, by the specialists to whom one has delegated authority). I am urging, on the contrary, that anybody who professes that a particular point of moral conviction is not discussable, not debatable, not negotiable, simply because it is the word of God, or because the Bible says so, or because "that is what all Muslims [Hindus, Sikhs . . .] believe, and I am a Muslim [Hindu, Sikh . . .]," should be seen to be making it impossible for the rest of us to take their views seriously, excusing themselves from the moral conversation, inadvertently acknowledging that their own views are *not* conscientiously maintained and deserve no further hearing.

The argument for this is straightforward. Suppose I have a friend, Fred, who is (in my carefully considered opinion) *always right*. If I tell you I'm against stem-cell research because "my friend Fred says it's wrong and that's all there is to it," you will just look at me as if I was missing the point of the discussion. This is supposed to be a consideration of reasons, and I have not given you a reason that I in good faith could expect you to appreciate. Suppose you believe that stem-cell research is wrong because that is what God has told you. *Even if you are right*—that is, even if God does indeed exist and has, personally, told you that stem-cell research is wrong—you cannot reasonably expect others who do not share your faith or experience to accept this as a reason. You are being unreasonable in taking your stand. The fact that your faith is so strong that you cannot do otherwise just shows (if you *really* can't) that you are *disabled* for moral persuasion, a sort of robotic slave to a meme that you are unable to evaluate. And if you reply that you *can* but you *won't* consider reasons for and against your conviction (because it is God's word, and it would be sacrilegious even to consider whether it might be in error), you avow your willful refusal to abide by the minimal conditions of rational discussion. Either way, your declarations of your deeply held views are posturings that are out of place, part of the problem, not part of the solution, and we others will just have to work around you as best we can.

Notice that this stand involves no disrespect and no prejudging of the possibility that God has told you. If God has told you, then part of *your* problem is convincing others, to whom God has not (yet) spoken, that this is what we ought to believe. If you refuse or are unable to attempt this, you are actually letting your God down, in the guise of demonstrating your helpless love. You can withdraw from the discussion if you must—that is your right—but then don't expect us to give your view any particular weight that we cannot discover by other means—and don't blame us if we don't "get it."

Many deeply religious people have all along been eager to defend their convictions in the court of reasonable inquiry and persuasion. They will have no difficulty at all with these observations—aside from confronting the diplomatic decision of whether they will join me in trying to convince their less reasonable coreligionists that they are making matters worse for their religion by their intransigence. And here is one of the most intractable moral problems confronting the world today. Every religion—aside from a negligible scattering of truly toxic cults—has a healthy population of ecumenical-minded people who are eager to reach out to people of other faiths, or no faith at all, and consider the moral quandaries of the world on a rational basis. In July 2004, the fourth Parliament of World Religions was held in Barcelona,[3] and it brought thousands of people of different religions together for a week of workshops, symposia, plenary sessions, performances, and worship services, all enjoined to observe the same principles:

> *listen and be listened to so that all speakers can be heard*
> *speak and be spoken to in a respectful manner*
> *develop or deepen mutual understanding*
> *learn about the perspective of others and reflect on one's own views,*
> *and*
> *discover new insights.* [*Pathways to Peace*, the Parliament program]

Colorful flocks of differently robed priests and gurus, nuns and monks, choirs and dancers, all holding hands and listening respectfully to one another—it was all very heartwarming, but these well-intentioned and energetic people are singularly ineffective in dealing with the more radical members of their own faiths. In many instances they are, rightly, terrified of them. Moderate Muslims have so far been utterly unable to turn the tide of Islamic opinion against Wahhabists and other extremists, but moderate Christians and Jews and Hindus have been equally feckless in countering the outrageous demands and acts of their own radical elements.

It is time for the reasonable adherents of all faiths to find the courage and stamina to reverse the tradition that honors helpless love of God—in any tradition. Far from being honorable, it is not even excusable. It is shameful. And most shameful are the priests, rabbis, imams, and other experts whose response to the sincere requests from their flock for moral guidance is to conceal their own inability to give reasons for their views about the tough issues by hiding behind some "inerrant" (read "above criticism") interpretation of the sacred texts. It is one thing for a well-meaning layperson with a deep allegiance to a religious tradition to delegate authority to his or her religious leaders, but it is quite another for those leaders to *pretend* to discover (thanks to their expertise) the right answers in their tradition by a process that has to be taken on faith and is inaccessible to even the most well-meant criticism.

As so often before, we should grant that it is entirely possible that this evasive question-ducking rationale is entirely free-floating. In other words, it is surely possible for people to *believe in all innocence* that their love of God absolves them from the responsibility to figure out reasons for these hard-to-fathom commands from their beloved God. We need make no accusations of insincerity or guile, but respecting someone's innocence does not oblige us to respect his belief. Here is what we should say to such a person: There is only one way to respect the *substance* of any purported God-given moral edict: consider it conscientiously in the full light of reason, using all the evidence at our command. No God that was pleased by displays of unreasoning love would be worthy of worship.

Here is a riddle: how is your religion like a swimming pool? And here is the answer: it is what is known in the law as an *attractive nuisance*. The doctrine of attractive nuisance is the principle that people who maintain on their property a dangerous condition that is likely to attract children are under a duty to post a warning or to take stronger affirmative action to protect children from the dangers of that attraction. It is an exception to the general rule that no particular care is required of property owners to safeguard trespassers from harm. Unenclosed swimming pools are the best-known example, but old refrigerators with their doors not removed, machinery or stacks of building materials, or other eminently climbable objects that could be an irresistible lure to young children have also been deemed attractive nuisances. Property owners are held responsible for harms that result when they maintain something that can lure innocent people into harm.

Those who maintain religions, and take steps to make them more attractive, must be held similarly responsible for the harms produced by some of those whom they attract and provide with a cloak of respectability. Defenders of religion are quick to point out that terrorists typically have political, not religious agendas, which may well be true in many or most cases, or even in all cases, but that is not the end of it. The political agendas of violent fanatics often lead them to adopt a religious guise, and to exploit the organizational infrastructure and tradition of unquestioning loyalty of whichever religion is handy. And it is true that these fanatics are rarely if ever inspired by, or guided by, the deepest and best tenets in those religious traditions. So what? Al Qaeda and Hamas terrorism is still Islam's responsibility, and abortion-clinic bombing is still Christianity's responsibility, and the murderous activities of Hindu extremists are still Hinduism's responsibility.

As Sam Harris argues in his brave book *The End of Faith* (2004), there is a cruel Catch-22 in the worthy efforts of the moderates and ecumenicists in all religions: by their good works they provide protective coloration for their fanatical coreligionists, who quietly condemn their open-mindedness and willingness to change while reaping the benefits of the good public relations they thereby obtain.

In short, the moderates in all religions *are being used* by the fanatics, and should not only resent this; they should take whatever steps they can find to curtail it in their own tradition. Probably nobody else can do it, a sobering thought:

> If a stable peace is ever to be achieved between Islam and the West, Islam must undergo a radical transformation. This transformation, to be palatable to Muslims, must also appear to come from Muslims themselves. It does not seem much of an exaggeration to say that the fate of civilization lies largely in the hands of "moderate" Muslims. [Harris, 2004, p. 154]

We must hold these moderate Muslims responsible for reshaping their own religion—but that means we must equally hold moderate Christians and Jews and others responsible for all the excesses in their own traditions. And, as George Lakoff has noted, we need to prove to those Islamic leaders that we hear their moral voices, and not just our own:

> We depend on the goodwill and courage of moderate Islamic leaders. To gain it, we must show our goodwill by beginning in a serious way to address the social and political conditions that lead to despair. [2004, p. 61]

How can we *all* keep the cloak of religious respectability from being used to shelter the lunatic excesses? Part of the solution would be to make religion *in general* less of a "sacred cow" and more of a "worthy alternative." This is the course somewhat haplessly followed by some of us brights—atheists, agnostics, freethinkers, secular humanists, and others who have liberated themselves from specifically religious allegiances. We brights are quite aware of all the good that religions accomplish, but we prefer to channel our charity and good deeds through secular organizations, precisely because we don't want to be complicit in giving a good name to religion! This keeps our hands clean, but that is not enough—any more than it is enough for moderate Christians to avoid giving funds to anti-Semitic organizations within Christianity, or for moderate Jews to restrict their charity to organizations that are working to secure peaceful coexistence for Palestinians and Israelis. That is a start, but there is more work to be done, and it is the unpleasant and even dangerous work of desanctifying the excesses in each tradition *from the inside*. Any religious person who is not actively and publicly involved in that effort is shirking a duty—and the fact that you don't belong to a congregation or denomination that is offending doesn't excuse you: it is Christianity and Islam and Judaism and Hinduism (for example) that are attractive nuisances, not just their offshoot sects.

Any vicious cult that uses Christian imagery or texts as its protective coloration should lie heavily on the conscience of all who call themselves Christians, for instance. Until the priests and rabbis and imams and their flocks explicitly condemn *by name* the dangerous individuals and congregations within their ranks, they are *all* complicit. I know many Christians who are privately sickened by many of the words and deeds done "in the name of Jesus," but expressions of dismay to close friends are not enough. In *Darwin's Dangerous Idea*, I wrote about the brave Muslims who dared to speak out publicly against the obscene travesty of the fatwa pronounced on Salman Rushdie, author of *The Satanic Verses*, condemned to death for his heresies, and urged, "Let us all distribute the danger by joining hands with them" (p. 517n). But here is the truly distressing Catch-22: if we non-Muslims join hands with them, we thereby mark them as "puppets of the enemies of Islam" in the eyes of many Muslims. Only those within the religious community can effectively start to dismantle this deeply immoral attitude, and multiculturalists who urge us to go easy on them are exacerbating the problem.

4 Bless my soul: spirituality and selfishness

He who has the most toys when he dies wins.
  —Well-known materialist slogan

Yes, we have a soul; but it's made of lots of tiny robots.
  —My materialist slogan[4]

Consider the two utterly different meanings of the word "material-istic." In its most common everyday sense, it refers to somebody who cares only about "material" possessions, wealth, and all its trappings. In its scientific or philosophical sense, it refers to a theory that aspires to explain all the phenomena without recourse to anything immaterial—like a Cartesian soul, or "ectoplasm"—or God. The standard negation of *materialistic* in the scientific sense is *dualistic,* which maintains that there are two entirely different kinds of substance, matter and . . . whatever minds are supposedly made of. The *apparent* bridge tying the two meanings together is obvious enough: if you don't think you have an immortal soul, then you don't believe you'll get your reward in heaven, so . . . you might as well go for whatever you can get in this material world. If we asked people what term was the negation of *materialistic* in the everyday sense, they might very well settle on *spiritual.*

In the course of my research on this book, I found one opinion expressed in slightly different ways by people across the spectrum of religious views: "man" has a "deep need" for "spirituality," a need that is fulfilled for some by traditional organized religion, for others by New Age cults or movements or hobbies, and for still others by the intense pursuit of art or music, pottery or environmental activism—or football! What fascinates me about this delightfully versatile craving for "spirituality" is that people think they know what they are talking about, even though—or perhaps because—nobody bothers to explain just what they mean. It is supposed to be obvious, I guess. But it really isn't. When I've asked people to explain themselves, they typically beg off, along the lines of Louis Armstrong's oft-quoted reply when asked what jazz was: "If you gotta ask, you ain't never gonna get to know." This will not do. To see for yourself just how hard it is to say what spirituality is, take a stab at improving on this parody, boiled down from many frustrating encounters: "Spirituality is, you know, like, it's like paying attention to your soul or having deep thoughts that really move you, and not just thinking about who's got nicer clothes and whether to buy a new car and what's for dinner and stuff like that. Spirituality is *really caring* and not being just, you know, *materialistic*." Along with this common and unreflective view of spirituality goes a stereotype of the atheist: atheists lack "values"; they are careless, self-centered, shallow, overconfident. They think they know it all, and yet they completely miss out on the spirit. (You really can't be a good person unless you have a spiritual life.)

Now let me try to put better words in their mouths. What these people have realized is one of the best secrets of life: let your *self* go. If you can approach the world's complexities, both its glories and its horrors, with an attitude of humble curiosity, acknowledging that however deeply you have seen, you have only just scratched the surface, you will find worlds within worlds, beauties you could not heretofore imagine, and your own mundane preoccupations will shrink to *proper* size, not all that important in the greater scheme of things. Keeping that awestruck vision of the world ready to hand while dealing with the demands of daily living is no easy exercise, but it is definitely worth the effort, for if you can stay *centered*, and *engaged*, you will find the hard choices easier, the right words will come to you when you need them, and you will indeed be a better person. That, I propose, is the secret to spirituality, and it has nothing at all to do with believing in an immortal soul, or in anything supernatural.

The psychologist Nicholas Humphrey has explored in some depth the relationship between belief in "psychic forces" and the everyday sense of morality. He notes that almost all stories of the paranormal, of extrasensory perception and clairvoyance and talking to deceased friends and relatives at séances, have a "somewhat self-righteous aura to them—a tag of holiness, a certain touch-me-not feel" (1995, p. 186). And although this may be due in part to the fact that so often the stories deal with the most emotionally sensitive areas of people's lives, he has another explanation:

> ... it originates with what is, arguably, one of the most remarkable confidence tricks our culture has played on us. This has been to persuade people that there is a deep connection between believing in the possibility of psychic forces and being a gracious, honest, upright, trustworthy member of society. ...

He deftly enunciates the free-floating rationale:

> Whether or not people have had any explicit religious education, they have all been exposed to the idea that some kind of supernatural parent figure watches over them and cares for them. It may easily follow therefore that people's sense of justice and propriety persuades them that, if such a figure does exist, then *not to believe in him* would be ungrateful in the extreme—and only wicked children could possibly be so ungrateful. But, if unbelievers are generally wicked, it is natural (though hardly logical) to assume that believers are generally good. So whether or not a person believes in this supernatural parent becomes in itself a measure of his moral virtue. ... The absurd, but quite widely accepted result has been that every paranormal story we hear is supposed to be automatically worthy of attention and respect. [pp. 186–87]

I have come to accept that this alignment of moral goodness with "spirituality" and moral evil with "materialism" is just a frustrating fact of life, so deeply rooted in our contemporary conceptual scheme that it amounts to a prevailing wind against which materialistic science has to strain. We materialists are the bad guys, and those who believe in anything supernatural, however goofy and gullible the particular belief, have at least this much going for them: they're "on the side of the angels."

This familiar phrase was born, by the way, in the Oxford Union, a debating society at Oxford University, in a speech by Benjamin Disraeli in 1864, in response to the challenge of Darwinism: "What is the question now placed before society with a glib assurance the most astounding? The question is this—Is man an ape or an angel. My Lord, I am on the side of the angels." The misalignment of goodness with the denial of scientific materialism has a long history, but it *is* a misalignment.[5] There is *no reason at all* why a disbelief in the immateriality or immortality of the soul should make a person less caring, less moral, less committed to the well-being of everybody on Earth than somebody who believes in "the spirit." But won't such a materialist care only about the *material* well-being of the people? If that means only their housing, their car, their food, their "physical" as opposed to "mental" health, no. After all, a good scientific materialist believes that mental health—spiritual health, if you like—is just as physical, just as material, as "physical" health. A good scientific materialist can be just as concerned about whether there is plenty of justice, love, joy, beauty, political freedom, and, yes, even religious freedom as about whether there is plenty of food and clothing, for instance, since *all* of these are material benefits, and some are more important than others. (But for goodness' sake, let's try to get food and clothing to everybody who needs them as soon as possible, since without them justice and art and music and civil rights and the rest are something of a mockery.)

That should correct the understandable logical confusion. There is also the factual misconception to correct: plenty of "deeply spiritual" people—and everybody knows this—are cruel, arrogant, self-centered, and utterly unconcerned about the moral problems of the world. Indeed, one of the truly nauseating side effects of the common confusion of moral goodness with "spirituality" is that it permits untold numbers of people to slack off on the sacrifice and good works and hide behind their unutterably sacred (and impenetrable) mask of piety and moral depth. It's not just the hypocrites, though there are always plenty of them around. There are many people who quite innocently and sincerely believe that if they are earnest in attending to their own personal "spiritual" needs, this amounts to living a morally good life. I know many activists, both religious and secular, who agree with me: these people are deluding themselves. Auden's sardonic quip may shake our faith in the obviousness of the imperative to help others, but it certainly does nothing to suggest that just taking care of one's own "soul" is anything but selfish. Consider, for instance, those contemplative monks, primarily in Christian and Buddhist traditions, who, unlike hardworking nuns in schools and hospitals, devote most of their waking hours to the purification of their souls, and the rest to the maintenance of the contemplative lifestyle to which they have become accustomed. In what way, exactly, are they morally superior to people who devote their lives to improving their stamp collections or their golf swing? It seems to me that the best that can be said of them is that they manage to stay out of trouble, which is not nothing.

I am under no illusions about how hard it will be to undo the centuries of presumption that tend to merge "spirit" and "goodness." Since "team spirit" is obviously good, how can the denial of "spirit" be anything but bad? Even deep in the trenches of cognitive neuroscience, I find annoying echoes and shadows of this prejudice, with us "hardheaded" materialists forever on the defensive against the now practically extinct species of "tenderhearted" dualists, who seem (to the laypeople at least) to occupy the moral high ground simply because they still believe in the immateriality of souls. It's an uphill battle, but perhaps it will go better for us when it is fought in broad daylight.

But what about that hunger for spirituality that so many of my informants think is the mainspring of religious allegiance? The good news is that people really do want to be good. Believers and brights alike deplore the crass materialism (everyday sense) of popular culture and yearn not just to enjoy the beauty of genuine love but to bring that joy to others. It may often have been true in the past that for most people the only available road to that fulfillment involved a commitment to the supernatural, and more particularly to a specific institutional version of the supernatural, but today we can see that there is a bounty of alternative highways and footpaths to consider.

\\\

Chapter 10 The widely prevailing opinion that religion is the bulwark of morality is problematic at best. The idea that heavenly reward is what motivates good people is demeaning and unnecessary; the idea that religion at its best gives meaning to a life is jeopardized by the hypocrisy trap into which we have fallen; the idea that religious authority grounds our moral judgments is useless in genuine ecumenical exploration; and the presumed relation between spirituality and moral goodness is an illusion.

\\\

Chapter 11 The research described in this book is just the beginning. Further research is needed, on both the evolutionary history of religion and on its contemporary phenomena, as they appear to different disciplines. The most pressing questions concern how we should deal with the excesses of religious upbringing and the recruitment of terrorists, but these can only be understood against a background of wider theories of religious conviction and practice. We need to secure our democratic society, the home base for this research, against the subversions of those who would use democracy as a ladder to theocracy and then throw it away, and we need to spread the knowledge that is the fruit of free inquiry.

||| ||| |||

CHAPTER ELEVEN

Now What Do We Do?

1 Just a theory

You philosophers are fortunate people. You write on paper—I, poor empress, am forced to write on the ticklish skins of human beings.
—Catherine the Great, to Diderot (who had advised her about land reforms)

Since 2002, schools in Cobb County, Georgia, have put stickers in some of their biology textbooks saying "Evolution is a theory, not a fact," but a judge recently ruled that these must be removed, since they may convey the message of endorsement of religion "in violation of First Amendment separation of church and state and the Georgia Constitution's prohibition against using public money to aid religion" (*New York Times*, January 14, 2005). This makes sense, since the only motivations for singling out evolution for this treatment are religious. Nobody is putting stickers in chemistry or geology books saying that the theories explained therein are theories, not facts. There are still plenty of controversies in chemistry and geology, but these rival theories are contested within the securely established background theories of each field, which are not just theory but fact. There are lots of controversial theories within biology, too, but the background theory that is not contested is evolution. There are rival theories of vertebrate flight, and the role of migration in speciation, and, closer to human home, theories about the evolution of language, bipedality, concealed ovulation, and schizophrenia, to name just a few particularly vigorous controversies. Eventually, these will all get sorted out, and some of the theories will prove to be not just theories but facts.

My description of the evolution of various features of religion in chapters 4–8 is definitely "just a theory"—or, rather, a family of proto-theories, in need of further development. In a nutshell, this is what it says: Religion evolved, but it doesn't have to be good for us in order to evolve. (Tobacco isn't good for us, but it survives just fine.) We don't all learn language because we think it's good for us; we all learn language because we cannot do otherwise (if we have normal nervous systems). In the case of religion, there is a lot more teaching and drill, a lot more deliberate social pressure, than there is in language learning. In this regard, religion is more like reading than talking. There are tremendous benefits to being able to read, and perhaps there are similar or greater benefits to being religious. But people may well love religion independently of any benefits it provides them. (I am delighted to learn that red wine in moderation is good for my health, since, *whether or not* it is good for me, I like it, and I want to go on drinking it. Religion could be like that.) It is not surprising that religion survives. It has been pruned and revised and edited for thousands of years, with millions of variants extinguished in the process, so it has plenty of features that appeal to people, and plenty of features that preserve the identity of its recipes for these very features, features that ward off or confound enemies and competitors, and secure allegiance. Only gradually have people come to have any appreciation of the reasons—the heretofore free-floating rationales—for these features. Religion is many things to many people. For some, the memes of religion are mutualists, providing undeniable benefits of sorts that cannot be found elsewhere. These people may well depend for their very lives on religion, the way we all depend on the bacteria in our guts that help us digest our food. Religion provides some people with a moti-vated organization for doing great things—working for social jus-tice, education, political action, economic reform, and so forth. For others, the memes of religion are more toxic, exploiting less savory aspects of their psychology, playing on guilt, loneliness, the longing for self-esteem and importance. Only when we can frame a com-prehensive view of the many aspects of religion can we formulate defensible policies for how to respond to religions in the future.

Some aspects of this theory sketch are pretty well established, but getting down to specifics and generating further testable hy-potheses is work for the future. I wanted to give readers a good idea of what a testable theory would be like, what sorts of questions it would raise, and what sorts of explanatory principles it could in-voke. My theory sketch may well be false in many regards, but if so, this will be shown by confirming some alternative theory of the same sort. In science, the tactic is to put forward something that can be either fixed or refuted by something better. A century ago, it was just a theory that powered fixed-wing flight was possible; now it is fact. A few decades ago, it was just a theory that the cause of AIDS was a virus, but the reality of HIV is not just a theory today.

Since my proto-theory is not yet established and may prove to be wrong, it shouldn't be used yet to guide our policies. Having in-sisted at the outset that we need to do much more research so that we can make well-informed decisions, I would be contradicting myself if I now proceeded to prescribe courses of action on the basis of my initial foray. Recall, from chapter 3, the moral that Taubes drew in his history of the misguided activism that led us on the low-fat crusade: "It's a story of what can happen when the de-mands of public health policy—and the demands of the public for simple advice—run up against the confusing ambiguity of real sci-ence." There is pressure on us all to act decisively today, on the basis of the little we already (think we) know, but I am counseling patience. The current situation is scary—one religious fanaticism or another could produce a global catastrophe, after all—but we should resist rash "remedies" and other overreactions. It is possible, however, to discuss *options* today, and to think *hypothetically* of what the sound policies *would be* if something like my account of religion is correct. Such a consideration of possible policies can help motivate the further research, giving us pressing reasons for finding out which hypotheses are really true.

If somebody wants to put a sticker in this book, saying that it presents a theory, not a fact, I would happily concur. *Caution!* it should say. *Assuming that these propositions are true without further research could lead to calamitous results.* But I would insist that we also put the stickers on any books or articles that maintain or presuppose that religion is the lifeboat of the world, which we dare not upset. The proposition that God exists is *not even* a theory, as we saw in chapter 8. That assertion is so prodigiously ambiguous that it expresses, at best, an unorganized set of dozens or hundreds—or billions—of quite *different* possible theories, most of them disqualified as theories in any case, because they are systematically immune to confirmation or disconfirmation. The refutable versions of the claim that God exists have life cycles like mayflies, being born and dying within a matter of weeks, if not minutes, as predictions fail to come true. (Every athlete who prays to God for victory in the big game and then wins is happy to thank God for taking his side, and chalks up some "evidence" in favor of his theory of God—but quietly revises his theory of God whenever he loses in spite of his prayers.) Even the secular and nonpartisan proposition that religion *in general* does more good than harm, either to the individual believer or to society as a whole, has hardly begun to be properly tested, as we saw in chapters 9 and 10.

So here is the only prescription I will make categorically and without reservation: Do more research. There is an alternative, and I am sure it is still hugely appealing to many people: Let's just close our eyes, trust to tradition, and wing it. Let's just *take it on faith* that religion is the key—or one of the keys—to our salvation. How can I quarrel with faith (for heaven's sake)? *Blind* faith? Please. Think.

This is where we began. My task was to demonstrate that there was enough reason to question the tradition of faith so that you could not in good conscience turn your back on the available or discoverable relevant facts. I am quite prepared to roll up my sleeves and get down to examining the evidence and considering alternative scientific theories of religion, but I think I have already made my case that it would be indefensibly reckless *not* to do this research.

My survey has highlighted a small fraction of the work that has already been done, using it to tell one of the possible stories of how religion became what it is today, leaving other stories unmentioned. I told what I think is the best current version, but perhaps I have overlooked some contributions that will eventually be recognized retrospectively to be more important. This is a risk that a project like mine takes: if, by drawing attention to one avenue of research, it helps bury some better avenue in oblivion, I will have done a disservice. I am acutely aware of this prospect, so I have shared drafts of this book with researchers who have their own vision of how to make progress in the field. My network of informants inevitably has its own bias, however, and I would like nothing better than for this book to provoke a challenge—a reasoned and evidence-rich scientific challenge—from researchers with opposing viewpoints.[1]

I anticipate that one of the challenges will come from those in academia who are unmoved by my discussion of the "academic smoke screen" in chapter 9, and who firmly believe that the only researchers qualified to do the research are those who enter into an exploration of religion with a "proper respect" for the sacred, with a deep commitment to hallowing the traditions if not converting to them. They will want to maintain that the sort of empirical, biology-based inquiries I have championed, what with their mathematical models and use of statistics and the rest, are bound to be woefully superficial, insensitive, and uncomprehending.

Recent history shows that this is a concern to take seriously. A few decades ago, the field of "science studies" was born, when historians of science and philosophers of science were joined by sociologists and anthropologists who decided to apply their techniques, honed on the exploration of tribal cultures isolated in distant jungles and archipelagoes, to science itself, such as the subcultures of particle physicists, or molecular biologists, or mathematicians. Some of the early attempts by well-intentioned teams of social scientists to study these phenomena "in the wild" (of the laboratory and seminar room) led to the publication of studies that were met with—and deserved—the derision of the scientists who were the topic of the research. However sophisticated the researchers may have been as anthropologists, they were still naïve observers, largely clueless about the technicalities of the science they were witnessing, so they often came up with comically bad interpretations of what they had observed. If you don't understand in some detail the enterprise of the people you are studying, you have scant chance of understanding their interactions and reactions at the human level. The same maxim should apply to the study of religious discourse and practices.

People in science studies have had to work hard to overcome the bad reputation the field garnered in its early days, and there are still many scientists who do not bother suppressing their contempt for it, but the misguided work has by now been more than balanced by deeply informed and comprehending work that has actually managed to open scientists' eyes to patterns and foibles in their own practice. The key to this more recent success is simple: do your homework. Anybody hoping to make sense of any highly sophisticated and difficult field of human effort needs to become a near-expert in that field *in addition* to having the training of his or her home field. Applied to the study of religion, the prescription is clear: scientists intent on explaining religious phenomena are going to have to delve deeply and conscientiously into the lore and practices, the texts and contexts, the daily lives and problems of the people they are studying.

How could this be guaranteed? Religious experts—priests,

imams, rabbis, ministers, theologians, historians of religion—who are skeptical of the qualifications of those scientists who would study them could create and administer an entrance examination! Anybody who could not pass the entrance exam that they devised would be quite appropriately judged not sufficiently knowledgeable to comprehend the phenomena under investigation, and could be denied access and cooperation. Let the experts make the entrance examination as demanding as they like, and give them total authority on grading it, but require some of their own experts to take the exam as well, and require that the examination be blind-graded, so the graders couldn't know the identity of the candidates. That would give the religious experts a way of confirming their mutual esteem while weeding out the clueless from their own ranks and certifying any qualified investigators.[2]

2   Some avenues to explore: how can we home in on religious conviction?

Thou shalt not answer questionnaires
Or quizzes upon World-Affairs,
  Nor with compliance
Take any test. Thou shalt not sit
With statisticians nor commit
  A social science.
  —W. H. Auden, "A Reactionary Tract for the Times"

What research is needed? Consider some of the unanswered empirical questions already raised by me so far in this book:

Chapter 4: What were our ancestors like before there was anything like religion? Were they like bands of chimpanzees? What, if anything, did they talk about, aside from food and predators and the mating game? Do the burial practices of Neanderthals show that they must have had fully articulate language?

Chapter 5: Could an ape (without language) *concoct* the counter-intuitive combination of a walking tree or an invisible banana? Why don't other species have art? Why do we human beings so consistently focus our fantasies on our ancestors? Does impromptu hypnosis work as effectively when the hypnotist is not the parent? How well have nonliterate cultures preserved their rituals and creeds over the generations? How did healing rituals arise? Does there have to be some*one* to prime the pump? (What is the role of charismatic innovators in the origin of religious groups?)

Chapter 6: For how long could folk religion be carried along by our ancestors before reflection began to transform it? How and why did folk religions metamorphose into organized religions?

Chapter 7: Why do people join groups? Is the robustness of a religion like the robustness of an ant colony or a corporation? Is religion the product of blind evolutionary instinct or rational choice? Or is there some other possibility? Are Stark and Finke right about the principal reason for the precipitous decline after Vatican II in Catholics seeking a vocation in the church?

Chapter 8: Of all the people who believe in belief in God, what percentage (roughly) also actually believe in God? At first it looks as if we could simply give people a questionnaire with a multiple-choice question on it:

I believe in God: _____ *Yes* _____ *No* _____ *I don't know*

Or should the question be:

God exists: _____ *Yes* _____ *No* _____ *I don't know*

Would it make any difference how we framed the questions?

You will notice that hardly any of these questions deal even indirectly with either brains or genes. Why not? Because having religious convictions is not very much like having either epileptic seizures or blue eyes. We can already be quite sure there isn't going to be a "God gene," or even a "spirituality" gene, and there isn't going to be a Catholicism center in the brain of Catholics, or even a "religious experience" center. Yes, certainly, whenever you think of *Jesus* some parts of your brain are going to be more active than others, but whenever you think of *anything* this is going to be true. Before we start coloring in your particular brain-maps for thinking about *jesting* and *Jet Skis* and *jewels* (and *Jews*), we should note the evidence that suggests that such hot spots are both mobile and multiple, heavily dependent on context—and of course not arrayed in alphabetical order across the cortex! In fact, the likelihood that the places that light up *today* when you think about Jesus are the same places that will light up *next week* when you think about Jesus is not very high. It is still possible that we will find dedicated neural mechanisms for some aspects of religious experience and conviction, but the early forays into such research have not been persuasive.[3]

Until we develop better *general* theories of cognitive architecture for the representation of content in the brain, using neuro-imaging to study religious beliefs is almost as hapless as using a voltmeter to study a chess-playing computer. In due course, we should be able to relate everything we discover by other means to what is going on among the billions of neurons in our brains, but the more fruitful paths emphasize the methods of psychology and the other social sciences.[4]

As for genes, compare the story I have told in the earlier chapters with this simplified version, from *Time* magazine's recent cover article "Is God in our Genes?":

> Humans who developed a spiritual sense thrived and bequeathed that trait to their offspring. Those who didn't risked dying out in chaos and killing. The evolutionary equation is a simple but powerful one. [Kluger, 2004, p. 65]

The idea that lurks in this bold passage is that religion is "good for you" because it was endorsed by evolution. This is just the sort of simpleminded Darwinism that rightly gives the subtle scholars and theorists of religion the heebie-jeebies. Actually, as we have seen, it isn't that simple, and there are more powerful evolutionary "equations." The hypothesis that there is a (genetically) heritable "spiritual sense" that boosts human genetic fitness is one of the less likely and less interesting of the evolutionary possibilities. In place of a single spiritual sense we have considered a convergence of several different overactive dispositions, sensitivities, and other co-opted adaptations that have nothing to do with God or religion. We did consider one of the relatively straightforward genetic possibilities, a gene for heightened hypnotizability. This might have provided major health benefits in earlier times, and would be one way of taking Hamer's "God gene" hypothesis seriously. Or we could put it together with William James's old speculation that there are two kinds of people, those who require "acute" religion and those whose needs are "chronic" and milder. We can try to discover if there really are substantial organic differences between those who are highly religious and those whose enthusiasm for religion is moderate to nonexistent.

Suppose we struck paydirt and found just such a pattern. What would be the implications—if any—for policy? We could consider the parallel with the genetic differences that help to account for some Asians' and some Native Americans' difficulty with alcohol. As with variation in lactose tolerance, there is genetically transmitted variation in the ability to metabolize alcohol, due to variation in the presence of enzymes, mainly alcohol dehydrogenase and aldehyde dehydrogenase.[5] Needless to say, since, through no fault of their own, alcohol is poisonous to people with these genes—or it turns them into alcoholics—they are well advised to forgo alcohol. A different parallel is with the genetically transmitted distaste for broccoli and cauliflower and cilantro that many people discover in themselves; they have no difficulty metabolizing these foods, but find them unpalatable, because of identifiable differences in the many genes that code for olfactory receptors. They don't have to be advised to avoid these foods. Might there be either "spiritual-experience intolerance" or "spiritual-experience distaste"? There might be. There might be psychological features with genetic bases that are made manifest in different reactions by people to religious stimuli (however we find it useful to classify these). William James offers informal observations that give us some reason to suspect this. Some people seem impervious to religious ritual and all other manifestations of religion, whereas others—like me—are deeply moved by the ceremonies, the music, and the art—but utterly unpersuaded by the doctrines. It may be that still others hunger for these stimuli, and feel a deep need to integrate them into their lives, but would be well advised to steer clear of them, since they can't "metabolize" them the way other people can. (They become manic and out of control, or depressed, or hysterical, or confused, or addicted.)

These are hypotheses that are definitely worth formulating in detail and testing *if* we can identify patterns of individual variation, whether or not they are genetic (they might be culturally transmitted, after all). To take a fanciful example, it could turn out that people whose native language was Finnish (whatever their genetic heritage) were well advised to moderate their intake of religion!

A "spiritual sense" (whatever that is) might prove to be a genetic adaptation in the simplest sense, but more specific hypotheses about patterns in human tendencies to respond to religion are apt to be more plausible, more readily tested, and more likely to prove useful in disentangling some of the vexing policy questions that we have to face. For instance, it would be particularly useful to know more about how secular beliefs differ from religious beliefs (and as we saw in chapter 8, "belief" is a misnomer here; we might better call them religious *convictions* to mark the difference). How do religious convictions differ from secular beliefs in the manner of their acquisition, persistence, and extinction, and in the roles they play in people's motivation and behavior? There has been a substantial research industry devoted to conducting surveys on all aspects of religious attitude.[6] We regularly see the highlights of the latest results in the media, but the theoretical underpinnings and enabling assumptions of the survey methodologies are in need of careful analysis. Alan Wolfe (2003, p. 152), for one, thinks that the surveys are unreliable: "The results are inconsistent and puzzling, depending, as is often the case with such research, on the wording of the questions in surveys or the samples chosen for analysis." But is Wolfe right? This should not just be a matter of personal opinion. We need to find out.

Consider one of the more striking recent reports. According to ARIS (American Religious Identification Survey) in 2001, the three categories with the *largest gain* in membership since the previous survey of 1990 were evangelical/born-again (42 percent), nondenominational (37 percent), and no religion (23 percent). These data support the view that evangelicalism is growing in the U.S.A., but they also support the view that secularism is on the rise. We are apparently becoming polarized, as many informal observers have recently maintained. Why? Is it because, as supply-siders such as Stark and Finke think, only the most costly religions can compete with no religion at all in the marketplace for our time and resources? Or is it that the more we learn about nature, the more science strikes many people as leaving something out, something that only an antiscience perspective can seem to supply? Or is there some other explanation?

Before we jump in to explain the data, we should ask how sure we are of the assumptions used in gathering them. Just how reliable are the data, and how were they gathered? (Telephone inquiry, in the case of ARIS, not written questionnaire.) What checks were used to avoid biasing context? What other questions were people asked? How long did it take to conduct the interview? And then there are offbeat questions that might have answers that mattered: What had happened in the news on the day the poll was conducted? Did the interviewer have an accent? And so on.[7] Large-scale surveys are expensive to conduct, and nobody spends thousands of dollars gathering data using a casually designed "instrument" (questionnaire). Much research has been devoted to identifying the sources of bias and artifact in survey research. When should you use a simple yes/no question (and don't forget to include the important "I don't know" option), and when should you use a five-point Likert scale (such as the familiar *strongly agree, tend to agree, uncertain, tend to disagree, strongly disagree*)? When ARIS did its survey in 1990, the first question was: "What is your religion?" In 2001, the question was amended: "What is your religion, if any?" How much of the increase in *Non-denominational* and *No religion* was due to the change in wording? Why was the "if any" phrase added?

In the course of writing *How We Believe: Science, Skepticism and the Search for God* (2nd ed., 2003), Michael Shermer, the director of the Skeptic Society, conducted an ambitious survey of religious convictions. The results are fascinating, in part because they differ so strikingly from the results found in other, similar surveys. Most recent surveys find approximately 90 percent of Americans believe in God—and not just an "essence" God, but a God who answers prayers. In Shermer's survey, only 64 percent said they believed in God—and 25 percent said they disbelieved in God (p. 79). That's a huge discrepancy, and it is not due to any *simple* sampling error (such as sending the questionnaires to known skeptics!).[8] Shermer speculates that education is the key. His survey asked people to respond in their own words to "an open-ended essay question" explaining why they believed in God:

> As it turns out, the people who completed our survey were significantly more educated than the average American, and higher education is associated with lower religiosity. According to the U.S. Census Bureau for 1998, one-quarter of Americans over twenty-five years old have completed their bachelor's degree, whereas in our sample the corresponding rate was almost two-thirds. (It is hard to say why this was the case, but one possibility

is that educated people are more likely to complete a moderately complicated survey.) [p. 79]

But (as my student David Polk pointed out) once self-selection is acknowledged as a serious factor, we should ask the further question: who would take time to fill out such a questionnaire? Probably only those with the strongest beliefs. People who just don't think religion is important are unlikely to fill out a questionnaire that involves composing answers to questions. Only one out of ten of the people who received the mailed-out survey returned it, a relatively low rate of return, so we can't draw any interesting conclusions from his 64 percent figure, as he acknowledges (Shermer and Sulloway, in press).[9]

3 What shall we tell the children?

It was the schoolboy who said, "Faith is believing what you know ain't so."
   —Mark Twain

A research topic of particular urgency, but also particular ethical and political sensitivity, is the effect of religious upbringing and education on young children. There is an ocean of research, some good, some bad, on early-childhood development, on language learning and nutrition and parental behavior and the effect of peers and just about every other imaginable variable that can be measured in the first dozen years of a person's life, but almost all of this—so far as I can determine—carefully sidesteps religion, which is still largely *terra incognita*. Sometimes there are very good— indeed, unimpeachable—ethical reasons for this. All the carefully erected and protected barriers to injurious medical research with human subjects apply with equal force to any research we might imagine conducting on variations in religious upbringing. We aren't going to do placebo studies in which group A memorizes one catechism while group B memorizes a different catechism and group C memorizes nonsense syllables. We aren't going to do cross-fostering studies in which babies of Islamic parents are switched with babies of Catholic parents. These are clearly off limits, and should remain so. But what *are* the limits? The question is important, because, as we try to design indirect and noninvasive ways of getting at the evidence we seek, we will confront the sort of trade-offs that regularly confront researchers looking for medical cures. Perfectly risk-free research on these topics is probably impossible. What counts as informed consent, and how much risk may even those who consent be permitted to tolerate? And whose consent? The parents' or the children's?

All these policy questions lie unexamined in the shadows cast by the first spell, the one that says that religion is out of bounds, period. We should not pretend that this is benign neglect on our part, since we know full well that under the protective umbrellas of personal privacy and religious freedom there are widespread practices in which parents subject their own children to treatments that would send any researcher, clinical or otherwise, to jail. What are the rights of parents in such circumstances, and "where do we draw the line"? This is a political question that can be settled not by *discovering* "the answer" but by working out *an* answer that is acceptable to as many informed people as possible.

It will not please everybody, any more than our current laws and practices regarding the consumption of alcoholic beverages please everybody. Prohibition was tried, and by general consensus—far from unanimous—it was determined to be a failure. The current understanding is quite stable; we are unlikely to go back to Prohibition anytime soon. But there are still laws forbidding the sale of alcoholic beverages to minors (with age varying by country). And there are plenty of gray areas: what should we do if we find parents giving alcohol to their children? At the ball game, the parents may get in trouble, but what about in the privacy of their own homes? And there is a difference between a glass of champagne at big sister's wedding, and a six-pack of beer every evening while trying to do homework. When do the authorities have not just the right but the obligation to step in and prevent abuse? Tough questions, and they don't get easier when the topic is religion, not alcohol. In the case of alcohol, our political wisdom is importantly informed by what we have learned about the short-term and long-term effects of imbibing it, but in the case of religion we're still flying blind.

Some people will scoff at the very idea that a religious upbringing *could* be harmful to a child—until they reflect on some of the more severe religious regimens to be found around the world, and recognize that in the United States we already prohibit religious practices that are widespread in other parts of the world. Richard Dawkins goes further. He has proposed that no child should ever be identified as a Catholic child or a Muslim child (or an atheist child), since this identification in itself prejudges decisions that have yet to be properly considered.

> We'd be aghast to be told of a Leninist child or a neo-conservative child or a Hayekian monetarist child. So isn't it a kind of child abuse to speak of a Catholic child or a Protestant child? Especially in Northern Ireland and Glasgow where such labels, handed down over generations, have divided neighbourhoods for centuries and can even amount to a death warrant? [2003b]

Or imagine if we identified children from birth as young *smokers* or *drinking* children because their parents smoked or drank. In this regard (and no other) Dawkins reminds me of my grandfather, a physician who was way ahead of his time back in the 1950s, writing impassioned letters to the editors of the Boston newspapers, railing against the secondhand smoke that was endangering the health of children whose parents smoked at home—and we all laughed at him, and went on smoking. How much harm could that little bit of smoke do anyone? We've found out.

Everybody quotes (or misquotes) the Jesuits, "Give me a child until he is seven, and I will show you the man," but nobody—not the Jesuits or anybody else—really knows how resilient children are. There is plenty of anecdotal evidence of young people turning their backs on their religious traditions after years of immersion and walking away with a shrug and a smile and no visible ill effects. On the other hand, some children are raised in such an ideological prison that they willingly become their own jailers, as Nicholas Humphrey (1999) has put it, forbidding themselves any contact with the liberating ideas that might well change their minds. In his deeply thoughtful essay, "What Shall We Tell the Children?," Humphrey pioneers the consideration of the ethical issues involved in deciding how to decide "when and whether the teaching of a belief system to children is morally defensible" (p. 68). He proposes a general test based on the principle of informed consent, but applied—as it must be—hypothetically: what *would* these children choose if they were, later in life, somehow given the information they would need in order to make an informed choice? Against the objection that we cannot answer such hypothetical questions, he argues that there is in fact plenty of empirical evidence, and general principles, from which clear conclusions can be conscientiously derived. We take ourselves to be sometimes permitted, and even obligated, to make such conscientious decisions on behalf of people who cannot, for one reason or another, make an informed decision for themselves, and this set of problems can be addressed using the understanding that we have already hammered out in the workshop of political consensus on these other topics.

The resolution of these dilemmas is not (yet) obvious, to say the least. Compare it with the closely related issue of what we, on the outside, should do about the Sentinelese and the Jarawas and the other peoples who still live a stone-age existence in remarkable isolation on the Andaman and Nicobar Islands, far out in the Indian Ocean. These people have managed to keep even the most intrepid explorers and traders at bay for centuries by their ferocious bow-and-arrow defense of their island territories, so little is known about them, and for some time now the government of India, of which the islands form a distant part, has prohibited all contact with them. Now that they have been drawn to the world's attention in the wake of the great tsunami of December 2004, it is hard to imagine that this isolation can be maintained, but even if it could be, should it be? Who has the right to decide the matter? Certainly not the anthropologists, although they have worked hard to protect these people from contact—even with themselves—for decades. Who are they to "protect" these human beings? The anthropologists do not own them as if they were laboratory specimens carefully gathered and shielded from contamination, and the idea that these islands should be treated as a human zoo or preserve is offensive—even when we contemplate the even more offensive alternative of opening the doors to missionaries of all faiths, who would no doubt eagerly rush in to save their souls.

It is tempting, but illusory, to think that they have solved the ethical problem for us, by *their* adult decision to drive away all outsiders without asking if they are protectors, exploiters, investigators, or soul-savers. They clearly want to be left alone, so we should leave them alone! There are two problems with this convenient proposal: Their decision is so manifestly ill informed that if we let it trump all other considerations are we not as culpable as somebody who lets a person drink a poisoned cocktail "of his own free will" without deigning to warn him? And in any case, although the adults may have reached the age of consent, are their children not being victimized by the ignorance of their parents? We would never permit a neighbor's child to be kept so deluded, so shouldn't we cross the ocean and step in to rescue these children, however painful the shock?

Do you feel a slight adrenaline surge at this moment? I find that this issue of parental rights versus children's rights has no clear rivals for triggering emotional responses in place of reasoned responses, and I suspect that this is one place where a genetic factor is playing a quite direct role. In mammals and birds who must care for their offspring the instinct to protect one's young from all outside interference is universal and extremely potent; we will risk our lives unhesitatingly—unthinkingly—to fend off threats, real or imagined. It's like a reflex. And in this case, we can "feel in our bones" that parents *do* have the right to raise their children the way they see fit. Never make the mistake of wandering in between a mother bear and her cub, and *nothing* should come between parents and their children. That's the core of "family values." At the same time, we do have to admit that parents don't literally *own* their children (the way slaveowners once owned slaves), but are, rather, their stewards or guardians and ought to be held accountable by outsiders for their guardianship, which does imply that outsiders have a right to interfere—which sets off that adrenaline alarm again. When we find that what we feel in our bones is hard to defend in the court of reason, we get defensive and testy, and start looking around for something to hide behind. How about a sacred and (hence) unquestionable bond? Ah, that's the ticket!

There is an obvious (but seldom discussed) tension between the supposedly sacred principles invoked at this point. On the one hand, many declare, there is the sacred and inviolable right to life: every unborn child has a right to life, and no prospective parent has the right to terminate a pregnancy (except maybe if the mother's life is itself in jeopardy). On the other hand, many of the same people declare that, once born, the child loses its right not to be indoctrinated or brainwashed or otherwise psychologically abused by those parents, who have the right to raise the child with any upbringing they choose, short of physical torture. Let us spread the value of freedom throughout the world—but not to children, apparently. No child has a right to freedom from indoctrination. Shouldn't we change that? What, and let *outsiders* have a say in how I raise *my kids*? (Now do you feel the adrenaline rush?)

While we wrestle with the questions about the Andaman Islanders, we can see that we are laying the political foundations for similar questions about religious upbringing in general. We shouldn't assume, while worrying over the likely effects, that the seductions of Western culture will automatically swamp all the fragile treasures of other cultures. It is worth noting that many Muslim women, raised under conditions that many non-Muslim women would consider intolerable, when given informed opportunities to abandon their veils and many of their other traditions, choose instead to maintain them.

Maybe people everywhere can be trusted, and hence allowed to make their own informed choices. Informed choice! What an amazing and revolutionary idea! Maybe people should be trusted to make choices, not necessarily the choices *we* would recommend to them, but the choices that have the best chance of satisfying *their* considered goals. But what do we teach them until they are informed enough and mature enough to decide for themselves? We teach them about *all* the world's religions, in a matter-of-fact, historically and biologically informed way, the same way we teach them about geography and history and arithmetic. Let's get *more* education about religion into our schools, not less. We should teach our children creeds and customs, prohibitions and rituals, texts and music, and when we cover the history of religion, we should include both the positive—the role of the churches in the civil-rights movement of the 1960s, the flourishing of science and the arts in early Islam, and the role of the Black Muslims in bringing hope, honor, and self-respect to the otherwise shattered lives of many inmates in our prisons, for instance—and the negative—the Inquisition, anti-Semitism over the ages, the role of the Catholic Church in spreading AIDS in Africa through its opposition to condoms. No religion should be favored, and none ignored. And as we discover more and more about the biological and psychological bases of religious practices and attitudes, these discoveries should be added to the curriculum, the same way we update our education about science, health, and current events. This should all be part of the mandated curriculum for both public schools and home-schooling.

Here's a proposal, then: as long as parents don't teach their children anything that is likely to close their minds

1. through fear or hatred or
2. by disabling them from inquiry (by denying them an education, for instance, or keeping them entirely isolated from the world)

then they may teach their children whatever religious doctrines they like. It's just an idea, and perhaps there are better ones to consider, but it should appeal to freedom lovers everywhere: the idea of insisting that the devout of all faiths should face the challenge of making sure their creed is worthy enough, attractive and plausible and meaningful enough, to withstand the temptations of its competitors. If you have to hoodwink—or blindfold—your children to ensure that they confirm their faith when they are adults, your faith *ought* to go extinct.

4 Toxic memes

Any creative encounter with evil requires that we not distance ourselves from it by simply demonizing those who commit evil acts. In order to write about evil, a writer has to try to comprehend it, from the inside out; to understand the perpetrators and not necessarily sympathize with them. But Americans seem to have a very difficult time recognizing that there is a distinction between understanding and sympathizing. Somehow we believe that an attempt to inform ourselves about what leads to evil is an attempt to explain it away. I believe that just the opposite is true, and that when it comes to coping with evil, ignorance is our worst enemy. —Kathleen Norris, "Native Evil"[10]

Writing this book has helped me to understand that religion is a kind of technology. It is terribly seductive in its ability to soothe and explain, but it is also dangerous.

—Jessica Stern, *Terror in the Name of God: Why Religious Militants Kill*

Have you heard about the Yahuuz, a people who think that what we call child pornography is just good clean fun? They smoke marijuana daily, make a public ceremony of defecation (with hilarious competition to see who gets to do the ritual wiping), and, whenever an elder reaches the age of eighty, have a special feast day on which the person ceremonially kills himself or herself—and is then eaten by all. Disgusted? Then you know how many Muslims feel about our contemporary culture, with its alcohol, provocative clothing, and casual attitudes toward familial authority. Part of my effort in this book is to get you to *think* and not just *feel.* In this instance, you need to see that your disgust, however strong, is only a *datum,* a fact about you and a very important fact about you, but not an inerrant sign of moral truth—it's just like the Muslim's disgust at some of our cultural practices. We should respect the Muslims, empathize with them, take their disgust seriously—but then propose that they join us in a discussion about the perspectives on which we differ. The price you should be willing to pay for this is your own willingness to consider the (imaginary!) Yahuuz' way of life calmly, and ask if it is so clearly indefensible. If they enter into these traditions wholeheartedly, with no apparent coercion, perhaps we should say, "Live and let live."

And perhaps not. The burden should be on us to demonstrate *to the Yahuuz* that their way of life includes traditions they should be ashamed of, and should banish. Perhaps, if we engaged in this exercise conscientiously, we would discover that *some* of our disgust with their ways was parochial and unjustifiable. They would teach us something. And we would teach them something. And perhaps the gulf of difference between us would never be crossed, but we shouldn't assume this worst-case prospect.

In the meantime, the way to prepare for this utopian global conversation is to study, as compassionately and dispassionately as we can, both their ways and our own ways. Consider the brave self-observation of Raja Shehadeh, writing about the grip of modern

Palestine: "Most of your energy is spent extending feelers to detect public perception of your actions, because your survival is contingent on remaining on good terms with your society."[11] When we can share similar observations about the problems in our own society, we will be on a good path to mutual understanding. Palestinian society, if Shehadeh is right, is beset with a virulent case of the "punish those who won't punish" meme, for which there are models (beginning with Boyd and Richerson, 1992) that predict other properties we should look for. It may be that this particular feature would foil well-intentioned projects that would work in societies that lack it. In particular, we mustn't assume that policies that are benign in our own culture will not be malignant in others. As Jessica Stern puts it:

> I have come to see terrorism as a kind of virus, which spreads as a result of risk factors at various levels: global, interstate, national, and personal. But identifying these factors precisely is difficult. The same variables (political, religious, social, or all of the above) that seem to have caused one person to become a terrorist might cause another to become a saint. [2003, p. 283]

As communications technology makes it harder and harder for leaders to shield their people from outside information, and as the economic realities of the twenty-first century make it clearer and clearer that education is the most important investment any parent can make in a child, the floodgates will open all over the world, with tumultuous effects. All the flotsam and jetsam of popular culture, all the trash and scum that accumulates in the corners of a free society, will inundate these relatively pristine regions along with the treasures of modern education, equal rights for women, better health care, workers' rights, democratic ideals, and openness to the cultures of others. As the experience in the former Soviet Union shows only too clearly, the worst features of capitalism and high tech are among the most robust replicators in this population explosion of memes, and there will be plenty of grounds for xenophobia, Luddism, and the tempting "hygiene" of backward-looking fundamentalism. At the same time, we shouldn't rush to be apologetic about American pop culture. It has its excesses, but in many instances it is not the excesses that offend so much as the egalitarianism and tolerance. The hatred of this potent American export is often driven by racism—because of the strong Afro-American presence in American pop culture—and sexism—because of the status of women we celebrate and our (relatively) benign treatment of homosexuality. (See, e.g., Stern, 2003, p. 99.)

As Jared Diamond shows in *Guns, Germs, and Steel*, it was European germs that brought Western Hemisphere populations to the brink of extinction in the sixteenth century, since those people had had no history in which to develop tolerance for them. In this century it will be our memes, both tonic and toxic, that will wreak havoc on the unprepared world. Our capacity to tolerate the toxic excesses of freedom cannot be assumed in others, or simply exported as one more commodity. The practically unlimited educability of any human being gives us hope of success, but designing and implementing the cultural inoculations necessary to fend off disaster, while respecting the rights of those in need of inoculation, will be an urgent task of great complexity, requiring not just better social science but also sensitivity, imagination, and courage. The field of public health expanded to include cultural health will be the greatest challenge of the next century.[12]

Jessica Stern, an intrepid pioneer in this endeavor, notes that individual observations such as hers are just the beginning:

> A rigorous, statistically unbiased study of the root causes of terrorism at the level of individuals would require identifying controls, youth exposed to the same environment, who felt the same humiliation, human rights abuse, and relative deprivation, but who chose nonviolent means to express their grievances or chose not to express them at all. A team of researchers, including psychiatrists, medical doctors, and a variety of social scientists,

would develop a questionnaire and a list of medical tests to be administered to a random sample of operatives and their families. [2003, p. xxx]

In chapter 10, I argued that researchers don't have to be believers to be understanders, and we had better hope I was right, since we want our researchers to understand Islamic terrorism from the inside without having to become Muslims—and certainly not terrorists—in the process.[13] But we also won't understand Islamic terrorism unless we can see how it is like and unlike other brands of terrorism, including Hindu and Christian terrorism, ecoterrorism, and antiglobalist terrorism, to round up the usual suspects. And we won't understand Islamic and Hindu and Christian terrorism without understanding the dynamics of the transitions that lead from benign sect to cult to the sort of disastrous phenomenon we witnessed in Jonestown, Guyana, in Waco, Texas, and in the Aum Shinrikyo cult in Japan.

One of the most tempting hypotheses is that these particularly toxic mutations tend to arise when charismatic leaders miscalculate in their attempts to be memetic engineers, unleashing memetic adaptations that they find, like the Sorcerer's Apprentice, they can no longer control. They then become somewhat desperate, and keep reinventing the same bad wheels to carry them over their excesses. The anthropologist Harvey Whitehouse (1995) offers an account of the debacle that overtook the leaders of Pomio Kivung, the new religion in Papua New Guinea mentioned at the outset of chapter 4, that suggests (to me) that something like runaway sexual selection took over. The leaders responded to the pressure from the people—*Prove that you mean it!*—with ever-inflated versions of the claims and promises that had brought them to power, leading inevitably to a crash. It's reminiscent of the accelerated burst of creativity you see in pathological liars when they can sense that their exposure is imminent. Once you've talked the people into killing all the pigs in anticipation of the great Period of the Companies, you have nowhere to go but down. Or out: *It's them—the infidels—who are the cause of all our misery!*

There are so many complexities, so many variables—can we ever hope to make predictions that we can act on? Yes, in fact, we can. Here is just one: in every place where terrorism has blossomed, those it has attracted are almost all young men who have learned enough about the world to see that their futures look otherwise bleak and uninspiring (like the futures of those who were preyed upon by Marjoe Gortner).

> What seems to be most appealing about militant religious groups—whatever combination of reasons an individual may cite for joining—is the way life is simplified. Good and evil are brought out in stark relief. Life is transformed through action. Martyrdom—the supreme act of heroism and worship—provides the ultimate escape from life's dilemmas, especially for individuals who feel deeply alienated and confused, humiliated or desperate. [Stern, 2003, pp. 5–6]

Where are we going to find an overabundance of such young men in the very near future? In many countries, but especially in China, where the draconian one-child-per-family measures that have slowed the population explosion so dramatically (and turned China into a blooming economic force of unsettling magnitude) have had the side effect of creating a massive imbalance between male and female children. Everybody wanted to have a son (a superannuated meme that had evolved to thrive in an earlier economic environment), so daughters have been aborted (or killed at birth) in huge numbers, so now there are not going to be anywhere near enough wives to go around. What are all those young men going to do with themselves? We have a few years to figure out benign channels into which their hormone-soaked energies can be directed.

5 Patience and politics

Congress shall make no law respecting an establishment of religion, or prohibiting the free exercise thereof; or abridging the freedom of speech, or of the press; or the right of the people peaceably to assemble, and to petition the Government for a redress of grievances.

—First Amendment to the Constitution of the United States of America

Traditions deserve to be respected only insofar as they are respectable—that is, exactly insofar as they themselves respect the fundamental rights of men and women.

—Amin Maalouf, *In the Name of Identity: Violence and the Need to Belong*

Praise Allah for the Internet. With the Web making self-censorship irrelevant—someone else is bound to say what you won't—it became a place where intellectual risk-takers finally exhaled.

—Irshad Manji, *The Trouble with Islam*[14]

Eternal vigilance is the price of liberty.

—Either Thomas Jefferson (date unknown) or Wendell Phillips (1852)

There's such a thing as growing up too fast. We all have to make the awkward transition from childhood through adolescence to adulthood, and sometimes the major changes come way too early, with lamentable results. But we cannot maintain our childhood innocence forever. It is time for us all to grow up. We must help one another, and be patient. It is overreaction that again and again has lost us ground. Give growing up some time, encourage it, and it will come about. We must have faith in our open society, in knowledge, in continuing pressure to make the world a better place for people to live, and we must recognize that people need to see their lives as having meaning. The thirst for a quest, a goal, a meaning, is unquenchable, and if we don't provide benign or at least nonmalignant avenues, we will always face toxic religions.

Instead of trying to destroy the madrassahs that close the minds of thousands of young Muslim boys, we should create alternative schools—for Muslim boys *and girls*[15]—that will better serve their real and pressing needs, and let these schools compete openly with the madrassahs for clientele. And how can we hope to compete with the promise of salvation and the glories of martyrdom? We could lie, and make promises of our own that could never be fulfilled in this life or anywhere else, or we could try something more honest: we could suggest to them that the claims of *any* religion should, of course, be taken with a grain of salt. We could start to change the climate of opinion that holds religion to be above discussion, above criticism, above challenge. False advertising is false advertising, and if we start holding religious organizations accountable for their claims—not by taking them to court but just by pointing out, often and in a matter-of-fact tone of voice, that of course these claims are ludicrous—perhaps we can slowly get the culture of credulity to evaporate. We have mastered the technology for creating doubt through the mass media ("Are you sure your breath is sweet?" "Are you getting enough iron?" "What has your insurance company done for you lately?"), and now we can think about applying it, gently but firmly, to topics that have heretofore been off limits. Let the honest religions thrive because their members are getting what they want, as informed choosers.

But we can also start campaigns to adjust specific aspects of the landscape in which this competition takes place. A bottomless pit in that landscape that strikes me as particularly deserving of paving over is the tradition of "holy soil." Here is Yoel Lerner, an Israeli and a former terrorist, quoted by Stern:

"There are six hundred thirteen commandments in the Torah. The temple service accounts for about two hundred and forty of these. For nearly two millennia, since the destruction of the Temple, the Jewish people, contrary to their wishes, have been unable to maintain the temple service. They've been unable to comply with those commandments. The Temple constituted a kind of telephone line to God," Lerner summarizes. "That link has been destroyed. We want to rebuild it." [2003, p. 88]

Nonsense, say I. Here is an imaginary case: Suppose it turned out that Liberty Island (formerly Bedloe's Island, on which the Statue of Liberty stands) was once a burial ground of the Mohawks—say the Matinecock Tribe of nearby Long Island. And suppose the Mohawks came forward with the claim that it should be restored to pristine purity (no gambling casinos, but also no Statue of Liberty, just one big holy cemetery). Nonsense. And shame on any Mohawks who had the chutzpah(!) to rile up their fellow braves on the issue. This would be ancient history—a lot *less* ancient than the history of the Temple—and it should be allowed to recede gracefully into the past.

We don't let religions declare that their holy traditions require that left-handed people be enslaved, or that people who live in Norway should be killed. We similarly cannot let religions declare that "infidels" who have been innocently living on their "holy" turf for generations have no right to live there. There is also, of course, culpable hypocrisy in the policy of deliberately building new settlements in order to *create* just such "innocent" dwellers and foreclose the claims of the previous dwellers on that land. This is a practice that goes back centuries; the Spaniards who conquered most of the Western Hemisphere often took care to build their Christian churches on the destroyed foundations of the temples of the indigenous people. Out of sight, out of mind. Neither side of these disputes is above criticism. If we could just devalue the whole tradition of holy soil, and its occupation, we could address the residual injustices with clearer heads.

Perhaps you disagree with me about this. Fine. Let's discuss it calmly and openly, with no untrumpable appeals to the sacred, which have no place in such a discussion. If we should continue to honor claims about holy soil, it will be because, all things considered, this is the course of action that is just, and life-enabling, and a better path to peace than any other we can find. Any policy that cannot pass that test doesn't deserve respect.

Such open discussions are underwritten by the security of a free society, and if they are to continue unmolested, we must be vigilant in protecting the institutions and principles of democracy from subversion. Remember Marxism? It used to be a sour sort of fun to tease Marxists about the contradictions in some of their pet ideas. The revolution of the proletariat was inevitable, good Marxists believed, but if so, why were they so eager to enlist us in their cause? If it was going to happen anyway, it was going to happen with or without our help. But of course the inevitability that Marxists believe in is one that depends on the growth of the movement and all its political action. There were Marxists working very hard to bring about the revolution, and it was comforting to them to believe that their success was guaranteed in the long run. And some of them, the only ones that were really dangerous, believed so firmly in the rightness of their cause that they believed it was permissible to lie and deceive in order to further it. They even taught this to their children, from infancy. These are the "red-diaper babies," children of hardline members of the Communist Party of America, and some of them can still be found infecting the atmosphere of political action in left-wing circles, to the extreme frustration and annoyance of honest socialists and others on the left.

Today we have a similar phenomenon brewing on the religious right: the inevitability of the End Days, or the Rapture, the coming Armageddon that will separate the blessed from the damned in the final Day of Judgment. Cults and prophets proclaiming the imminent end of the world have been with us for several millennia, and it has been another sour sort of fun to ridicule them the morning after, when they discover that their calculations were a little off. But, just as with the Marxists, there are some among them who are working hard to "hasten the inevitable," not merely anticipating the End Days with joy in their hearts, but taking political action to bring about the conditions they think are the prerequisites for that occasion. And these people are not funny at all. They are dangerous, for the same reason that red-diaper babies are dangerous: they put their allegiance to their creed ahead of their commitment to democracy, to peace, to (earthly) justice—and to truth. If push comes to shove, some of them are prepared to lie and even to kill, to do whatever it takes to help bring what they consider celestial justice to those they consider the sinners. Are they a lunatic fringe? They are certainly dangerously out of touch with reality, but it is hard to know how many they are.[16] Are their numbers growing? Apparently. Are they attempting to gain positions of power and influence in the governments of the world? Apparently. Should we know all about this phenomenon? We certainly should.

Hundreds of Web sites purport to deal with this phenomenon, but I am not in a position to endorse any of them as accurate, so I will not list any. This in itself is worrisome, and constitutes an excellent reason to conduct an objective investigation of the whole End Times movement, and particularly the possible presence of fanatical adherents in positions of power in the government and the military. What can we do about this? I suggest that the political leaders who are in the best position to call for a full exposure of this disturbing trend are those whose credentials could hardly be impugned by those who are fearful of atheists or brights: the eleven senators and congressmen who are members of the "Family" (or the "Fellowship Foundation"), a secretive Christian organization that has been influential in Washington, D.C., for decades: Senators Charles Grassley (R., Iowa), Pete Domenici (R., N.Mex.), John Ensign (R., Nev.), James Inhofe (R., Okla.), Bill Nelson (D., Fla.), Conrad Burns (R., Mont.), and Representatives Jim DeMint (R., S.C.), Frank Wolf (R., Va.), Joseph Pitts (R., Pa.), Zach Wamp (R., Tenn.), and Bart Stupak (D., Mich.).[17] Like the nonfanatical Muslim leaders in the Islamic world on whom the world is counting to cleanse

Islam of toxic excess, these nonfanatical Christians have the influence, the knowledge, and the responsibility to help the nation protect itself from those who would betray our democracy in pursuit of their religious agendas. Since we certainly don't want to relive McCarthyism in the twenty-first century, we should approach this task with maximal public accountability and disclosure, in a bipartisan spirit, and in the full light of public attention. But of course this will require that we break the traditional taboo against inquiring so openly and searchingly about religious affiliations and convictions.

So, in the end, my central policy recommendation is that we gently, firmly educate the people of the world, so that they can make truly informed choices about their lives.[18] Ignorance is nothing shameful; *imposing* ignorance is shameful. Most people are not to blame for their own ignorance, but if they willfully pass it on, they *are* to blame. One might think this is so obvious that it hardly needs proposing, but in many quarters there is substantial resistance to it. People are afraid of being more ignorant than their children— especially, apparently, their daughters. We are going to have to persuade them that there are few pleasures more honorable and joyful than being instructed by your own children. It will be fascinating to see what institutions and projects our children will devise, building on the foundations earlier generations have built and preserved for them, to carry us all safely into the future.

APPENDIX A

The New Replicators

[For context, see p. 81. Reprinted, with permission, from *The Encyclopedia of Evolution* (Oxford: Oxford University Press, 2002).]

It has long been clear that in principle the process of natural selection is *substrate-neutral*—evolution will occur whenever and wherever three conditions are met:

1. replication
2. variation (mutation)
3. differential fitness (competition)

In Darwin's own terms, if there is "descent [1] with modification [2]" and "a severe struggle for life" [3], better-equipped descendants will prosper at the expense of the competition. We know that a single material substrate, DNA (with its surrounding systems of gene expression and development), secures the first two conditions for life on earth and the third condition is secured by the finitude of the planet as well as more directly by uncounted environmental challenges. But we also know that DNA won out over early variations that have left their traces and ongoing exemplars, such as the RNA viruses and prions. Are there on this planet any other completely different evolutionary substrates that have arisen? The best candidates are the brainchildren, planned or unplanned, of one species, *Homo sapiens*.

Darwin himself proposed *words* as an example: "The survival or preservation of certain favoured words in the struggle for existence is natural selection" (*Descent of Man*, 1871, p. 61). Billions of words are uttered (or inscribed) every day, and almost all of them are replicas—in a sense to be discussed below—of earlier words perceived by their utterers. Replication is not perfect, and there are many opportunities for variation or mutation in pronunciation, inflection, or meaning (or spelling, in the case of written words). Moreover, words are roughly segregated into lineages of replication chains; for instance, we can trace a word's descendants from Latin to French to Cajun. Words compete for airtime and print space in many media, with words going obsolete and dropping out of the word pool, while other words spring up and flourish. We discover *conTROVersy* going to fixation in some regions and *CONtroversy* going to fixation in others, while the original meaning of "begs the question" is supplanted in some quarters by a variant. The detectable historical changes in languages have been studied from one Darwinian perspective or another since Darwin's own day, and a great deal is known about patterns of replication, variation, and competition in the processes that have yielded the diverse languages of today. Indeed, some of the investigative methods of modern evolutionary biology, in bio-informatics, for instance, are themselves descended from pre-Darwinian researches conducted by paleographers and other early scholars of historical linguistics. As Darwin noted, "The formation of different languages and of distinct species, and the proofs that both have been developed through a gradual process, are curiously the same" (1871, p. 59).

Words, and the languages they populate, are not the only culturally transmitted variants that have been proposed, however. Other human acts and practices that spread by imitation have been identified as potential replicators, as have some of the habits of nonhuman animals. The physical substrates of these media are various indeed, including sounds and all manner of visible, tangible patterns in the behavior of the vector organisms. Moreover, behaviors often produce artifacts (paths, shelters, tools, weapons, . . . signs or symbols) that may serve as better exemplars for the purposes of replication than the behaviors that produce them, being relatively stable over time, and hence in some regards easier to copy, as well as being independently movable and storeable—like seeds in this regard. One human artifact, the computer, with its prolific copying ability, has recently provided a distinctly new substrate, in which both deliberate and inadvertent experiments in artificial evolution are now burgeoning, taking advantage especially of the emergence of gigantic networks of linked computers that permit the swift dispersal of propagules made of nothing but bits of information. These *computer viruses* are simply sequences of binary digits that can have an effect on their own replication. Like macromolecular viruses, they travel light, being nothing more than information packets including a phenotypic overcoat that tends to gain them access to replication machinery wherever they encounter it. And, finally, researchers in the new field of Artificial Life aspire to generate both virtual (simulated, abstract) and real (robotic) self-replicating agents that can take advantage of evolutionary algorithms to explore the adaptive landscapes they are situated in, generating improved designs by processes that meet the three defining conditions while differing from carbon-based life-forms in striking ways. While at first glance these phenomena may appear to be only *models* of evolving entities, thriving in modeled environments, the boundary between an abstract demonstration and an application in the real world is more easily crossed by these evolutionary phenomena than by others, precisely because of the substrate-neutrality of the underlying evolutionary algorithms. Artificial self-replicators can escape from their original environments on researchers' computers and take on a "life" of their own in the rich new medium of the Internet.

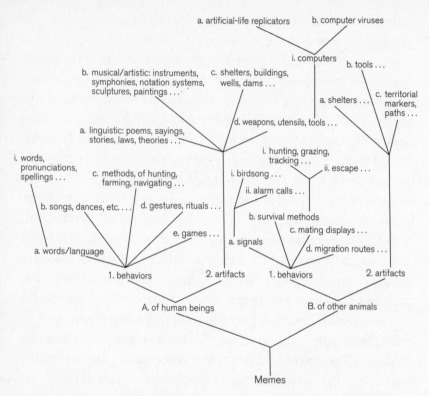

A Simple Taxonomy of the New Replicators

It can be seen that all these categories of new replicators are dependent, like viruses, on replicative machinery that is built and maintained directly or indirectly by the parent process of biological evolution. Were all DNA life-forms to go extinct, all their habits and metahabits, their artifacts and meta-artifacts, would soon die with them, lacking the wherewithal (both the machinery and the energy to run the machinery) to reproduce on their own. This might not be a permanent feature of the planet. For the time being, our computer networks and robot fabrication and repair facilities require massive supervision and maintenance by us, but it has been suggested by the roboticist Hans Moravec (1988) that silicon-based electronic (or photonic) artifacts could become entirely self-sustaining and self-replicating, weaning themselves from their dependence on their carbon-based creators. This improbable and distant eventuality is not a requirement for evolution, however, or for life itself. After all, our own self-replication and self-maintenance is entirely dependent on the billions of bacteria without which our metabolisms would fail, and if our artifactual descendants similarly have to enslave armies of our biological descendants to keep their systems up and running, this would not detract from their claim to be a new branch on the tree of life.

As with many taxonomies in evolutionary theory, there are controversies and puzzles about how to draw the branchings, and how to name them. Some of these puzzles are substantive and some are merely disagreements about which terms to use. The zoologist Richard Dawkins coined the term "meme" in a chapter of his 1976 book, *The Selfish Gene,* and the term has caught on. He opened his discussion of these "new replicators" with a discussion of birdsong, but others who have adopted the term have wanted to restrict memes to human culture. Should such evolving animal traditions as alarm calls, nest-building methods, and chimpanzee tools also be called memes? Researchers concentrating on cultural transmission in animals, such as John Tyler Bonner (1980) and Eytan Avital and Eva Jablonka (2000), have resisted the term, and others writing on human cultural evolution, such as Luigi Luca Cavalli-Sforza and Marcus Feldman (1981), and Robert Boyd and Peter Richerson (1985), have also chosen to use alternative terms. But since the word "meme" has secured a foothold in the English language, appearing in the most recent edition of the *Oxford English Dictionary* with the definition "an element of culture that may be considered to be passed on by non-genetic means," we may conveniently settle on it as the general term for any culturally based replicator—if such there are. Those who are squeamish about using a term whose identity conditions are still so embattled should remind themselves that similar controversies continue to swirl around how to define its counterpart, "gene," a term that few would recommend abandoning altogether.

Memes include not just animal traditions, then, but also computer-based replicators, for two reasons: not only do computers and their maintenance and operation depend on human culture, but the boundaries between computer viruses and more traditional human memes have already been blurred. Simple computer viruses in effect carry the instruction *copy me*, which is directed to the computer, in machine language, and is entirely invisible to the computer's user. Like the toxins unwittingly ingested by people who catch and eat freshwater fish, such a computer virus, while an element of the users' environment, is arguably not part of their *cultural* environment. However, at least as widespread and virulent as such "proper" computer viruses are bogus computer-virus warnings, directed to the computer-user, in natural language. These, which depend directly on a comprehending (but duped) human vector to get themselves replicated on the Internet, are definitely within the intended understanding of memes, and intermediate cases are the computer viruses that depend on enticing human users to open attachments (thereby triggering the invisible copying instruction) by promising some amusing or titillating contents. These, too, depend on human comprehension; one written in German will not spread readily to the computers of monoglot English speakers. (This pattern may change if users avail themselves regularly of on-line translation services.) In the arms race between virus and antivirus, ever more elaborate exploitations of human interests are to be expected, so it seems best to include all these replicators under the rubric of memes, noting that some of them make only indirect use of human vectors, and hence are only indirectly elements of human culture. We are beginning to see this porous boundary crossed in the other direction as well: it used to be true that the differential replication of such classic memes as songs, poems, and recipes depended on their winning the competition for residence in human brains, but now that a multitude of search engines on the Web have interposed themselves between authors and their (human) audiences, competing with one another for reputation as high-quality sources of cultural items, significant fitness differences between memes can accumulate independently of any human appreciation or cognizance at all. The day may soon come when a cleverly turned phrase in a book gets indexed by many search engines, and thereupon enters the language as a new cliché, without anybody human having read the original book.

Problems of classification and individuation

Some problems of classification are substantive, depending in part on historical facts that are not well established, and others are tactical problems for the theorist: what divisions of the phenomena will prove most perspicuous? Are all computer viruses properly descended from the earliest forays into Artificial Life, or should at least some of them be shown as arising independently of that intellectual movement? Not all computer hackers are A-Life hackers, but there is also the unanswered tactical question of how to characterize what is copied. If one hacker gets the *general idea* of a computer virus from somebody else and then goes on to make an entirely new kind of computer virus, is that new virus properly a descendant with modifications of the virus that inspired its creation? What if the hacker adapts elements of the original virus's design in the new type? How much sheer mindless copying must there be, or, alternatively, how much comprehending inspiration *may* there be, in an instance of replication? (More on this question below.) Is there cross-species meme-copying in the animal world? Polar bears build a den that includes a raised snow shelf that permits cold air to drain out the depressed opening of the den. Is this wise trend in arctic technology entirely innate (now) or do bear cubs have to copy their mother's example? The same snow shelf is found in an Inuit's igloo or quincy. Did the Inuit copy this tradition from the polar bear, or was it an independent invention? Does it ever happen that one species begins attending to the alarm calls of another and then develops an alarm-call tradition of its own? Does the *alarm-call* meme spread from species to species, or should we consider the intraspecific alarm calls and their variants as entirely independent lineages?

Exacerbating these problems are other problems of meme individuation. Should the (English) *word* "windsurfing" be seen as distinct from the (language-neutral) windsurfing *meme?* Are these two memes or one? Do styles, such as *punk* and *grunge,* count as memes before they have names? Why not? Joining forces with a name-meme is no doubt an excellent fitness advantage for almost any meme. (An exception could be a meme that depends on spreading insidiously; the coining of a name-meme, such as *male chauvinism,* may actually hinder the spread of male chauvinism by sensitizing something like an immune reaction in potential vectors.) It is probably true that as soon as any human meme becomes salient enough in the environment to be discerned, it will thereupon be named by one of its discerners, tightly linking the two memes thereafter: the name and the named, which typically have a shared fate, but not always. (The musical characteristics identifiable as the blues include many robust instances that are not *called* the blues by those who play and listen to them.) Undiscerned memes can also flourish. For instance, changes in the pronunciation or meaning of a word can move to fixation in a large community before any sharp-eared linguist or other cultural observer takes note. There are more than a few people—comedians as well as anthropologists and other social scientists—who earn their living detecting and commenting on evolving trends in cultural patterns that have heretofore been at best dimly appreciated.

Until these and other problems of initial theoretical orientation are resolved, skepticism about memes will continue to be widespread and heartfelt. Many commentators are deeply opposed to any proposals to recast questions in the social sciences and humanities in terms of cultural evolution, and this opposition is often expressed in terms of a challenge to prove that "memes exist":

> Genes exist [these critics grant] but what are memes? What are they made of? Genes are made of DNA. Are memes made of neuron-patterns in the brains of encultured people? What is the material substrate for memes?

There are some proponents of memes who have argued in favor of an attempt to *identify* memes with specific brain structures—a project still entirely uncharted, of course. But on current understandings of how the brain might store cultural information, it is unlikely that any *independently identifiable* common brain structures, in different brains, could ever be isolated as the material substrate for a particular meme. While some genes for making eyes do turn out to be identifiable whether they occur in the genome of a fly, a fish, or an elephant, there is no good reason to anticipate that the memes for wearing bifocals might be similarly isolatable in neuronal patterns in brains. It is vanishingly unlikely, that is, that the brain of Benjamin Franklin, who invented bifocals, and the brains of those of us who wear them, should "spell" the *idea* of bifocals in a common brain-code. Besides, this imagined path to scientific respectability is based on a mistaken analogy. In his 1966 book, *Adaptation and Natural Selection*, the evolutionary theorist George Williams offered an influential definition of a *gene* as "any hereditary information for which there is a favorable or unfavorable selection bias equal to several or many times its rate of endogenous change," and as he went on to stress in his 1992 book, *Natural Selection: Domains, Levels, and Challenges*, "A gene is not a DNA molecule; it is the transcribable information coded by the molecule" (p. 11).

Genes, genetic recipes, are all written in the physical medium of DNA, using a single canonical language, the nucleotide alphabet of Adenine, Cytosine, Guanine, and Thymine, triplets of which code for amino acids. Let every strand of smallpox DNA in the world be destroyed; if the smallpox genome is preserved (translated from *nucleotides* into the *letters* A, C, G, and T and stored on hard disks on computers, for instance), smallpox is not truly extinct; it could have descendants someday, because its genes *still exist* on those hard disks, as what Williams calls "packages of information" (1992, p. 13).

Memes, cultural recipes, similarly depend on one physical medium or another for their continued existence (they aren't magic), but they can leap around from medium to medium, being translated from language to language, from language to diagram, from diagram to rehearsed practice, and so forth. A recipe for chocolate cake, whether written in English in ink on paper, or spoken in Italian on videotape, or stored in a diagrammatic data structure on a computer's hard disk, can be preserved, transmitted, translated, and copied. Since the proof of the pudding is in the eating, the likelihood of a recipe getting *any* of its physical copies replicated depends (mainly) on how successful the cake is. How successful the cake is at doing what? At getting a host to make another cake? Usually, but even more important is getting the host to make another copy of the recipe and passing it on. That's all that matters, in the end. The cake may not enhance the fitness of those who eat it; it may even poison them, but if it first somehow provokes them to pass on the recipe, the meme will flourish.

This is perhaps the most important innovation in outlook permitted by recasting investigations in terms of memes: they have their own fitness as replicators, independently of any contribution they may or may not make to the genetic fitness of their hosts, the human vectors. Dawkins (1976) put it this way: "What we have not previously considered is that the cultural trait may have evolved in the way that it has, simply because it is *advantageous to itself*" (p. 200 of rev. ed.). The anthropologist F. T. Cloak (1975) put it this way: "The survival value of a cultural instruction is the same as its function; it is its value for the survival/replication of itself or its replica."

Those who question whether "memes exist" because they cannot see what material thing a meme could be should ask themselves if they are equally dubious about whether words exist. What is the word "cat" made of? Words are recognizable, reidentifiable products of human activity; they come in many media, and can leap from substrate to substrate in the process of being replicated. Their standing as real things is not in the slightest impugned by their abstractness. In the proposed taxonomy, words are but one species of memes, and the other species of memes are the same kind of things that words are—you just can't pronounce or spell them. Some of them you can dance, and some of them you can sing, or play, and others you can follow by making something out of the various building materials the world provides. The word "cat" isn't made out of some of the ink on this page, and a recipe for chocolate cake isn't made of flour and chocolate.

There is no single proprietary code, parallel to the four-element code of DNA, that can be used to anchor meme-identity the way gene-identity can be anchored for most practical purposes. This is an important difference, but one of degree. If the current trend of language extinctions continues at its present pace, in the not-so-distant future every person on earth will speak the same language, and it will then be difficult to resist the temptation (which should still be resisted!) to *identify* memes with their (now practically unique) verbal labels. But so long as there are multiple languages, to say nothing of the multiple media in which nonlinguistic cultural items can be replicated, we are better off to keep strictly to the abstract, code-neutral understanding of a meme as a "package of information," bearing in mind that, for high-fidelity replication to occur, there must always be some "code" or other. Codes play a crucial role in all systems of high-fidelity replication, since they provide finite, practical sets of norms against which relatively mindless editing or proofreading can be done. But even in the clearest cases of codes, there are often multiple levels of norms. Suppose Tommy writes the letters "SePERaTE" on the blackboard, and Billy "copies" it by writing "seperate." Is this really copying? The normalization to all lowercase letters shows that Billy is not slavishly copying Tommy's chalkmarks but, rather, being triggered to execute a series of canonical, normalized acts: *make an "s," make an "e,"* etc. It is thanks to these letter-norms that Billy can "copy" Tommy's word at all. But he *does* copy Tommy's spelling error, unlike Molly, who "copies" Tommy by writing "separate," responding to a higher norm, at the level of word spelling. Sally then goes a step higher, "copying" the phrase "separate butt equal"—all words in good standing in the dictionary—as "separate but equal," responding to a recognized norm at the phrase level. Can we go higher? Yes. Anybody who, when "copying" the line in the recipe "Separate three eggs and beat the yolks until they form stiff white cones," would replace "yolks" with "whites," knows enough about cooking to recognize the error and correct it. Above spelling and syntactic norms are a host of semantic norms as well.

Norms can both hinder and help replication. The anthropologist Dan Sperber (2000) has distinguished copying from what he calls "triggered production" and has noted that in cultural transmission "the information provided by the stimulus is complemented with information already in the system." This complementing tends to absorb mutations instead of passing them on. Evolution depends on the existence of mutations that can survive the proofreading processes of replication intact, but it does not specify the level at which this survival must occur. A brilliant cooking innovation might indeed get corrected away by an all-too-knowing chef in the course of passing on the recipe, but other "errors" might get through and replicate indefinitely. Meanwhile, the correction of other varieties of noise at other levels, responding to spelling norms or others, must be ongoing, in order to keep the copying process faithful enough so that multiple exemplars of each innovation can be tested against the environment. As Williams puts it, "A given package of information (codex) must proliferate faster than it changes, so as to produce a genealogy recognizable by some diagnostic effects" (1992, p. 13). Recognizable, that is, to the unfocused, independently varying environment, so it can yield probabilistic verdicts of natural selection that have some likelihood of identifying adaptations of projectible fitness.

Just how big or small can a meme be? A single musical tone is not a meme, but a memorable melody is. Is a symphony a single meme or is it a system of memes? A parallel question can be asked about genes, of course. No single nucleotide or codon is a gene. How many notes or letters or codons does it take? The answer in both cases tolerates blurred boundaries: a meme, or a gene, must be large enough to carry information worth copying. There is no fixed measure of this, but the bountiful system of case law on copyright and patent infringements indicates that verdicts on particular cases form a relatively trustworthy equilibrium that is stable enough for most purposes.

Other objections to memes seem to exhibit an inverse relationship between popularity and soundness: the more enthusiastically they are championed, the more ill informed they are. They have been patiently rebutted again and again by proponents, but those who are appalled by the prospect of an evolutionary account of anything in human culture don't seem to notice. A common mistake is for critics to imagine that memes must be more like genes than they need to be for the three conditions to be met. It has been observed, for instance, that when an individual first acquires some encountered cultural item, this is typically *not* a case of imitating a single instance of it. (If I take up the practice of wearing my baseball cap backward, or add a new word to my working vocabulary, am I copying the first instance of it I ever noticed, or the most recent instance, or am I somehow averaging over all of them?) This embarrassment of riches in the search for *the* parent of the new offspring does complicate the model of cultural replication, but it does not in itself disqualify the process as one of replication. For instance, the ultra-high-fidelity copying of computer files depends in many instances on error-correcting code-reading systems that in effect let "majority rule" determine which of several candidate exemplars should count as canonical. In such cases, no single vehicle of the information can be identified as the source, but it is an instance of replication if anything is. Darwin's trio of requirements is both substrate-neutral and implementation-neutral to a degree that is not always appreciated.

Is cultural evolution Darwinian?

Marking these unresolved problems of nomenclature and individuation, we can turn to the more fundamental and important question: Do any of these candidates for Darwinian replicator actually fulfill the three requirements in ways that permit evolutionary theory to explain phenomena not already explicable by the methods and theories of the traditional social sciences? Or does this Darwinian perspective provide only a relatively trivial unification? It would still be important to conclude that cultural evolution obeys Darwinian principles in the modest sense that nothing that happens in it *contradicts* evolutionary theory, even if cultural phenomena are best accounted for in other terms. In *The Origin of Species*, Darwin himself identified three processes of selection: "methodical" selection by the foresighted, deliberate acts of farmers and others intent on artificial selection, "unconscious" selection, in which human beings have engaged in activities that have unwittingly contributed to the differential survival and reproduction of species, mostly on their way to domestication, and "natural" selection, in which human intentions have played no role at all. To this list we can add a fourth phenomenon, genetic engineering, in which the intention and foresight of human designers plays a still more prominent role. All four of these phenomena are Darwinian in the modest sense. Genetic engineers do not produce counterexamples to the theory of evolution by natural selection, any more than plant breeders over the eons have done; they produce novel fruits of the fruits of the fruits of evolution by natural selection. The idea of memes promises similarly to unify under a single perspective such diverse cultural phenomena as deliberate, foresighted scientific and cultural inventions (memetic engineering), such authorless productions as folklore, and even such unwittingly redesigned phenomena as languages and social customs themselves. As we enter the age of deliberate, purportedly foresighted tinkering with our own genomes and the genomes of other species, we face the prospect of strong interactions between genetic and memetic evolution, including many that may take off without having been foreseen at all. It behooves us to investigate these possibilities with the same vigor and attention to detail we devote to the investigation of the evolution of carbon-based pathogens and the swift disappearance of natural barriers that have structured the biosphere until very recently.

We should also remind ourselves that, just as population genetics is no substitute for ecology, which investigates the complex interactions between phenotypes and environments that ultimately yield the fitness differences presupposed by genetics, no one should anticipate that a new science of memetics would overturn or replace all the existing models and explanations of cultural phenomena developed by the social sciences. It might, however, recast them in significant ways, and provoke new inquiries in much the way genetics has inspired a flood of investigations in ecology. The books listed under Further Reading explore these prospects in some detail, but still at a very programmatic and speculative level. At this time there are still only a few works that might be listed as pioneering empirical investigations in specialized branches of memetics: Hull (1988), Pocklington and Best (1997), Gray and Jordan (2000).

Further Reading

Aunger, Robert, [June 2002], *The Electric Meme: A New Theory of How We Think and Communicate*. New York: Free Press.

———, ed., 2000, *Darwinizing Culture: The Status of Memetics as a Science*. Oxford: Oxford University Press.

Avital, Eytan, and Eva Jablonka, 2000, *Animal Traditions: Behavioural Inheritance in Evolution*. Cambridge: Cambridge University Press.

Blackmore, Susan, 1999, *The Meme Machine*. Oxford: Oxford University Press.

Bonner, John Tyler, 1980, *The Evolution of Culture in Animals*. Princeton: Princeton University Press.

Boyd, Robert, and Peter Richerson, 1985, *Culture and the Evolutionary Process*. Chicago: University of Chicago Press.

Brodie, Richard, 1996, *Virus of the Mind: The New Science of the Meme*. Seattle: Integral Press.

Cavalli-Sforza, Luigi Luca, and Marcus Feldman, 1981, *Cultural Transmission and Evolution: A Quantitative Approach*. Princeton: Princeton University Press.

Dawkins, Richard, 1976, *The Selfish Gene*. Oxford: Oxford University Press. Rev. ed., 1989.

Dennett, Daniel, 1995, *Darwin's Dangerous Idea*. New York: Simon & Schuster.

———, 2001, "The Evolution of Culture." *Monist*, vol. 84, no. 3, pp. 305–24.

———, 2005, "From Typo to Thinko: When Evolution Graduated to Semantic Norms." In S. Levinson and P. Jaisson, eds., *Culture and Evolution*. Cambridge, Mass.: MIT Press.

Durham, William, 1992, *Coevolution: Genes, Culture and Human Diversity*. Stanford, Calif.: Stanford University Press.

Hull, David, 1988, *Science as a Process*. Chicago: University of Chicago Press.

Laland, Kevin, and Gillian Brown, 2002, *Sense and Nonsense: Evolutionary Perspectives on Human Behaviour*. Oxford: Oxford University Press.

Lynch, Aaron, 1996, *Thought Contagion: How Belief Spreads Through Society*. New York: Basic Books.

Pocklington, Richard, in press, "Memes and Cultural Viruses." In *Encyclopedia of the Social and Behavioral Sciences*.

Journal

Artificial Life

Web Journal

Journal of Memetics. Available at http://www.cpm.mmu.ac.uk/jom-emit/.

Other References

Cloak, F. T., 1975, "Is a Cultural Ethology Possible?" *Human Ecology*, vol. 3, pp. 161–82.

Gray, Russell D., and Fiona M. Jordan, 2000, "Language Trees Support the Express-Train Sequence of Austronesian Expansion." *Nature*, vol. 405 (June 29, 2000), pp. 1052–55.

Moravec, Hans, 1988, *Mind Children: The Future of Robot and Human Intelligence*. Cambridge, Mass.: Harvard University Press.

Pocklington, Richard, and Michael L. Best, 1997, "Cultural Evolution and Units of Selection in Replicating Text." *Journal of Theoretical Biology*, vol. 188, pp. 79–87.

Sperber, Dan, 2000, "An Objection to the Memetic Approach to Culture." In Robert Aunger, ed., *Darwinizing Culture*. Oxford: Oxford University Press.

Williams, George, 1966, *Adaptation and Natural Selection*. Princeton: Princeton University Press.

———, 1992, *Natural Selection: Domains, Levels, and Challenges*. Oxford: Oxford University Press.

APPENDIX B

Some More Questions About Science

[For context, see p. 93.]

1 An invitation to an investigation

In a democracy with freedom of religion, people are entitled to declare their religion to be the only true religion, and then to refuse all invitations to defend their declaration. In a democracy, we also let people be conscientious objectors, but we don't thereby give or imply any endorsement whatever to their claims. If you decline to put your beliefs on the line, then your beliefs, whatever they are, really cannot be given any *consideration* in the ongoing investigation, which has no use for one-sided declarations that will not be subjected to rigorous scrutiny and cross-examination. We'll definitely consider your (apparent) beliefs as *data*—there are people, and you are one of them, who make various avowals but cannot be enticed to place those avowals in the arena of investigation—but we will not make the mistake of counting your declaration as an opinion offered as a contribution to our inquiry.

It is sometimes held that such a refusal to submit one's creed to inquisitive probing is a commendable act of loyalty to one's religious group, an honorable declaration of faith. You may be among the many people who proudly assert that their religion is more important to them than their loyalty to family or friends or nation—or anything else. "*Don't even think about* alternatives!" could be your motto, except that its very articulation would be a self-violation. As we saw in chapter 1, that is one thing you could mean by saying your religion is *sacred* to you.

I want to put this attitude in a larger context. Even if you are convinced that your religion is a unique path to truth, you must be curious about why all the other religions are so popular around the world. And if you think it would be a good thing to bring these people—who constitute a majority of the world's people, whatever religion is yours—to see the truth as you do, then you should see the point of looking intently, as an outsider, at these religions, to "see what makes them tick." Considering how your own religion looks to an outsider would also be a valuable exercise, wouldn't it, since understanding how outsiders react to what they discover when they encounter you could hardly fail to improve your effectiveness in carrying your message to others.

As we look around the troubled world today, we see failed states, ethnic violence, and grotesque injustice arising on all sides, and a question we all have to face is *which lifeboats* we should strive to keep afloat. Some people believe that the world's democratic nations are the best hope of the world, that they provide the most secure and reliable—though hardly foolproof—platforms on the planet for improving human welfare and staving off nuclear chaos and genocide. If they capsize, we're all in deep trouble. Others believe that their transnational religions make better lifeboats, and if they had to choose between the welfare of their religion and the welfare of the nation of which they are citizens, they would unhesitatingly opt in favor of their religion. Perhaps you are among them. Since—if you are reading this book—you almost certainly live in a democratic nation with a principle of freedom of religion, you are then in a delicate position: you are enjoying the security of the democratic lifeboat while withholding your ultimate allegiance to it.

By availing yourself of the freedom granted you by a nation that honors the freedom of religion, you excuse yourself—as is your right (it's like "taking the Fifth Amendment" when called to testify in court)—from helping your fellow citizens explore a problem of national and international security of the utmost urgency. You are a free rider, putting your loyalty to your religion ahead of your duty to your fellow citizens. Fortunately for you, there are enough public-spirited citizens to make up the loss and keep the nation intact while you indulge yourself in your faith-based stand "on principle." In this regard, you are no different from the Shiite or Sunni who says in his heart: Let Iraq perish, if need be, so long as my religious tribe prospers. The main difference (and it is huge) is that the shaky state of Iraq is not (currently) anybody's idea of a seaworthy lifeboat, whereas the free society in which you live is manifestly the guarantor of such security and freedom as we now enjoy. So you have fewer grounds for withholding your allegiance to the nation and its laws than the Iraqis do.

For many of us, the price we pay—accepting the rule of secular law—is one of the best bargains on the planet. Those of us who therefore put our first allegiance—critically and tentatively and conditionally—with our secular systems of democracy recognize the wisdom of the principle of freedom of religion, and will defend it even when it interferes seriously with our particular interests. Those with other allegiances who refuse to make this commitment pose a problem—and not just a theoretical problem. In Turkey today, an Islamic party governs with a majority that would enable it to impose Islamic law on the whole nation, but it wisely refrains and even goes so far as to outlaw some practices of radical Muslims as inconsistent with religious liberty for all. The result is fragile, and fraught with problems, but it contrasts dramatically with the situation in Algeria, where violence and insecurity continue to blight the lives of everybody in the wake of a civil war that was triggered in 1990, when it became apparent that democratic elections would put in power an Islamic party intent on throwing away the ladder of democracy and creating a theocracy.

Fifty years ago, President Eisenhower nominated Charles E. Wilson, then president of General Motors, as his secretary of defense.

At the nomination hearing before the Senate Armed Services Committee, Wilson was asked to sell his shares in General Motors, but he objected. When asked if his continued stake in General Motors mightn't unduly sway his judgment, he replied, "For years, I thought what was good for the country was good for General Motors and vice versa." Some in the press, unsatisfied with this response, stressed only the second half of his response—"What's good for General Motors is good for the country"—and in response to the ensuing furor, Wilson was forced to sell his stock in order to win the nomination. This was a fine object lesson on the importance of being clear about priorities. Even if it were true, other things being equal, that what was good for General Motors was good for the country, people wanted to be clear about where Wilson's loyalties would lie in the rare event that there was a conflict. Whose benefit would Wilson further in those circumstances? That is what had people upset, and rightly so. They wanted the actual decision-making by the secretary of defense to be *directly* responsive to the *national* interest. If decisions reached under those benign circumstances benefited General Motors (and presumably most of them would, if Wilson's long-held homily is true), that would be just fine, but people were afraid that Wilson had his priorities backward. Imagine the furor that would have been provoked had Wilson said that for years, as a good Methodist, he had believed that what was good for the Methodist Church was good for the country.

Allegiance to the principles of a free and democratic society *only so long as they support the interests of your religion* is a start, but we can ask for more. If it is the best you can muster, then fair enough, but you should recognize that the rest of us are right to view you as part of the problem. Is this a fair judgment? This is controversial, and I have deliberately expressed it in stark terms to bring out the contrast. It is a view that deserves to be taken just as seriously as the more traditional, and more obviously biased, insistence that deep respect is due to all such exemptions from scrutiny. A similar impasse often arises during ecumenical attempts to resolve the different perspectives of science and religion, and it puts the scientifically minded discussants in a quandary: how should they respond? The polite tack is to acknowledge profound differences in viewpoint and paper over the cracks with some bland assurances of mutual respect. But this conceals and postpones indefinitely the consideration of an asymmetry: we wouldn't for one moment pay respectful attention to any scientist who retreated to "If you don't understand my theory, it's because you don't have *faith* in it!" or "Only official members of my lab have the ability to detect these effects," or "The contradiction you think you see in my arguments is simply a sign of the limitations of human comprehension. There are some things beyond all understanding." Any such declaration would be an intolerable abdication of responsibility as a scientific investigator, a confession of intellectual bankruptcy.

According to Avery Cardinal Dulles (2004), *apologetics* is "the rational defense of faith," and in the past it was often supposed to prove rigorously that God exists, that Jesus was divine, was born of a virgin, and so forth, but it fell into disrepute. "Apologetics fell under suspicion for promising more than it could deliver and for manipulating the evidence to support the desired conclusions. It did not always escape the vice that Paul Tillich labeled 'sacred dishonesty'" [p. 19]. Recognizing this problem, many of the devout have retreated to a less aggressive avowal of their creed, but Cardinal Dulles regrets this development, and calls for a renewal and reformation of apologetics.

> This withdrawal from controversy, though it seems to be kind and courteous, is insidious. Religion becomes marginalized to the degree that it no longer dares to raise its voice in public. . . . The reluctance of believers to defend their faith has produced all too many fuzzyminded and listless Christians, who care very little about what is to be believed. [p. 20]

Dulles urges that "apologetics needs to shift its ground":

> In a revealed religion such as Christianity, the key question is how God comes to us and opens up a world of meaning not accessible to human investigative powers. The answer, I suggest, is testimony. . . . Personal testimony calls for an epistemology quite distinct from the scientific, as commonly understood. The scientist treats the datum to be investigated as a passive object to be mastered and brought within the investigator's intellectual horizons. Interpretations proffered by others are not accepted on authority but are tested by critical probing. But when we proceed by testimony, the situation is very different. The event is an interpersonal encounter, in which the witness plays an active role, making an impact on us. Without in any way compelling us to believe, the witness calls for a free assent that involves personal respect and trust. To reject the message is to withhold confidence in the witness. To accept it is a trusting submission to the witness's authority. To the extent that we believe, we renounce our autonomy and willingly depend on the judgment of others. [p. 22]

This candid assessment articulates the free-floating rationale for the "witnessing" move, which deftly eludes the probing of the scientist by making it an affront to question the witness, a bit of impoliteness, and worse. This tactic exploits the widespread desire of people not to offend, a very effective way of disabling the critical apparatus of science. Dulles observes with equal candor that the scientific method does have a drawback, from his proselytizing perspective: "As philosophers or historians we treat the datum as something impersonal to be brought within the compass of our own world of thought. This method is useful for confirming certain doctrines and refuting certain errors, but it rarely leads to conversion" [p. 21]. In other words, use the scientific method when it helps, and use other methods when it doesn't. There is a name for this practice among scientists. It is known as *cherry-picking,* and it is a scientific sin.[1]

Nobody had to invent the witnessing practice; it just arises, and it *works* (it works better than the competition, in some circumstances), so it gets replicated. Cardinal Dulles commends the practice, and explains why it works, but is not responsible for it, and the basic rationale of witnessing is not by any means restricted to Catholicism. I vividly remember my great discomfort some years ago when a student of mine from India told me about the miracles she had witnessed her holy man perform during her vacation trip home. She made it indirectly but abundantly clear to me that if I challenged her account, even privately (outside of class), she would be deeply humiliated and dishonored. I mustn't do that to a student! What to do? When she, raising the stakes, told me about the photograph that she had in her dormitory room, with real honey flowing from the eyes of the guru, I eagerly requested to see for myself and taste the honey, but although she promptly agreed to arrange for me to examine the marvelous object myself, no further invitation to investigate was ever issued. I have often wondered whether she ever brought herself to reflect on what had happened, and if so, what conclusions she reached, but of course politeness bade me to let the matter drop there. Politeness also overwhelms the skeptical instincts of many a target of deliberate con men who know that just a touch of "hurt feelings" can deflect most if not all the questions any reasonable person would want to have answered. A tactic that works *can* be used deliberately and viciously, but it can also work—sometimes better—in the hands of an innocent enthusiast who would never dream of doing anything duplicitous.

Cardinal Dulles is interested in getting conversions; and so are scientists. They campaign with vigor and ingenuity for their pet theories. But they are constrained by the rules of science not to engage in practices that would tend to disable the critical faculties of potential hosts for the memes they want to spread. No such rules have yet evolved to govern the practice of religion.

2 What pays for science?

The religion that is afraid of science dishonors God and commits suicide.
—Ralph Waldo Emerson

What about science itself? What happens when we turn the harsh light of evolutionary theory on itself, for instance, and ask what conspiracy of conditions and payoffs led to its existence? Science in general is a very expensive human activity. What dark cravings might it be satisfying? Might it not have its share of ignoble ancestors, or be driven by embarrassing lusts? The practical benefits that have driven the scientific quest are often there, to be sure, but perhaps just as often science has proceeded by an arguably pathological excess of curiosity—knowledge for its own sake, at whatever cost. Might science turn out to be an irresistible bad habit? It might be. So might religion. Let's find out, with the scientific study of science itself, an investigation already well under way.

Why do we do science? Our brains certainly didn't evolve to do quantum physics or even long division. The standard answer, which may mask important complexities, begins with what we might call our native curiosity drive, which we share with almost all animals, and which focuses our attention on just about anything novel or complex, especially if it is in motion, and more or less compels us to examine it (cautiously). The free-floating rationale of this is obvious: as locomotors, we diminish the risks of damage and enhance our chances of finding what we need by looking where we are going. If we found that trees were also curious, we'd have to rethink this common wisdom, but the famous example of the sea squirt suggests that the principle is safe. The juvenile sea squirt wanders through the ocean looking for a good place to settle. For guidance in this task it needs a rudimentary nervous system. When it finds a suitable rock to cling to for the rest of its life (as a *sessile* filter-feeder), it no longer needs its nervous system, which it disassembles and assimilates, a vivid example in support of the hypothesis that curiosity is costly, and when it can't pay for itself by guiding locomotion, it is abandoned. As the joke has it, this is like tenure for a professor—once you have it, you are free to eat your own brain!

Curiosity must be tempered by caution, and by thrift as always, so it is not surprising that animals tend to exhibit curiosity only about the most immediately pressing ecological concerns. Herbivores check out the plants in the vicinity, whereas carnivores largely ignore them. Omnivores are busier investigators than herbivores, though both keep an eye out for predators, and so forth. Our closest relatives, the great apes, show a more catholic interest in almost all things, but even chimpanzees born in captivity are remarkably uninterested in all the human speech they hear all around them from the day they are born, ecologically relevant though it surely is to them in their evolutionarily novel circumstances. A human infant's intense interest in speech sounds may in fact be one of the most important genetic differences between us and chimpanzees. Nobody knows how differently an infant chimpanzee's brain might develop if it simply had the *urge to attend* to the torrent of overheard verbal input that its auditory system receives but regularly discards, the way ours discards the rustling of the leaves in the wind. We know of no organ of the body that pays greater homage than the brain does to the maxim "Use it or lose it," and it is conceivable that a tiny genetic change, turning up the competitive volume, in effect, for the category of speech sounds, might cascade into major anatomical changes in the developing brain.

It is extremely unlikely that such a small genetic change could be responsible for *all* the differences between chimpanzee brains and human brains, but there has been time in any case for a whole suite of genetic adjustments to make our brains more language-friendly than chimpanzee brains. Whatever the differences are, they mark a major innovation in evolutionary history, because once language evolved we became not just curious but *inquisitive:* we actually asked questions aloud, in articulated language. Questions became ubiquitous items in our perceptual worlds, and provoked reactions, which provoked more questions, and so forth, snowballing into an accumulation of lore that could be orally transmitted, and eventually written down. On one point at least, the Darwinian and Biblical accounts of how *we* got here agree: in the beginning was the Word.

But it was a long time before this accumulation of lore, of both wisdom and superstition, history and myth, practical facts and frozen lies, came to look at all like science. It was neither systematic nor self-conscious about its methods. It had not yet paid much attention *to itself.* This reflexive move, giving us the science of science, the history of history, the philosophy of philosophy, the logic of logic, and so forth, is one of the great enabling strokes of human civilization, refining the ore obtained by millennia of informal curiosity into the purified metal of investigation. Can you "pull yourself up by your own bootstraps"? Not without defying the law of gravity, but you can do something almost as good: you can use your existing, imperfect, ill-understood methods of inquiry to refine those very methods, pitting good ideas against better ideas, and using your *current* sense of what counts for a good idea as your temporary, defeasible guide to improvement. In this regard it is like the strategy, when moving to a foreign country, of picking a few informants and trusting them—until you learn otherwise. If you have really bad luck with your initial choices, you may end up almost helplessly misinformed and victimized. If your informants are *somewhat* reliable, on the other hand, you can soon discover some of the limits of their reliability and begin making targeted adjustments. It isn't *logically guaranteed* to work, but so what? It is much more likely to work than flipping a coin, and the odds get better over time.

Consider the curious problem of drawing a straight line. A *really* straight line. How do we do it? We use a straight-edge, of course.

And where did we get it? Over the centuries we refined our techniques for making straighter and straighter so-called straight-edges, pitting them against one another in supervised trials and mutual adjustments that have kept raising the threshold of accuracy. We now have large machines that are accurate to within a millionth of an inch over their entire length, and we have no difficulty in using our current vantage point to appreciate the practically unattainable but readily conceivable norm of a *really* straight edge. We discovered that norm, the eternal Platonic Form of the Straight, if you like, through our creative activity.[2]

Whether we date the beginning of science to early Egyptian geometry (literally, earth-measuring) or follow the transformation of religious fascination with "heavenly bodies" and calendar cycles into astronomy, science began to take on its self-critical concern for evidence and rigorous argumentation only a few thousand years ago. Religion is much older, of course, although *organized* religion— with creeds and hierarchies of ecclesiastical officials and codified systems of prohibitions and requirements—is roughly contemporaneous with organized science, and with writing. This is unlikely to be a coincidence. It takes a lot of record-keeping to overcome the memory limitations of the human brain—a topic considered in more detail in chapters 5 and 6.

Astronomers and mathematicians collaborated with priests at the outset, helping each other with difficult questions: How many days till we can have our winter-solstice ritual? When will the stars be in the right position for the most effective and proper sacrificial ceremony? So, without the question-*posing* by religion, science might never have found the funding it needed to get off the ground. More recently, of course, these specialists' perspectives have diverged into competing worldviews, a divorce made public and irrevocable at the dawn of modern science in the seventeenth century. The evolution of warfare also played a significant role in the development of science, as *literal* arms races paid for the R & D of new weapons, vehicles, maps, navigational devices, systems of human organization, and much more. Swords before plowshares, no doubt, and catalogues of plunder before bird lists and taxonomies of flowers. Agriculture, manufacturing, and trade—every project of human civilization has generated questions that needed answers, and over time the techniques for systematic and reliable question-answering evolved, by cultural, not genetic, evolution.

Thus was science born out of religion and civilization's other projects, a very recent cultural phenomenon but one that has transformed the planet like nothing else in the last sixty-five million years. The visionary engineer Paul MacCready has made an arresting calculation: Ten thousand years ago, human beings (plus their domestic animals) accounted for less than a tenth of 1 percent (by weight) of all vertebrate life on land and in the air. Back then, we were just another mammalian species, and not a particularly populous one (he estimates eighty million people worldwide). Today, that percentage, including livestock and pets, is in the neighborhood of 98! As MacCready (2004) puts it:

> Over billions of years, on a unique sphere, chance has painted a thin covering of life—complex, improbable, wonderful and fragile. Suddenly we humans (a recently arrived species no longer subject to the checks and balances inherent in nature), have grown in population, technology, and intelligence to a position of terrible power: we now wield the paintbrush.[3]

So science, and the technology it spawns, has been explosively practical, an amplifier of human powers in almost every imaginable dimension, making us stronger, faster, able to see farther in both space and time, healthier, more secure, more knowledgeable about just about everything, including our own origins—but that doesn't mean it can answer all questions or serve all needs.

Science doesn't have the monopoly on truth, and some of its critics have argued that it doesn't even live up to its advertisements as a reliable source of objective knowledge. I am going to deal swiftly with this bizarre claim, for two reasons: I and others have dealt with it at length elsewhere (Dennett, 1997; Gross and Levitt, 1998; Weinberg, 2003), and, besides, everybody knows better—whatever people may say in the throes of academic battle. They reveal this again and again in their daily lives. I have yet to meet a postmodern science critic who is afraid to fly in an airplane because he doesn't trust the calculations of the thousands of aeronautical engineers and physicists who have demonstrated and exploited the principles of flight, nor have I ever heard of a devout Wahhabi who prefers consulting his favorite imam about the proven oil reserves in Saudi Arabia over the calculations of geologists. If you buy and install a new battery in your mobile phone, you expect it to work, and will be mightily surprised *and angry* if it doesn't. You are quite ready to *bet your life* on the extraordinary reliability of the technology that surrounds you, and you don't even give it a second thought. Every church trusts arithmetic to keep track accurately of the receipts in the collection plate, and we all calmly ingest drugs from aspirin to Zocor, confident that there is ample scientific evidence in support of the hypothesis that these are safe and effective.

But what about all the controversies in science? New theories are trumpeted one week and discredited the next. When Nobel laureates disagree over a scientific claim, at least one of them is just wrong, in spite of being an anointed prince or princess of the church of science. And what about the occasional scandals of fraudulent data and suppression of results? Scientists are not infallible, nor are they, as a rule, more virtuous than laypeople, but they do submit to a remarkable discipline that *keeps them* honest in spite of themselves, imposing elaborate systems of self-restraint and review, and to a remarkable degree *depersonalizing* their individual contributions. So, although it is true that there have been eminent scientists who were racists, or sexists or drug addicts or just plain crazy, their contributions almost always stand or fall independently of these personal failings, thanks to the filters, checks, and balances that weed out the unreliable work. (Occasionally, a scientist or a whole school of scientific research will fall into dishonor or political disrepute, and since serious scientists don't want to cite those pariahs in their own work, this blocks perfectly good research for a generation or more. In psychology, for example, research on eidetic imagery—"photographic memory"—was stalled for a long time because some of the early work was done by Nazis.)

Through a microscope, the cutting edge of a beautifully sharpened ax looks like the Rocky Mountains, all jagged and irregular, but it is the dull heft of the steel behind the edge that gives the ax its power. Similarly, the cutting edge of science seen up close looks ragged and chaotic, a bunch of big egos engaging in shouting matches, their judgment distorted by jealousy, ambition, and greed, but behind them, agreed upon by all the disputants, is the massive routine weight of accumulated results, the facts that give science its power. Not surprisingly, those who want to puncture the reputation of science and drain off its immense prestige and influence tend to ignore the wide-angle perspective and concentrate on the clashes of schools and their not-so-hidden agendas. But, ironically, when they set out to make their case for the prosecution (using all the finely polished tools of logic and statistics), all their *good* evidence of the failings and biases of science comes from science's own highly vigorous exercises in self-policing and self-correction. The critics have no choice: there is no better source of truth on any topic than well-conducted science, and they know it.

What about the distinction between the "hard" sciences—physics, chemistry, mathematics, molecular biology, geology, and their kin among the *Naturwissenschaften*—and the "soft" social sciences (along with history and the other disciplines in the humanities), the *Geisteswissenschaften?* It is widely believed that the social sciences aren't really science at all, but, rather, just gussied-up political propaganda of one sort or another. Or at best they are a kind of science (*hermeneutical* or *interpretive* science) that plays by different rules, with different goals and methodologies. There is no denying that ideological battles rage within the social sciences over just these issues. What chance is there that the work that passes muster with one camp or another will be worthy of the respectful attention we give to results in the hard sciences? The discipline of anthropology, notoriously, is divided in two, with the physical anthropologists siding with the biologists and other hard scientists and typically unable to conceal their contempt for the cultural anthropologists, who side with the literary theorists and other folks in the humanities and typically express an equally withering contempt for their "reductionist" colleagues in the other camp. This is deplorable. A few hardy anthropologists, such as Atran (2002), Boyer (2001), Cronk et al. (2000), Dunbar (2004), Durham (1992), and—leaping to late in the alphabet—Sperber (1996), try to bridge the gap between evolutionary biology and culture, and they have to deal with an incessant swarm of ideologically driven critics.

Similar if less extreme divisions can be found in psychology, economics, political science, and sociology. With Freudians and Marxists and Skinnerians and Gibsonians and Piagetians and Chomskians and Foucauldians—and structuralists and deconstructionists and computationalists and functionalists—waging their campaigns, it is undeniable that ideology plays a large role in how these putatively scientific investigations are carried out. Is it all just ideology? While the earthquakes of controversy rage on the jagged peaks, do valuable objective results accumulate down in the valleys that can be used by any school of thought? Yes, and it is quite obvious. Researchers in one school routinely avail themselves of the hard-won results of their opponents, since, *if the science is done right,* everybody has to accept the results—but not the interpretations put on them. A lot of the valuable work done in these fields consists in confirming the well-gathered data (and replicating the experiments), and then showing that a better interpretation of the results follows from a rival theoretical perspective.

3 Putting ideology in its place

Ideology is like halitosis—it is what the other fellow has.
   —Terry Eagleton, *Ideology*

That's the practical answer, but I want to consider a deeper challenge as well. (A philosopher is someone who says, "We know it's possible in practice; we're trying to work out if it's possible in principle!") In 1998, the Yale legal scholar J. M. Balkin published *Cultural Software: A Theory of Ideology*, a fascinating book that looks at these controversies from a biologically informed perspective. In particular, he attempts to resolve what he calls Mannheim's paradox: "If all discourse is ideological, how is it possible to have anything other than an ideological discourse on ideology?"(p. 125). Is there—could there be—any ideology-free, neutral standpoint from which to judge these issues objectively? Just what *is* ideology? Not just any mistaken thinking, but thinking that is pathological or bad for us in some way. After reviewing a variety of representative (and of course highly ideological!) definitions of ideology, Balkin proposes that ideology be identified with ways of thinking that *help maintain unjust social conditions.*

> To understand what is ideological, we need a notion not only of what is true but also of what is just. False beliefs about other people, no matter how mistaken or unflattering, are not ideological until we can demonstrate that they have ideological effects in the social world. [p. 105]

This brings into the open a major difference between goals and methods in the social sciences and the hard sciences: social sciences are not just about people (so is the molecular biology of HIV and the chemistry of human nutrition) but about *how people should live.* There are *moral* judgments implicit in the very setting of the research agendas of these fields, and although these are *like* the value judgments implicit in such questions as "How can we interfere with HIV replication?" (why would we want to do this?) and "How can we improve human nutrition?" (what standard do we use to measure *good* nutrition?), the value judgments implicit in the social sciences are less obviously judgments that every sane person would agree on. To call somebody's thinking ideological is thus to condemn it *from a moral perspective that the target may not accept*. Much of the controversy is fueled, Balkin observes, by the quite justifiable fear of what he calls *imperialist universalism*:

> . . . the view that there are universal concrete standards of justice and human rights that apply to every society, whether pre- or postindustrial, whether secular or religious, and that it is the duty of right-minded people to change the positive norms and institutions of all societies so that they conform with these universal norms of justice and universal human rights. [p. 150]

Certainly many people in the United States are blithely confident that this is true, and hold that it is our duty to spread the American Way to all the peoples of the world. They think that any culture that finds our message repugnant is just deeply misinformed about how things are and how they ought to be. The only alternative they can see to this is truly shocking, a *moral relativism* that holds that *whatever* a particular culture approves of—polygamy, slavery, infanticide, cliteridectomy, you name it—is beyond rational criticism. Since such relativism is intolerable, in their eyes, imperialist universalism must be endorsed. Either we're right and they're wrong, or "right" and "wrong" have no meaning!

Meanwhile, many Muslims—for instance—would agree that moral relativism is beneath contempt, while insisting that *they* have the only true insight into what ought to be done in the world. Many Hindus think likewise, of course. The more one learns of the different passionately held convictions of peoples around the world, the more tempting it becomes to decide that there really couldn't be a standpoint from which truly universal moral judgments could be constructed and defended. So it is not surprising that cultural anthropologists tend to take one variety of moral relativism or another as one of their enabling assumptions. Moral relativism is also rampant in other groves of academia, but not all. It is decidedly a minority position among ethicists and other philosophers, for example, and it is by no means a necessary presupposition of scientific open-mindedness.

We don't have to assume that there are no moral truths in order to study other cultures fairly and objectively; we just have to set aside, for the time being, the assumption that we already know what they are. Imperialist universalism (of any variety) is not a good way to start. Even if "we" are right, insisting on it from the outset is ultimately neither diplomatic *nor* scientific. Science is not supposed to have all the moral answers and shouldn't be advertised as providing them. We may appeal to science to clarify or confirm factual presuppositions of our moral discussions, but it doesn't provide or establish the values that our ethical judgments and arguments are based on. We who put our faith in science should be no more reluctant to acknowledge this than those who put their faith in one religion or another. *Everybody* should consider adopting the stable middle ground that Balkin provides: an open-minded ("ambivalent") stance that permits a rational *dialogue* to engage the issues between people, no matter how radically different their cultural backgrounds. We can engage in this conversation with some reasonable hope of resolution that isn't simply a matter of one culture overwhelming the other by brute force. We cannot expect, Balkin argues, to persuade others if we leave no room and opportunity for them to persuade us. Success does depend on the participants' sharing, and knowing that they share, two *transcendent* values of truth and justice. What this means is only that both parties accept that these values are inescapably presupposed by human projects that *we all* participate in, simply by being alive: the projects of *staying* alive, and staying *secure*. Nothing more parochial need be assumed, and even "Martians" should be able to agree on this.

The idea of a transcendent value is rather like the idea of a per-

fectly straight line—not achievable in practice, but readily comprehended as an ideal that can be approximated even if it can't be fully
articulated. At first this may look like a dubious dodge—an ideal
that we all somehow accept even if nobody can say what it is! But in
fact, just such ideals are accepted and inescapable even in the most
rigorous and formalistic of investigations. Consider the ideal of rationality itself. When logicians disagree about whether classical
logic is to be preferred to intuitionistic logic, for instance, they have
to have in mind a prior standard of rationality, by appeal to which
one logic could be seen (by all) as better than another, and they have
to presume that they share this ideal, but they don't have to be able
to formulate this standard explicitly—that's what they're working
on. And in just the same spirit, people with radically different ideas
about which policies or laws would best serve humanity can—
indeed, must—presuppose *some* shared ideal if there is to be any
point in talking it over at all.

Balkin provides an imaginary dialogue that illustrates the appeal
to transcendent values in its simplest form. A marauding army
massacres the people and we call them war criminals. They object,
saying that their culture permits what they have done, but we can
turn their point back on them.

> . . . we can say to them: "If standards of justice and truth are in-
> ternal to each culture, you can have no objection to our character-
> ization of you as war criminals. For just as our standards can
> have no application to you, your standards can have no applica-
> tion to us. We are as correct in proclaiming your evil in our cul-
> ture as you are correct in proclaiming your uprightness in yours.
> But your very assertion that we have misunderstood you under-
> mines this claim. It presupposes common values of truth and
> justice that we are somehow obligated to recognize. And on that
> ground we are prepared to argue for your wickedness." [p. 148]

This plea may fall on deaf ears, but if so, then there really are
objective grounds for a verdict of irrationality: they are making a

mistake that they themselves have no grounds to defend *to themselves,* and that we need not respect in deference.

Cultural evolution has given us the thinking tools to create our societies and all their edifices and perspectives, and Balkin sees that these thinking tools—which he calls cultural software—are inevitably both liberating and constraining, both empowering and limiting. When our brains come to be inhabited by memes that have evolved under earlier selection pressures, our ways of thinking are restricted just as surely as our ways of talking and hearing are restricted when we learn our mother tongue. But the *reflexivity* that has evolved in human culture, the trick of *thinking* about *thinking* and *representing* our *representations,* makes all the restrictions temporary and revisable. As soon as we recognize that, we are ready to adopt what Balkin calls the ambivalent conception of ideology which avoids Mannheim's paradox: "A subject constituted by cultural software is thinking about the cultural software that constitutes her. It is important to recognize that this recursion in and of itself involves no contradiction, anomaly, or logical difficulty" (pp. 127–28). Balkin insists, "Ideological critique does not stand above other forms of knowledge creation or acquisition. It is not a master form of knowing" (p. 134). This book is intended to be an instance of just such an ecumenical effort, relying on the respect for truth and the tools of truth-finding to provide a shared pool of knowledge from which we can work *together* toward mutually comprehended and accepted visions of what is good and what is just. The idea is not to bulldoze people with science, but to get them to see that things they already know, or could know, have implications for how they should want to respond to the issues under discussion.

The Bellboy and the
Lady Named Tuck

[For context, see note 11 to chapter 5.]

For years, Dan Sperber and his colleagues Scott Atran and Pascal Boyer have expressed their skepticism about the utility of the meme's-eye perspective. First, let me try to give their main objections a clear expression, before saying why they have not persuaded me, in spite of what I have learned from them. This is my own summary of their position:

It is obvious that cultural items (ideas, designs, methods, behaviors . . .) have population explosions and extinctions, and that there are large noncoincidental family resemblances between such items and the models that inspire them or from which they are otherwise descended. But the phenomenon of transmission in most but not quite all these cases is not the sort of high-fidelity copying that the gene model requires. The cause of a new instance is not *copying* at all: "The cause may merely trigger the production of a similar effect" (Sperber, 2000, p. 169). Thus produced, the similarities between instances are not like the similarities between genes and hence require a different sort of Darwinian explanation. Culture evolves, but not strictly by descent with modification. And it is true that there are some few memes meeting Dawkins's specifications, such as chain letters, but such true memes play a relatively negligible role in the dynamics of cultural evolution (Sperber, 2000, p. 163). It is better to concentrate *instead* on the constraints and biases discernible in the psychological mechanisms that people share (Atran, 2002, pp. 237–38; Boyer, 2001, pp. 35–40).

My main reply to this objection is to be found in appendix A, "The New Replicators." Here I will expand on that reply by concentrating on the word italicized above: *"instead."* I want to challenge the Sperberians' conviction that they need to turn their backs on memes in order to study the constraints and biases of psychology. Atran, for instance, complains that the memetic approach is "mind-blind" (2002, pp. 241ff.) in that it ignores the detailed role of specific psychological mechanisms in shaping cultural items that proliferate. This is not at all an obvious point of disagreement, since Atran agrees that there is differential proliferation of cultural items. It is tempting to see the dispute as an artifact of miscommunication, with (some) memeticists promising too much and antimemeticists taking them at their word. As I note at the end of appendix A, memetics does not replace or pre-empt psychology any more than population genetics replaces or pre-empts ecology. (Is population genetics *environmentblind?* Yes, in general, and none the worse for it, since its models typically don't go into the details of how and why there are selective pressures in the environment; they just show how the effects of those selective forces, whatever they are, will be manifested in populations over time as migrations and births and deaths take their toll. To get a whole biological explanation, you still need the ecology, and for a whole cultural explanation, memeticists still need psychology—though they may deny it in the throes of partisanship.)

Boyer expresses the Sperberian objection in similar terms, but in spite of his stated opposition to memes, he often cannot resist couching his points in terms of differential replication. Indeed, his theory has been summarized by one *sympathetic* commentator as the thesis that "religion can primarily be understood as the systematic exploitation of mundane psychological systems by *especially virulent strains* of cultural concepts" (Bering, 2004, p. 126).

"Virulent" is not quite the word Bering is seeking, since its (dictionary) connotations are all negative; "prolific" or "fit" would be a more accurate summary of Boyer's thesis, since Boyer is careful to be neutral on the issue of whether religion is a good or bad accompaniment to human life, but, leaving that aside, it seems that Bering would include Boyer among the memeticists in spite of his disclaimers. So why couldn't we just encourage Boyer and Atran and Sperber to concentrate on the selective forces provided by psychology, which they do so well, leaving the (trivial?) unificatory work to the memeticists down the corridor?

But there is more to be said. We want to conceive of cultural evolution in terms of memes *and* in terms of the constraints of psychology—*and* the further constraints that emerge from the earlier interaction of memes and those very constraints! Consider an experiment we might do inspired by the research on "urban legends" by Heath, Bell, and Sternberg (2001). Did you hear about the bellboy who was caught on surveillance video putting the hotel guests' toothbrushes up his . . . ? How about the driver who heard a thump and when he stopped his car, many miles later, found the body of a baby embedded in the grillework of his car? Noting that many of the most popular urban legends involve shockingly disgusting tales, these researchers investigated the role of disgust in heightening the likelihood of transmission of a wide variety of urban legends. They provided competing "alleles" (alternative tellings) of each story and found that, sure enough, the more disgusting versions traveled better. Alas, they didn't measure actual transmission, just their subjects' convictions about how likely they were to repeat the stories. Research is expensive. But thought experiments are cheap, so let us *imagine* an experiment that would nicely illustrate the Sperberians' point—and why I don't find it a good argument against the memes approach.

Suppose we concoct a thousand different urban legends—new ones, not yet circulating on the World Wide Web—and carefully plant them in ten thousand different hearers, one to a customer, each story going to ten hearers. We *try* to give these meme candidates "radioactive tags" by including telltale details in each planted version, along the lines of "Did you hear about the *Brazilian taxi-driver* who . . ." And suppose we also spend *lots* of money tracking these trajectories, by hiring armies of private detectives to eavesdrop on our initial subjects, tapping their phones, and so forth (another virtue of thought experiments—you don't have to clear them with your university's internal review board or the police!), so that we get quite a lot of good data about which stories evaporate after a single telling, which actually get transmitted, in what words. The Sperberians' dream result would be that we came up with . . . *zilch!* Almost all our radioactive tags would disappear, and all that would remain of the thousand different stories would be *seven* (say) stories that kept getting reinvented, time and again, because these seven stories were the only ones that tickled all the innate psychological constraints. When we looked at the lineages, we would see that, say, a hundred initially very different stories had all converged eventually on a single tale, the closest "attractor" in urban-legend space. Sometimes a story would be gradually modified in the direction of the favored attractor, but if the hearer already knew that tale, a new story might end abruptly in a cul-de-sac: "Hey, interesting. That reminds me—have you heard about the guy who . . . ?"

If this were the result, we would see that all the *content* in the urban legends that prevailed over time was already implicit in the psychology of the hearers and tellers, and virtually none was replicated faithfully from the initial stories. Here is Atran's way of expressing the point:

> In genetic evolution there is only "weak selection" in the sense that there are no strong determinants of directional change. As a result, the cumulative effects of small mutations (on the order of one in a million) can lead to stable directional change. By contrast, in cultural evolution, there is very "strong selection" in the

sense that modularized expectations can powerfully constrain transmitted information into certain channels and not others. As a result, despite frequent "error," "noise," and "mutation" in socially transmitted information, the messages tend to be steered (snapped back or launched forward) into cognitively stable paths. Cognitive modules, not memes themselves, enable the cultural canalization of beliefs and practices. [2002, p. 248]

It would be almost as if we each have a CD in our brains with a few (dozen? hundred?) urban legends recorded on it; whenever we hear a close approximation to one of these urban legends, this triggers the CD to go to that track and play *it*—"triggered production," not imitation of what we've heard. (This is suggested by Sperber's "theoretical example" of the sound recorders [2000, p. 169].) That extreme null result is unlikely, of course, and if some content *did* get replicated from host to host, those who were infected by it would set up a *new constraint* on the fate of whatever urban legends they heard next. Cultural canalization can be due as much to prior cultural exposure as to one's underlying cognitive modules. Perhaps, if you haven't heard the one about the Chinese midget, you replicate the one about the boy with the pet gerbil and pass it along more or less intact, and if you have, you tend to merge them into something that eventually emerges as the one about the policewoman and the gerbil, and so forth. To investigate the interaction between contents culturally transmitted and constraints that are shared independently of culture, you really have to track the replication of memes—as best you can. Nobody said it was a practical research program in most instances.

A remarkable instance of this occurred in the preparation of this book. One of the readers of the penultimate draft noticed a typographical error in chapter 2, and since it was repeated in the bibliography, it occurred to him that I might miss it: Gould's 1999 book is *Rocks of Ages*, he told me, but I had written it *Rock of Ages*. My first reaction was frank disbelief. I thought my reader was making the mistake; the first word of Gould's book *couldn't* be "Rocks," could it? I had read the book, and noted his plays on words (the paleontologist studies the ages of rocks, while . . .) but had completely missed his putting the mutation in his title, because the hymn title was so well branded into my memory! I had to check the book for myself, and, sure enough, the title is *Rocks of Ages*, but then I hopped on the Web to see if I was alone in making the error. On March 23, 2005, there were approximately as many Google citations for "Gould 'Rock of Ages'" (3,860) as for "Gould 'Rocks of Ages'" (3,950), and although many of the former entries proved to have both the correct title of Gould's book and the hymn title, among the entries with the title misspelled were reviews of the book, and discussions of the book, both positive and negative. To casual inspection, there didn't seem to be any obvious pattern to the errors, but here is a fine elementary project in computational memetics for anybody who wants to dig deeper. There is sure to be an interesting story to be told about how often this error has crept in by mutation and who has copied whose error. (See Dawkins's discussion [1989, pp. 325–29] of a similar transcription error of a title, and an introduction to the methods of memetics using the resources of the Scientific Citation Index.)

In addition to having the genetically evolved mechanisms or modules beloved by evolutionary psychologists, our brains are packed with culturally transmitted mechanisms of every imaginable sort, and the presence or absence of these sets up immunities and receptivities in hosts just as powerful as—or even more powerful than—the constraints exhibited by the underlying machinery. In his chapter against memes, Atran quotes me on this topic, but misses the point I was trying to make. I had said that the structure of Chinese and Korean minds is "dramatically different" from that of American or French minds (Dennett, 1995b, p. 365), and Atran supposes (2002, p. 258) that I am trying to make a subtle point about how people with different native tongues will interpret drawings or attribute causation or blame in different scenarios. He cites experiments in which people from different cultural groups respond quite similarly in a variety of circumstances designed by psychologists to elicit such differences. But I had something much simpler and more obvious in mind: People with Chinese minds won't laugh at, or remember, or repeat, jokes told in English! (A few years ago, the brilliant songwriter and singer Lyle Lovett released an album entitled *Joshua Judges Ruth*. I found that in general my friends didn't get it; I'd ask them what Lovett's *next* album might be entitled and none of them replied, *"First and Second Samuel?"*—which was the first thing that would pop into my head, thanks to Sunday-school drill more than half a century before.) And just as we can be quite sure that jokes told in French have a hard time getting spread in Anglophone neighborhoods, we can be quite sure that a person's political views, and knowledge of art (or quantum physics, or sexual practices), would provide strong constraints or biases on his or her receptivity and eagerness to transmit various candidate memes. For instance, to my way of thinking, one of the funniest limericks I have ever heard is the following, which you will find funny only if you've heard lots of limericks:

> There was a young lady named Tuck
> Who had the most terrible luck:
>   She went out in a punt,
>   And fell over the front,
> And was bit on the leg by a duck.

I couldn't resist transmitting it to you. Who will transmit it further? That depends a lot on what other memes infect your brain, and the brains of those you talk to. In the complex world of cultural transmission, the patterns that are *directly* due to fixed features of human psychology will perhaps not loom particularly large. So it seems to me that those who follow Sperber in his opposition to memes are making points that can better be made in the language of memes: one of the things they are saying, for instance, is that *convergent evolution* plays such a dominating role in cultural evolution that the transmission of design by actual descent through cultural lineages is much less of a factor in accounting for observed similarities than the shaping of design by selective forces. This is often very plausible, and can be investigated in any case. But we should also be alert to the possibility that many of the similarities between, say, Islam and Christianity *may* be due to their common Abrahamic ancestor religion rather than to their each having adjusted to similar found conditions in their human adherents.

Kim Philby as a Real Case of Indeterminacy of Radical Interpretation

[For context, see note 14 to chapter 8.]

Philosophers have spent decades dreaming up thought experiments designed to prove or disprove W.V.O. Quine's (1960) principle of the *indeterminacy of radical translation:* the surprising claim that *in principle* there could be two different ways of translating one natural language into another natural language and *no evidence at all* about which one was the *right* way to translate the language. (Quine insisted that in that case there wouldn't *be* a right way; each way would be as good as the other, and there would be no further fact of the matter.) It seems deeply unlikely at first that this would be possible. Couldn't a well-informed bilingual, for instance, always tell which of two competing translations of a sentence in one of his languages was the better translation into the other? How could there *not* be plenty of evidence in favor of one of the two translations?

If you think the resolution is obvious, you haven't read, or understood, the voluminous philosophical literature on this curious puzzle. A good place to start, after reading Quine's masterpiece, *Word and Object* (1960), would be the special 1974 issue of *Synthese* devoted to a University of Connecticut conference on intentionality, language, and translation, in which Quine took on his most distinguished opponents and left them, and the issue, unresolved, which is where the issue stands today (Quine, 1974a, b).

In the case of Philby's ultimate beliefs, we have a tantalizing glimpse of how close we might come in the real world (as contrasted with the strange world of many philosophers' thought experiments) to a case of indeterminacy of radical *interpretation* (see David Lewis's essay "Radical Interpretation," 1974, in the *Synthese* issue). We may imagine two indefatigable observers, following Philby's every move, recording his every utterance, reading his most secret papers, listening to him talk in his sleep, and even (now we're back in philosophy land) recording his every brainwave; and we can see that they might, *on the same evidence,* come up with staunchly held opposing verdicts: he's a loyal Brit after all; no, he's a loyal Soviet.

It would be no use just asking Philby, of course; both observers are well aware of how he's going to respond to such a query, and their opposing theories account for it about equally well. (For a related argument, see my discussion of the beliefs of "Ella" in "Real Patterns," 1991b.) No doubt it is hugely unlikely that in such a case neither of the two interpretations would ever unravel, but (Quine insisted) not impossible. That was his point. In every real-world situation, probably, two such radically different interpretations of a whole life history would balance on the knife edge of *no verdict* for only a short time, and eventually one interpretation would collapse, leaving the other victorious, but we shouldn't make the mistake of supposing that this was a metaphysical certainty, guaranteed by some special inner fact that settled the issue. We can even come to see, from this perspective, that Philby might himself come to wonder which view was the truth—and not be able to tell! This problem would also face the imagined bilingual who is asked which translation manual is right. He might be astonished to discover that he himself had no resources to say which was "right," and in that case, Quine insisted, there would be no fact of the matter about which was right. They would be equally good translations, and that's all one could say.

If the point still eludes you, it may help to consider a simpler case of the same phenomenon, my "Quinian Crossword Puzzle." It is not easy concocting a crossword puzzle with two equally good solutions, but here is one. Which is the *real* solution? Neither, for I deliberately set out to make it that way. In principle, it is possible to make a higher-dimension crossword puzzle, a Philby, whose whole structure and history and current set of proclivities are equally amenable to two different intentional interpretations. In practice, impossible, but we should not, for that reason, imagine a category of inner facts that would settle any case.

A Quinian Crossword Puzzle

1	2	3	4
2			
3			
4			

Across
1. Dirty stuff
2. A great human need
3. To make smooth
4. Movie actor

Down
1. Vehicle dependent on H_2O
2. We usually want this
3. Just above
4. U.S. state (abbrev.)

Notes

|||\\\|||

1 Breaking Which Spell?

1. I discussed the example of *Dicrocelium dendriticum* in Dennett, 2003c; for more on its fascinating life cycle, see Ridley, 1995, and Sober and Wilson, 1998. For a striking case of a fish parasite, see LoBue and Bell (1993). A parasite of mice, *Toxoplasma gondii*, will be discussed in more detail in chapter 3. The epigram from Hugh Pyper is found in Blackmore (1999), as well as in Pyper (1998). All references can be found in the bibliography at the end of the book, and in general will be inserted in the text, not footnoted. Notes such as this will be used to expand on the points in the text in ways that may be of interest only to specialists.

2. Why the potential for breeding loyalty was present in dogs but not in cats is itself an interesting chapter of biology, but it would take us far afield. For more on the limits of domestication, see Diamond, 1997.

3. Here are two of the best-known definitions of religion with which to compare mine:

> . . . a unified system of beliefs and practices relative to sacred things, that is to say, things set apart and forbidden—beliefs and practices which unite into one single moral community called a Church. [Emil Durkheim, *The Elementary Forms of the Religious Life*]

> (1) A system of symbols which acts to (2) establish powerful, pervasive, and long-lasting moods and motivations in men by (3) formulating conceptions of a general order of existence and (4) clothing these conceptions with such an aura of factuality that (5) the moods and motivations seem uniquely realistic. [Clifford Geertz, *The Interpretation of Cultures*]

4. These transformations typically happen gradually. Doesn't there have to have been a Prime Mammal, the first mammal whose mother was not a mammal? Not really. There doesn't have to be a principled way of drawing the boundary between the therapsids, those descendants of reptiles whose descendants include all the mammals, and the mammals (for a discussion of this perennially puzzling point, see Dennett, *Freedom Evolves*, 2003c, pp. 126–28). A religion of long standing could turn into a former religion gradually, as its participants gradually shed the doctrines and practices that mark the genuine

article. No value judgment is implied by such a description; mammals are former therapsids and birds are former dinosaurs, and none the worse for it. Of course the *legal* implications of whether or not the boundary had been crossed would have to be settled, but this is a political issue, like the moral status of the octopus, not a theoretical issue.

5. May the Force Be With You! Is Luke Skywalker religious? Think how differently we would react to this incantation if the Force were presented by George Lucas as satanic. The recent popularity of cinematic sagas with fictional religions—*The Lord of the Rings* and *The Matrix* offer two other examples—is an interesting phenomenon in its own right. It is hard to imagine such delicate topics being tolerated in earlier times. Our growing self-consciousness about religion and religions is a good thing, I think, for all its excesses. Like science fiction generally, it can open our eyes to other possibilities, and put the actual world in better perspective.

6. During the 1950s and 1960s, when Freudian psychoanalysis was riding high, critics who tried to point out to its devotees the many weaknesses and mistakes of Freud's theory were typically stymied by an infuriatingly bland wall of psychoanalytic deflection, along the lines of "Let's see if we can figure out why you're so *hostile* to psychoanalysis, and why you feel this emotional need to 'refute' its claims. Why don't you start by telling us about your relations with your mother. . . ." This was question-begging (or circular reasoning) even when it was sincerely meant, and it was often simply dishonest. I recognize that my *postponement* of consideration of the issue of whether God exists may be seen by those who are armed with arguments as a similarly un-principled evasion of intellectual responsibility. But if I began this book with their issue, framed as it traditionally is, it would take hundreds of pages of plowing over familiar terrain before I could ever get to a novel contribution. Bear with me, please. I will not forget my obligation to treat this topic!

2   Some Questions About Science

1. For more on the role of science in avoidance, and the explosion of "evitability" that human civilization has achieved, see my *Freedom Evolves*, 2003c.

2. Following recent practice, I use the term "Islamist" to refer to those radical or fundamentalist strains of Islamic thought that in general condemn democracy, women's rights, and the freedom of inquiry in which science and technology can flourish. Many, probably most, Islamic thinkers and leaders are deeply opposed to the Islamist position.

3. The only study I have found is Anderson and Prentice, 1994.

3 Why Good Things Happen

1. In a few cases I met with small groups of coreligionists, and the occasional discovery by my informants of differences among them was particularly telling, perhaps even life-changing in a few instances.

2. Current thinking is that the various coyote calls serve different purposes. The bloodcurdling "group yip-howl" is most plausibly "important in announcing territorial occupancy and preventing visual contact between groups of coyotes" (Lehner, 1978a, p. 144; see also Lehner, 1978b). If you can *avoid* an actual battle over territory by engaging in impressive saber-rattling, this may be the thrifty way of preserving energy and health for another day's hunting. On this hypothesis, the signal's impressive volume is a hard-to-fake sign of its veracity, a common phenomenon in animal communication. (See Hauser, 1996, chapter 6, for an excellent discussion of the theoretical and experimental investigations of the evolution of honest signaling.) It also suggests some interesting experiments to be conducted in using high-quality playbacks of recorded coyote howls to regulate population densities. Would the coyotes catch on? How long would it take?

3. An opening survey of the voluminous literature on creationism and Intelligent Design should include: Pennock, 1999, *Tower of Babel: The Evidence Against the New Creationism*; Perakh, 2003, *Unintelligent Design*; Shanks, 2004, *God, the Devil, and Darwin: A Critique of Intelligent Design Theory*; Young and Edis, 2004, *Why Intelligent Design Fails: A Scientific Critique of the New Creationism*; and National Academy of Sciences, 1999, *Science and Creationism*. The May–August 2004 issue of *Reports of the National Center for Science Education* reviewed several dozen recent books on the topic, including more than a dozen (of varying quality) written from a Christian or Jewish perspective. For excellent surveys of contemporary evolutionary biology, I highly recommend the anthology of current work edited by Moya and Font, 2004, *Evolution: From Molecules to Ecosystems;* the two-volume *Encyclopedia of Evolution*, ed. Pagel, 2002; and the seventh edition of the textbook *Life: The Science of Biology*, by Purves, et al., 2004. There are also dozens of good Web sites on which one can find authoritative and fair refutations of the work of the most prominent critics of evolution, such as William Dembski and Michael Behe. The National Center for Science Education is one of the best, at http://www.ncseweb.org.

There are also plenty of Web sites devoted to Intelligent Design, of course, but no serious peer-reviewed journals. Why might that be? If Intelligent Design were an idea whose time has come, you would think that young scientists would be dashing around their labs, glued to their computers, vying to win the Nobel Prizes that surely are in store for anybody who can overturn any significant proposition of contemporary evolutionary biology. Intelligent Design fans insist that the scientific establishment has a bias against their work that makes it impossible for them to break into the mainstream journals, but this is simply not credible. The Discovery Institute and other well-funded havens for Intelligent Design research could easily afford to produce a high-quality, peer-reviewed journal if there were anything to publish in it, and if they could find credible scientists to do the peer reviewing. Literally thousands of peer-reviewed scientific articles are published every year elaborating and extending the basic theory of evolution, and most of the authors of these articles never become famous, in spite of their proven expertise. Surely a few of them would happily jump ship and risk ridicule from the establishment for the chance to become world-famous as the Scientist Who Refuted Darwin. But the backers of creationism don't even bother offering the lure. They know better. They know that all they have going for them is propaganda, so that is what they spend their endowment on.

William Dembski (2003) has made available a list of four (count 'em!) peer-reviewed scientific articles that, he says, support Intelligent Design themes. (He also lists his own 1998 book, which is indeed published in a peer-reviewed series by Cambridge University Press.) But Dembski's own comments on these essays make it clear that their arguments are at best, as he puts it, "non-Darwinian" (they are conducted without any specifically Darwinian premises), and hence might be put to use in support of an Intelligent Design argument. None of them actually advances an argument for Intelligent Design.

4. This standard way of talking masks a complication. When we speak of "half your genes" here, we mean half of those of your genes that are idiosyncratic, that distinguish you, genetically, from others in your species. In cloning, whatever genes you have that make you "special" (for better or for worse) get passed on in toto to your offspring. In sexual reproduction, only half of those genes show up in your offspring; your mate provides the balance of the idiosyncratic genes.

5. Does money emerge from pure barter systems by a series of gradual and scarcely noticed shifts in practice (the "commodity" theory), or does it always require some sort of "fiat" from some state authority or conscious agreement or compact (the "chartal" theory)? The origin of money has been debated for centuries. For a fascinating discussion of the history of the debate, together with some elegant economic models of the possible processes, see Awai, 2001. See also Burdett et al., 2001; and Seabright, 2004.

6. It is *possible,* of course, that in fact some one historical individual did *so much* of the early design work on money, or language, or music that he or she deserves the title of author, but this is extremely unlikely and entirely unnecessary. Evolution permits cultural design innovations to accumulate so gradually that authorship gets distributed over millions of clueless innovators over thousands of generations, just like the design innovations that revise genes.

7. The difference in the reproduction system makes a huge difference, of course. When the mint changes the year engraved on the die with which it stamps all the coins it makes, this is a *sort* of mutation, but such mutations don't accumulate, normally. If a nick or blemish in the die doesn't get repaired, it may mark all the coins for many years, and even get copied onto the successor die (if one of the coins it has made is chosen as the male from

which the new female die is made), and that is more like a genetic mutation that gets transmitted to offspring.

8. On the imagined "intrinsic" value of money, see "Consciousness: How Much Is That in *Real* Money?," in Dennett, 2005c.

9. On what it is like to be a turkey vulture, see Dennett, 1995a, reprinted in Dennett, 1998a.

10. Biologists can perhaps better understand the resistance of many social scientists to the *biologizing* of their disciplines by reflecting on their own discomfort with attempts to . . . *physicize* biology. Ernst Mayr, the legendary evolutionary biologist, recently published a book (shortly after his hundredth birthday) on the autonomy of biology, showing why it doesn't "reduce" to physics (Mayr, 2004). I agree with most of the claims he makes. He is *not* declaring that physics provides no constraints or principles that biologists must understand and may exploit. There are different brands of reductionism; only some of them—which I call greedy reductionisms (Dennett, 1995b)—are mistakes. When somebody declares that a view under attack is reductionistic, we have to look closely to see whether this is a bad thing.

11. Among those credited with this aphorism are the philosopher Ludwig Wittgenstein, the artist Paul Klee, and the critic Viktor Shklovsky.

12. Of course there are plenty of intermediate cases, in which boatbuilders have some idea or other, good or bad, dim or brilliant, behind the mutations that they introduce, which are thus not all slips of the adze. What seemed like a good idea at the time may prove worthless in fairly short order. This speeds up the design process, but in both directions—larger bad ideas get tried out in the trial-and-error process as well as good ideas. Richard Dawkins has proposed to call designs-without-designers "designoid" (1996, p. 4). The coinage is useful for marking the error people often make in supposing that anything that *appears* designed must have been produced by a deliberate conscious mind, but it shouldn't be taken to mark a bright line in nature. Are the short legs of dachshunds design or designoid? Human breeders set out to achieve the effect, and they had reasons for it. Are genetically engineered organisms design or designoid? Is the beaver's dam that ingeniously makes use of local and unprecedented opportunities for dam-building design or designoid? A beaver's dam requires considerably more cognitive talent to build than the ant lion's conical sand trap. The work of exploring the grand unity of Design Space is distributed between the slow ratcheting of natural selection of genes, and the swift trial-and-error explorations of individual brains (and their numerous artifactual exploration vehicles), so I will continue to use the umbrella term "design" to cover it all.

13. One of the main themes of *Darwin's Dangerous Idea* (Dennett, 1995b) is that what Darwin discovered is fundamentally an *algorithm,* an information-processing recipe that can be executed in many different media, just as the *long-division algorithm* can be done with pencil, pen, chalk, or scratching with a stick on the ground.

14. For more on memes, see also Dennett, 1995b, 2001b, 2001c, 2005c, and appendix C of this book.

15. For some of the details, see Dawkins, 2004a, pp. 31–32.

16. Group selection has had a controversial career in evolutionary theory, and technical disputes make it treacherous territory for the uninitiated. See Wilson and Sober, 1994 (and all the commentaries published in the same journal); Sober and Wilson, 1998; and Dennett, 2002a (and Sober and Wilson's reply in the same journal). Wilson's views will be discussed in more detail in a later chapter.

4   The Roots of Religion

1. There is no consensus among surveys about how to count religions (as contrasted with cults and other typically short-lived organizations), but by any benchmark there are many thousand distinct (independent, noncommunicating) religions. The almanacs have identified over thirty thousand distinct *Christian* churches. The more or less standard reference work for all religions is Barrett et al., *World Christian Encyclopedia* (2nd ed., 2001). Religions crop up so frequently that even Web sites have difficulty keeping their lists up to date. A few good ones are http://www.religioustolerance.org/worldrel.htm and http://www.watchman.org/cat95.htm—the latter indexes over a thousand *new* cults and religions. There are also journals and other organizations devoted to the study of new religions, easily found on the Web.

2. Dunbar (2004) calls these graves unequivocal evidence of religion, but they are in fact highly enigmatic. There is no doubt the bodies were deliberately placed in position with objects covered in red ocher, but the meaning of the tableau is highly contentious. See, e.g., http://home/earthlink.net/~ekerilaz/dolni.html.

3. A useful overview is Atran and Norenzayan, 2004, with its two dozen accompanying expert commentaries and a response from the authors. Other essential reading includes Sperber, 1975, 1996; Lawson and McCauley, 1990, 2002; Guthrie, 1993; Whitehouse, 1995; Barrett, 2000; Pyysièainen, 2001; Andresen, 2001; Shermer, 2003.

4. This theme has been developed by many authors in recent years. My own contributions to this literature include Dennett, 1991a, 1995b, 1996, and many articles.

5. The main reason I am opposed to speaking of animals—or even adult human beings—as "having a theory of mind" is that this typically conjures up entirely too intellectual an image of a theorem-deriving, proposition-consulting, hypothesis-testing little scientist, whereas I see adopters of the intentional stance—even virtuoso practitioners such as the most manipulative people you have ever encountered—as more like intuitive *artists* than sophisticated *theorists*. Craft is more in evidence than ideology, and the development of explicit, self-conscious *models* of the folk craft is a still more recent innovation— emerging first, really, in the wonderful novels of the eighteenth and nineteenth centuries and made more systematic (but arguably no more powerful)

by psychologists and sociologists and the like in the twentieth century (Dennett, 1990, 1991c). The "theory theorists" will retort that this wonderful craft or know-how has to be implemented *somehow* in the brains of those who have the competence, and that we should be trying to develop a computational neuroscience model of this competence. I entirely agree, but calling this a *theory* still pinches the imagination of the theorist in ways I think we should avoid. What else could it be but some sort of theory? That's a good question, I think, that we ought to try to answer, not a rhetorical question that forecloses the issue.

6. See, for instance, Tomasello and Call, 1997; Hauser, 2000; and Povinelli, 2003.

7. This is a delicate and controversial topic in theoretical cognitive science these days: just what is pleasure or pain, and what is addiction or habit or willpower? I have a little to say about the current state of the art in Dennett, 2003b, but more is in progress.

5 Religion, the Early Days

1. Do we *know* that other species don't have language or art? If so, how do we know? Among the many good recent books on these subjects, I recommend Hauser, 1996, 2000. The bowerbirds' bowers are perhaps the closest counterpart to human art, since they are nonfunctional or decorative artifacts whose manifest (if free-floating) purpose is to charm the opposite sex, which has often been hypothesized to be the original mainspring of our artistic impulses.

2. Dunbar (2004) defends the thesis that whereas our nearest relatives, the chimpanzees, can manage at most two orders of intentionality (beliefs about beliefs, say, or beliefs about desires) normal human beings can appreciate and respond to the complexities of fourth- or fifth-order intentionality, and argues that the virtuosi among us can go even higher, keeping track of sixth-order intentionality, as they maneuver their way among their conspecifics. "Religious leaders, like good novelists, are a rare breed" (p. 86). See also Tomasello, 1999.

3. Faber (2004) observes that human life begins with an infant crying for food, for comforting, for protection (out of fear), for help, and getting answered by a big warm wonderful thing. Thousands of times, the infant cries out; thousands of times, the cries are answered. "One would be hard-pressed to discover within the realm of nature another example of physiological and emotional conditioning to compare with this one in both depth and duration" (p. 18). This prepares the child, Faber argues, for religious stories:

He makes contact easily with the supernatural domain because in a manner of speaking *he has been there all along.* He has been living with or in the company of powerful, unseen, life-sustaining presences since he commenced the process of mind-body internalization, or interactional, physiological *imprinting*, as it naturally and persistently arose from his *affective interaction* with the all-powerful provider, the big one who appeared over and over again, ten thousand times, to rescue him from hunger and distress and to respond to his emotional and interpersonal needs, to his deep affective drive for *attachment*. [p. 20] ... The child's unconscious mind resonates to religious narratives *before* his rational faculties have ripened, *before* he can see and critically evaluate what it is that asks for his perceptual assent. [p. 25]

4. A list of over eighty different methods can be found at http://en.wiki pedia.org/wiki/Divination.

5. Dumbo's magic feather is discussed at some length in Dennett, 2003b.

6. Burkert (1996) offers a different speculative evolutionary scenario of a cascade of bottlenecks that could select for genes for susceptibility to religion: "Although religious obsession could be called a form of paranoia, it does offer a chance of survival in extreme and hopeless situations, when others, possibly the nonreligious individuals, would break down and give up. Mankind, in its long past, will have gone through many a desperate situation, with an ensuing breakthrough of *homines religiosi*" (p. 16). I cannot yet see how to test this hypothesis, but it is certainly a possibility to consider seriously, if we can find some way to do so.

7. My use of the term *folk religion* is at variance with the usage of some anthropologists and ethnomusicologists (e.g., Yoder, 1974; Titon, 1988), who use it to describe the contrast between "official" organized religion and what people of those denominations actually believe and practice in their daily lives (see Titon, 1988, pp. 144ff., for a discussion). See also the related concept of "theological incorrectness" (Slone, 2004). What I am calling folk religion is often called tribal or primitive religion.

8. Few folk music fans today are such purists as to turn up their noses at all composed "folk" songs, but for my purposes purism rules: those relatively ancient melodies and lyrics without authors are the folk music I am talking about. In every age, these songs get artfully adjusted and rearranged, with new lyrics and new rhythms, and sometimes new melodies as well, and along the way folk artists add songs of their own composition. To take just the recent past, Huddie Ledbetter and Woody Guthrie and Pete Seeger composed hundreds of "folk songs" that have joined the canon, even though in these cases we know who the author was. We tend to exclude from the canon the equally singable ballads of Gilbert and Sullivan and the Gershwins, but time may well erase the distinction. My point is that, although it is *possible in principle* that if we had perfect historical knowledge we could *always* identify a composer and a lyricist, it is also possible—and more likely—that in many cases the authorship was so distributed over the centuries that nobody deserves credit for either the melody or the lyrics of the "classic" folk song that now appears in the canon. Did Ravenscroft just *write down* "The Three Ravens" in 1611, or did he compose it? Or did he *adapt* it as he wrote it down—or did it adapt itself?

9. Some of this is too obvious to notice. Why should a ~~~~~
serial at all (just one word at a time)? Because we have ~~~~
which to speak, to put it crudely. The ideograms of j~~~
show that it is possible for written languages to untie their o~~~
not shed them altogether. Would a system of symbols that could ~~~
nounced," that was three-dimensional (a word sculpture of sorts), or ~~~
dependent on the use of color, count as a language? The very idea of silen~~
reading—let alone reading without moving your lips!—came along late in the
development of writing (in medieval times, historians aver—see, e.g., Saenger,
2000). Archaic spelling is also, of course, a trace of earlier pronunciations.

10. Blackmore (1999, p. 197) argues that "memetic drive" is possible and
likely in accounting for our love of rituals: that idiosyncrasies in *culturally*
transmitted rituals would be variably responded to by people, and this would
create a novel selective environment in which talent for and appreciation of
these idiosyncrasies were *genetically* selected—just as talent for language was
genetically selected once language got under way. What started as a more or
less undifferentiated sweet tooth for ritual, in other words, could evolve genetically into a sweet tooth for supernormal versions of the local idiosyncrasies, a case of gene-culture coevolution that was led by cultural exploration
of the space of possibilities, a possible extension of the Baldwin Effect, in
which innovations of behavior achieved by individuals in their lifetimes (innovations discovered or learned by them) can create and focus selection pressures that eventually lead to innate proclivities to perform these innovations, a
non-Lamarckian way that acquired characteristics can influence the evolution
of genetically determined characteristics (see Dennett, 1995b, 2003a, 2003d).

11. Some readers may be bothered by my persistent talk of memes in this
chapter, since the anthropologists whose work I am discussing so favorably,
Boyer and Atran and their mentor, Sperber, are united in their rejection of the
memes perspective, as they make quite clear in their books and articles. I have
been discussing this with them for some time, both in print (Dennett, 2000,
2001a, 2001b, 2002b [reprinted here as appendix A], and especially 2005b;
and see Sperber, 2000) and at conferences. I think they are making a mistake, but it is a bit of a technical disagreement that would be a distraction to
most readers. Still, a reply to their objections is in order, and is supplied in appendix C. See also the other essays in Aunger, ed., 2000, where Sperber,
2000, appears; and Laland and Brown, 2002, chapter 6.

12. Thanks to Dan Sperber for popping this balloon by drawing my attention to Mahadevan and Staal, 2003, from which the passage is quoted.

13. For a vivid but controversial introduction to the field, now somewhat
out of date, see Ruhlen, 1994. For an overview of the current state of the science, see Christiansen and Kirby, eds., 2003. Other thought-provoking studies
are Carstairs-McCarthy, 1999, and Cavalli-Sforza, 2001.

14. Swimming is an interestingly intermediate case: Unlike running and
walking, swimming strokes have quite a memetic history. In the late nineteenth century, an Englishman, Arthur Trudgen, carried the overarm Native
American way of swimming (soon called the "trudgeon" or "trudgen crawl"

this meme-vector) to England, but he miscopied the kick, using a breast-
roke "frog" kick instead of the flutter kick used by Native Americans. This
transmission error was corrected by Richard Cavill in 1902, and today's front
crawl is the descendant of that quite recent improvement. But versions of the
crawl have probably been invented and reinvented numerous times over the
eons, since it is so clearly superior to all other known ways of propelling oneself through the water at high speed. Not for nothing is this Good Trick known
as *freestyle* in competitive swimming. The only rule in freestyle is that you
must break the surface every now and then (and this rule was introduced to
prevent swimmers from experimenting with dangerous underwater strokes
that might drown them if they passed out). In freestyle, you are welcome to
improve on the front crawl if you can.

15. Note that today, thanks to writing and other storage media, this is not a
problem, so a religion no longer needs such regular rituals of unison to keep
the text pure. But a religion that makes the rituals optional is in danger of suc-
cumbing for other reasons.

16. Atran, 2002, and Lawson and McCauley, 2002, provide detailed critiques of the hypotheses of Whitehouse (1995, 2000) and others.

17. Orgel's Second Rule is "Evolution is cleverer than you are!" (Dennett,
1995b, p. 74). Stark and Finke (2000) argue that many religious "reforms" deliberately and consciously executed in recent times undo the wise design work
implicit in traditional religious practices. It is a serious design error, they
argue, to make religious ritual too easy, too inexpensive, too painless.

6 The Evolution of Stewardship

1. The ethnomusicologist Jeff Todd Titon introduced me to the music of
gospel preaching in his pioneering analysis of the art of John Sherfey (Titon,
1988); you can see and hear for yourself in his documentary video, *Powerhouse
for God* (Documentary Educational Resources, 101 Morse Street, Water-
town, MA 02472). Dozens of C. L. Franklin's sermons in Detroit and Mem-
phis were recorded and broadcast nationwide by Chess Records, and are
available through various Web sites.

2. It is also *possible* that some stable elements in leks are transmitted by
imitation, not through the genes—yet another instance of animal *tradition*,
not instinct (Avital and Jablonka, 2000). Cross-fostering studies, in which
birds' eggs from one lek tradition were hatched and raised by birds with a different lek, could shed light on this.

3. Pinker, 1994; Deacon, 1997; and Jackendoff, 2002, are the most accessible recent works on this topic.

4. And, yes, the pendulum is swinging back about tans. It now emerges
that sunlight is so good for you (in moderation) that the coverup recom-
mended by many dermatologists was going too far. It's hard to keep up with

all this information, and so mostly we just don't question "what everyone knows."

5. I should emphasize this, to keep well-meaning but misguided multicul- turalists at bay: the theoretical entities in which these tribal people frankly believe—the gods and other spirits—don't exist. These people are mistaken, and you know it as well as I do. It is possible for highly intelligent people to have a very useful but mistaken theory, and *we don't have to pretend otherwise* in order to show respect for these people and their ways.

6. In an important but underappreciated discussion, Sperber (1985, pp. 49ff.) proposes that we call such indeterminate cognitive states *semi- propositional representations*. These are the "half-understood ideas" that we all use every day, and that typically get turned into proper propositional represen- tations only under the pressure of systematic inquiry. This hypothesized *folie- à-deux* process of theology generation is similar to the generate-and-test model of dream production and hallucination generation described in Den- nett, 1991a, chapter 1.

7. I'm adopting here the active voice of "selfish meme" talk; it is the same shorthand we use when we say that HIV "attacks" and "hides" and "adjusts its strategy" in response to our efforts to eradicate it. Ideas don't have minds any more than viruses or bacteria do, but they can be usefully and predictively de- scribed *as if* they were selfish and clever.

8. Many years ago, I published a paper on pain (Dennett, 1975, reprinted in 1978) that included some shocking facts about the use of *amnestics* by anes- thesiologists to wipe out postsurgical memories of pain experienced by insuf- ficiently anesthetized patients during surgery. Several anesthesiologists who read my piece in draft implored me not to publish these details in a nonmedi- cal journal, since it would make their jobs more difficult. Anything that heightens the anxiety of patients presurgically makes the induction of safe anesthesia more difficult, and hence more dangerous to them, so it is best to keep this information where it belongs: restricted to the medical community. This is the strongest case I know of a fact that people might be better off not knowing—but it was not strong enough to dissuade me. You might want to ask yourself if you would approve of the policy of doctors' having secret knowl- edge that was systematically kept from their patients, at all costs.

9. The theory that all religion is just such *Priestertrug*, deception or ma- nipulation by priests for their own benefit, has a history going back to Diderot and the Enlightenment. "Yet in spite of suspicions both ancient and modern, in spite of the unimpeachable existence of cunning and trickery among hu- mans, the hypothesis of pure deception does not explain anything," Burkert avers (1996, p. 118), but this is too strong; it may not explain *everything*, but it explains many features of religion around the world, from psychic-healing frauds to the worst abuses of televangelism.

7   The Invention of Team Spirit

1. One tradition would speak here of "selfless" caring, but since this inevitably invites objections about the purported incoherence of true selflessness, I prefer to think of this as the possibility of extending the domain of the self. Here is one good reason: Supposedly "selfless" agents are not at all immune to the problems that bedevil the selfish agents described by economists. Say I am an agent in a bargaining situation, or in a prisoner's dilemma, or faced with a coercive offer, or an attempt at extortion. My problem is not resolved, or diminished, or even significantly adjusted, if the "self" I am protecting is other than my proper self—if I am not just trying to save my own skin, so to speak. An extortionist or a benefactor who knows what I care about is in a position to frame the situation to hit me where it matters to me, whatever matters to me. (Material in this note and the text paragraph to which it is keyed is drawn from Dennett, 2001b and 2003b.)

2. Manji provides a telling example: the deliberate squelching of *ijtihad*, the Muslim tradition of inquiry that flourished until the tenth century (and accounted for the glorious intellectual and artistic achievements of early Islam).

> In the guise of protecting the world-wide Muslim nation from disunity (known as *fitna* and considered a crime), Baghdad-approved scholars formed a consensus to freeze debate within Islam. These scholars benefited from patronage and weren't about to chirp an ode to openness when their masters wanted harsher lyrics. . . . The only thing this imperial strategy has achieved is to spawn the most dogged oppression of Muslims by Muslims: the incarceration of interpretation. [2003, p. 59]

What has been spread, Manji notes, is the "imitation of imitation," a copy-fidelity-enhancing mechanism like those discussed in chapter 5, but in this case deliberately designed by stewards, to edit out all exploratory mutations before they can spread.

3. Wilson's book is brimming with important evidence and analyses, but one of the disappointments for evolutionary theorists is that the machinery of multilevel-selection theory, so strenuously developed and defended by Sober and Wilson in *Unto Others* (1998), is not put to use here. We never see any analyses of empirical data showing populations of groups periodically dissolving into their constituents and re-forming into groups with higher proportions of altruists, for instance. We don't see differential *group* replication at all—except for some tantalizing informal remarks late in the book on the way established religions give birth to sects. An early endnote (n. 3 on p. 14) acknowledges these complications: "If the groups remain permanently isolated from each other, the local advantage of selfishness will run its course within each group and drive altruism extinct. There must be a sense in which the groups compete with each other in the formation of new groups, although the competition need not be direct . . ."(p. 235). But that is the only place these complications are treated in the book, aside from unargued claims such as

this one: "In general, social control mechanisms do not alter the basic conclusion that group-level adaptations require a corresponding process of group selection" (p. 19). This claim is in need of more careful defense, however, and depends critically on the definition of group selection used.

4. In his list of theories on p. 45, he defines the meme theory as "1.3. Religion as a cultural 'parasite' that often evolves at the expense of human individuals and groups."

5. It is not just that many of the points Wilson makes in support of his group-selection theory can be readily translated into meme talk and used to support the meme-selection theory. Wilson acknowledges that his theory of group selection *depends* on the existence of cultural evolution:

> . . . it is important to remember that moral communities larger than a few hundred individuals are "unnatural" as far as genetic evolution is concerned because to the best of our knowledge they never existed prior to the advent of agriculture. This means that culturally evolved mechanisms are absolutely required for human society to hang together above the level of face-to-face groups. [p. 119]

And since, as Wilson notes, excellent features of one religion often get copied by other, unrelated religions, he is already committed to tracing the ease of host-hopping by innovations quite independently of any "vertical" transmission of the features to descendant *groups*. Wilson makes a variety of important points that really cannot be understood except as a tacit reversion to the "meme's-eye view," so one could view my "mild memetic alternative" as a friendly amendment, though I expect that Wilson will go on carrying the torch for group selection. That is the meme that he has devoted his career to spreading, after all.

6. The fact that the supply-side theory offends them is not in itself an argument against it, of course. Neither is the claim (which many make) that they don't consider themselves to be making rational market choices about their religion. They may be deluding themselves about their actual thought processes. But, other things being equal (which they may not be), the fact that people respond with disbelief and outrage when considering the supply-siders' theories is some evidence that the reasonableness of these theories is not as obvious as Stark and his colleagues like to claim. See Bruce, 1999, for a detailed critique of rational choice theories of religion.

7. An introductory discussion of this recent literature is given in Dennett, 2003c, chapter 7, "The Evolution of Moral Agency."

8. Quoted in Armstrong, 1979, p. 249.

9. In my terminology, gods as conscious beings are higher-order intentional systems, rational agents with whom one can converse, bargain, argue, to whom promises can be made, and from whom promises can be solicited. It is hard to imagine the point of making a promise to the Ground of All Being.

10. The models of Bowles and Gintis are about the evolution of memes within communities, though they choose not to use the term: ". . . we adopt

the evolutionary view that key to the understanding of behaviors in the kinds of social interactions we are studying is *differential replication:* durable aspects of behavior, including norms, may be accounted for by the fact that they have been copied, retained, diffused, and hence replicated, while other traits have not" (p. 347). And they go on to point out that these effects are *not* the result of group-selection mechanisms (p. 349), even though they explain the organism-like adaptations that communities exhibit.

8   Belief in Belief

1. As Richard Lewontin recently observed, "To survive, science must expose dishonesty, but every such public exposure produces cynicism about the purity and disinterestedness of the institution and provides fuel for ideological anti-rationalism. The revelation that the paradoxical Piltdown Man fossil skull was, in fact, a hoax was a great relief to perplexed paleontologists but a cause for great exultation in Texas tabernacles" (2004, p. 39).

2. For a discussion of Nietzsche and his philosophical response to Darwin's theory of evolution by natural selection, see my *Darwin's Dangerous Idea* (1995b).

3. There are significant differences in breast cancer (Li and Daling, 2003), hypertension, diabetes, alcohol tolerance, and many other well-studied conditions. For an overview, see Health Sciences Policy (HSP) Board, 2003.

4. Thomas Kuhn, in *The Structure of Scientific Revolutions* (1962), is the godfather of all the subsequent discussions, and it should be noted that Kuhn's book is perhaps the all-time champion in the category of Enthusiastically Misunderstood Classic. It's a wonderful book, in spite of all the misuse to which it's been put.

5. Newberg, D'Aquili, and Rause entitle their 2001 book *Why God Won't Go Away: Brain Science and the Biology of Belief,* and claim to show by "careful conventional science" (p. 141) the "deeper, neurobiologically endorsed assurances that make God real" (p. 164), but the God that they claim to uncover by studying the "neurology of transcendence" is something they call Absolute Unitary Being, which is so undefinable that I myself have no idea whether I believe in it. (I believe that something exists—is *that* Absolute Unitary Being?) The authors acknowledge, "If Absolute Unitary Being is real, then God, in all the personified ways humans know him, can only be a metaphor" (p. 171). In other words, there's nothing in their neuroscience that an atheist would have to disagree with.

6. In Lee Siegel's delightful novel, *Love and Other Games of Chance* (2003), there is a character who has written a best-selling religious book entitled *He's Not Called God for Nothing.* Think about it.

7. The same reluctance poisons the debates about creationism and "Intelligent Design." At one extreme there are "Young Earth" creationists, who deny that our planet is billions of years old and defend hilarious hypotheses to

explain away the fossils and all the other evidence, and then there are the somewhat more reasonable Intelligent Design advocates, who readily acknowledge the age of the planet, the fossil record, and indeed the descent from a common single-celled ancestor of all plants and animals, but still think they can prove that there is work for an Intelligent Designer to do. When pressed in private, these more sophisticated thinkers sometimes acknowledge that the Young Earth nonsense is a mixture of fantasy and fraud, but they won't say it in public. And then they complain bitterly that the scientific community ignores them: "We're *serious* about this!" they insist—"but please don't ask us to acknowledge the falsehood of the sillier versions of our position!" No. Not if you want to play in the big leagues.

8. For a survey of the state of the art circa 1980 (along with some contentious proposals of my own), see Dennett, 1982, reprinted in 1987. I recently took a brief look at the literature that has piled up on the topic since then, and concluded that the intervening quarter century of effort had not produced anything that would change my 1982 opinions substantially, but of course many philosophers would disagree vehemently.

9. Cannon, 1957, is a classic exploration of the widespread lore claiming that evil spells have actually killed people. He concludes that it is by no means impossible to induce the death of somebody by fatally unnerving him, in effect. "In his terror [the victim] refuses both food and drink, a fact which many observers have noted and which, as we shall see later, is highly significant for a possible understanding of the slow onset of weakness. The victim 'pines away'; his strength runs out like water, to paraphrase words already quoted from one graphic account; and in the course of a day or two he succumbs" (p. 186).

10. In Dennett, 1978, I proposed a distinction between beliefs and "opinions," which are (roughly) sentences one would bet on as true (even if one didn't entirely understand them). Sperber (1975) made a similar division between intuitive and reflective beliefs, and has expanded and revised this analysis in Sperber, 1996.

11. See also Palmer and Steadman, 2004, on the adaptive tactic of literalization of metaphors.

12. My introduction to this somewhat depressing idea came in 1982, when I was told by the acquisitions editor of a major paperback publishing company that her company wasn't going to bid for the paperback rights for *The Mind's I*, the anthology of philosophy and science fiction that Douglas Hofstadter and I had edited, because it was "too clear to become a cult book." I could see what she meant: we actually explained things as carefully as we could. John Searle once told me about a conversation he had with the late Michel Foucault: "Michel, you're so clear in conversation; why is your written work so obscure?" To which Foucault replied, "That's because, in order to be taken seriously by French philosophers, twenty-five percent of what you write has to be impenetrable nonsense." I have coined a term for this tactic, in honor of Foucault's candor: *eumerdification* (Dennett, 2001a).

13. Professor Faith is the successor to Otto in *Consciousnesss Explained* (1991a), and Conrad in *Freedom Evolves* (2003c), not to be identified with any actual interlocutor of mine, but expressing, as best I can muster, the objections I have often heard.

14. Philosophers have spent decades dreaming up thought experiments designed to prove or disprove W.V.O. Quine's principle of the *indeterminacy of radical translation* (1960): the surprising claim that *in principle* there could be two different ways of translating one natural language into another natural language and no evidence at all about which one was the *right* way to translate the language. (Quine insisted that in that case there wouldn't *be* a right way; each way would be as good as the other, and there would be no further fact of the matter.) The Philby case can help us see that his claim is not so incredible as it first appears, and appendix D presents a brief discussion of this point (for philosophers only, probably).

15. Philosophers will recognize this as an application of Quine's theory of meaning (1960), and an extension of his observation that in the great "web of belief," *theoretical* statements far from the periphery of empirical confirmation and disconfirmation most readily exhibit inscrutability of reference.

16. Gödel's Theorem states that if you try to axiomatize *arithmetic* (the way plane geometry is axiomatized by Euclid—remember high-school geometry?) your system of axioms will be either inconsistent (which you certainly don't want, since anything at all, falsehoods as well as truths, can be proved from inconsistent axioms) or incomplete—there will be at least one truth of arithmetic, the system's Gödel sentence, that can never be proved from your axioms. Gödel's Theorem is provable *a priori*, but to make it have any real-world application (for instance, to describe limitations on actual, implemented Turing machines), you have to add an empirical premise or two, and this is where problems of interpretation arise to confound the would-be dualist, for instance. See "The Abilities of Men and Machines," in Dennett, 1978; and the chapter on Roger Penrose in Dennett, 1995b.

17. I may be wrong, of course. There are several worthy religious critics of my book (and many desperate misrepresenters). Christian metaphysician Alvin Plantinga's negative review (1996), which is available (along with other essays on these topics) on his Web site at http://id-www.ucsb.edu/fscf/library/plantinga/dennett.html, is a good place to start, since, although he can't resist misconstruing some of my arguments, he explains very clearly the power of the Darwinian challenge to his Christianity. He, for one, has no illusions about the two *"magisteria"* of Stephen Jay Gould discussed in chapter 2. If Darwinism is right, many cherished Christian doctrines are in trouble, which is why he—a metaphysician, not a philosopher of science—takes it upon himself to endorse some of the bad arguments of the Intelligent Design community. Plantinga, in his many books and articles, has also been an indefatigable and ingenious defender of the *a priori* arguments of theology, including attempts to rebut the atheists' favorite counterargument, the Argument from Evil, which has recently been given a good rehearing in the wake

of the tsunami in the Indian Ocean. To balance Plantinga, I recommend an older book, John Mackie's *The Miracle of Theism: Arguments For and Against the Existence of God* (1982), as patient and sympathetic—but also rigorous and relentless—a treatment as I have encountered.

18. Descartes had raised the question of whether God had created the truths of mathematics. His follower Nicolas de Malebranche (1638–1715) firmly expressed the view that they needed no inception, being as eternal as anything could be.

9   Toward a Buyer's Guide to Religions

1. For a recent example, see Dupré, 2001. I would have preferred to ignore it, as I recommend, but, asked to review it, I decided to use the occasion for a scolding (Dennett, 2004). On the lamentable excesses of postmodernism, see also Dennett, 1997.

2. According to Burkert, Diagoras made the same point several millennia earlier:

> "Look at all these votive gifts," Diagoras the atheist was told in the sanctuary of Samothrace, which houses the great gods who were famous for saving people from the dangers at sea. "There would be many more votives," the atheist unflinchingly retorted, "if all those who were actually drowned at sea had had the chance to set up monuments." [1996, p. 141]

3. As discussed in chapter 7, Stark and Finke (2000) argue that costly sacrifice is actually an important *attraction* of religion, but only because "you get what you pay for," and part of what you get can be health and prosperity.

4. There has been a huge amount of research on this topic. A few of the best surveys are Ellison and Levin, 1998; Chatters, 2000; Sloan and Bagiella, 2002; and Daaleman et al., 2004.

5. In 1996, Pope John Paul II declared that "new knowledge leads us to recognize in the theory of evolution more than a hypothesis," and though many biologists were cheered by this acknowledgment of the fundamental scientific theory that unifies all of biology, they noted with dismay that he went on to insist that the transition from ape to human being involved a "transition to the spiritual" that could not be accounted for by biology:

> Consequently, theories of evolution which, in accordance with the philosophies inspiring them, consider the spirit as emerging from the forces of living matter or as a mere epiphenomenon of this matter, are incompatible with the truth about man. Nor are they able to ground the dignity of the person.... The sciences of observation describe and measure the multiple manifestations of life with increasing precision and correlate them with the time line. The moment of transition

to the spiritual cannot be the object of this kind of observation, which nevertheless can discover at the experimental level a series of very valuable signs indicating what is specific to the human being. [John Paul II, 1996]

More recently, Christoph Schönborn, the Roman Catholic cardinal archbishop of Vienna, published an op-ed essay in the *New York Times* (July 7, 2005) deploring the misrepresentation of this letter as an endorsement of evolution and emphasizing that the official position of the Roman Catholic Church is actually opposed to the neo-Darwinian theory of evolution by natural selection. The spectacle of Roman Catholic bishops and cardinals instructing the faithful on the falsehood of neo-Darwinian biology would be comical if it weren't such a clear reminder of that church's sorry history of persecution of scientists whose theories were doctrinally inconvenient.

According to Archbishop Schönborn, Catholics may use "the light of reason" to arrive at the conclusion that "evolution in the neo-Darwinian sense—an unguided, unplanned process of random variation and natural selection" is not possible, a conclusion firmly refuted by thousands of observations, experiments, and calculations by experts in biology when they use their own light of reason. So, in spite of some important concessions over the years—and an official apology to Galileo centuries after the fact—the Roman Catholic Church is still in the awkward and indefensible position of trying to lean on scientific authority when Catholics like what it concludes while flatly rejecting it when it contradicts their traditions.

10   Morality and Religion

1. Some have cited the survey work by McCleary (2003) and McCleary and Barro (2003) as demonstrating a link between belief in heaven and hell and having a strong work ethic, but other interpretations of their work have not been ruled out. Econometrics is a field in which permitted rearrangements of the data often yield strikingly different "results," so one shouldn't be surprised when theorists of different persuasions find different readings.

2. Muslim scholars disagree on the interpretation of the relevant passages of the Koran (and hadith 2,562 in the Sunan al-Tirmidhi), but the scriptural passages definitely exist, and have not been mistranslated.

3. Earlier Parliaments were held in Chicago in 1893, at the Columbian Exposition; in 1993, in Chicago; and in 1999, in Cape Town.

4. This was the headline, in Italian, of an interview with me by Giulio Giorelli published in *Corriere della Serra* in Milan in 1997. Ever since then, I have adopted it as my slogan, opening my book *Freedom Evolves* (2003c) with it.

5. For a recent attempt to exploit it, see Johnson, 1996.

11   Now What Do We Do?

1. In *Darwin's Dangerous Idea*, I joined Ronald de Sousa in disparaging philosophical theology as "intellectual tennis without a net" (1995b, p. 154), and showed why an appeal to faith is out of bounds, quite literally, in the serious game of empirical research. That passage has drawn fire from Plantinga (1996) and others, but I stand by it. Let's play real intellectual tennis: this book is my serve, and I welcome serious returns—with the net of reason always up.

2. I am proposing this in advance, with scant hope of forestalling the usual reaction: defensive sneering. Consider some of the response to Jared Diamond's new book, *Collapse* (2005), as described in the *Boston Globe* by Christopher Shea (2005):

> "He is one of those people who—I don't want to sound catty, because he is an elegant writer—is not taken seriously by most historians," says Anthony Grafton, a professor of early European history at Princeton, who deems Diamond's work "superficial." Books like "Guns, Germs, and Steel," he says, are less important for their arguments than for "showing what historians have given up"—grand, sweeping history that connects the dots created by thousands of monographs.

To me, Professor Grafton doesn't sound catty; he sounds complacent. Perhaps he and his fellow historians are underestimating the force of Diamond's "superficial" arguments. We won't know until they take them seriously enough to dispose of them properly. As the saying goes, it's a dirty job but somebody's got to do it. We evolutionists don't all have to take the creationists seriously, because some of our folks have done that job well, and we've checked it out and approved it (see note 3 to chapter 3). Once the historians have duly rebutted Diamond's theses with the same care, they can go back to ignoring his arguments, if they haven't been persuaded. For another response to a response to Diamond, see Gregg Easterbrook's review (2005) and my reply (Dennett, 2005b).

3. The researchers who have made the headlines are Michael Persinger (1987), Vilayanur Ramachandran et al. (1997; for Ramachandran's popular account see Ramachandran and Blakeslee, 1998), and Andrew Newberg and Eugene D'Aquili (Newberg et al., 2001). The prospects and shortcomings connected with this work are discussed fairly by Atran (2002, chapter 7, "Waves of Passion: The Neuropsychology of Religion"). See also Churchland, 2002, and Shermer, 2003, for good reviews of religion and the brain. The more recent book by Dean Hamer (2004) was discussed in chapter 5. There are others working on such topics, and the best of the recent work is discussed by Atran.

4. The new field of neuro-economics (e.g., Montague and Berns, 2002; Glimcher, 2003) is making progress as much because of advances in economic

thinking as because of the new neuro-imaging technology. For a discussion, see chapter 8 of Ross, 2005.

5. An initial opening into this politically delicate but biologically secure research can be found in Ewing et al., 1974; Shriver, 1997; Gill et al., 1999; Wall et al., 2003. See Duster, 2005, for a thoughtful evaluation of the pitfalls to be avoided in studying the genetic factors in human diseases.

6. See, for instance, the encyclopedic Hill and Hood, 1999, *Measures of Religiosity*, which reviews hundreds of different surveys and instruments.

7. These questions may seem too fanciful to take seriously, but they are not. Research has shown striking effects of apparently trivial differences. The news of the day does matter in some conditions (Iyengar, 1987). In a survey about personal happiness (or subjective well-being), if the telephone caller asks subjects, "How's the weather where you are?," then how the weather is *doesn't* matter; if the telephone caller *doesn't* ask this innocuous question and the weather is sunny, people say they are significantly happier! Drawing attention to the local weather makes answerers less likely to be covertly influenced by it in their responses to questions on other topics (Schwarz and Clore, 1983). For other examples, see Kahnemann et al., eds., 2000.

8. Shermer designed the study in collaboration with Frank Sulloway, a former MIT statistician and Darwin scholar and author of *Born to Rebel: Birth Order, Family Dynamics, and Creative Lives* (1996). After extensive pretesting and refining of their questionnaire, they first sent it out to the five thousand Skeptic Society members and got over seventeen hundred replies, and then they sent the same survey to a *random sample* of ten thousand people across the country and got over a thousand respondents. The statistics above are for the random sample, not the skeptics. See Shermer, 2003, for some of the details. Shermer and Sulloway, in press, is the formal presentation of the results.

9. My own foray into questionnaire design has been exploring other possible sources of distortion, such as looking at how the same questions in two different contexts (challenging and supportive) get answered differently. There are definitely significant differences, but they are not what we initially expected, and are ambiguous between several different interpretations, so we are designing follow-up studies and have not yet submitted any of our results to a peer-reviewed journal. By the way, we have attempted to answer the question raised in chapter 8 and reviewed above, regarding whether it makes a difference whether a question reads "God exists" or "I believe that God exists" (strongly agree, agree somewhat . . .). Our preliminary results suggest that this minor difference in wording does not make a difference when, for instance, the test items are "Jesus walked on water" versus "I believe Jesus walked on water." But further studies may discover a context yielding a different result.

10. Quoted in Stern, 2003, p. xiii.

11. Quoted in Manji, 2003, p. 90.

12. This paragraph and its predecessor are drawn, with revisions, from Dennett, 1999b.

13. Scott Atran has begun studying future Hamas leaders in Palestine and Gaza. See his important editorial, "Hamas May Give Peace a Chance," *New York Times*, December 18, 2004.

14. No Arabic-language publisher would dare publish a translation of Manji's book, but an Arabic translation of it is available, free, on the Web. Young Muslims all over the Arab world can download it in discreet PDF files, to be read and shared and discussed, the beginnings of what Manji calls Operation Ijtihad. *Ijtihad* means "independent thinking," and it flourished as a tradition during the greatest period of Islam, the five hundred years beginning about A.D. 750 (Manji, 2003, p. 51).

15. Irshad Manji reports seeing a sign in a new school for girls in Afghanistan: "Educate a boy and you educate only that boy, educate a girl and you educate her entire family" (speech at Tufts University, March 30, 2005).

16. A recent poll in *Newsweek* (May 24, 2004) claimed that 55 percent of Americans think that the faithful will be taken up to heaven in the Rapture and 17 percent believe the world will end in their lifetimes. If this is even close to being accurate, it suggests that End Timers in the first decade of the twenty-first century outnumber the Marxists of the 1930s through the 1950s by a wide margin. But what percentage of these adherents are prepared to take any steps, overt or covert, to hasten the imagined Armageddon is anybody's guess, I fear to say.

17. Sharlet, 2003, provides a fascinating and unsettling introduction to this little-known organization, which includes this list of congressmen (including a few who are no longer in Congress), and also describes highlights of the history of its activities around the world, which include the National Prayer Breakfasts but also the covert support of political leaders and movements. Its current leader, Douglas Coe, is described by *Time* magazine (February 7, 2005, p. 41) as "the Stealth Persuader." Sharlet comments:

> At the 1990 National Prayer Breakfast, George H.W. Bush praised Doug Coe for what he described as "quiet diplomacy, I wouldn't say secret diplomacy," as an "ambassador of faith." Coe has visited nearly every world capital, often with congressmen at his side, "making friends" and inviting them back to the Family's unofficial headquarters, a mansion (just down the road from Ivanwald) that the Family bought in 1978 with $1.5 million donated by, among others, Tom Phillips, then the C.E.O. of arms manufacturer Raytheon, and Ken Olsen, the founder and president of Digital Equipment Corporation. [p. 55]

I think we need to know more about the activities of this quiet, nongovernmental diplomacy, since it may be pursuing policies that are antithetical to those of the democracy of which these congressmen are elected representatives.

18. We also need to keep *ourselves* informed, and this is becoming more difficult, oddly enough. We used to think that secrecy was perhaps the greatest enemy of democracy, and as long as there was no suppression or censorship, people could be trusted to make the informed decisions that would preserve

our free society, but we have learned in recent years that the techniques of misinformation and misdirection have become so refined that, even in an open society, a cleverly directed flood of misinformation can overwhelm the truth, even though the truth is out there, uncensored, quietly available to anyone who can find it. For instance, I do not fear that this book will be censored or suppressed, but I do anticipate that it (and I) will be subjected to ruthless misrepresentation when those who cannot honestly face its contents seek to poison the minds of readers to it or direct attention away from it. In my recent experience, even some respectable academics have been unable to resist the temptation to do this (Dennett, 2003e). Relying on that experience, I have made a list of the passages in this book most likely to be ripped out of context and used deliberately to misrepresent my position. This is not the first time I have done this. In *Consciousness Explained*, I provided a premonitory footnote to a passage on zombies (don't ask; you don't want to know), asserting, "It would be an act of desperate intellectual dishonesty to quote this assertion out of context!" (1991a, p. 407n), and, sure enough, several authors could not resist quoting it out of context—but at least they had to quote the footnote, too, not being *quite* that desperate or dishonest. In this case, stronger measures are called for, since the stakes are higher, so I am keeping my list of predicted deliberate misrepresentations sealed and ready to release. For instance, which of my little jokes, quite innocuous in context, will be brandished to demonstrate my "intolerance," my "disrespect," my anti-Christian, anti-Semitic, anti-Muslim "bias"? (As all you careful readers know full well, I am an equal-opportunity teaser, who refuses to tiptoe around for fear of offending people—because I want to take the "I'm mortally offended" card out of the game.) It will be interesting to see who, if anyone, falls into my trap. They won't be assiduous note readers, will they?

Appendix B   Some More Questions About Science

1. William Dembski, the author of numerous books and articles attacking evolutionary theory, often complains loudly that his "scientific" work is not treated with respect by working biologists. As the coeditor of *Unapologetic Apologetics: Meeting the Challenges of Theological Studies* (2001), he can find the reason for this in his own practices. For a detailed critique of Dembski's methods, see the Web site of Thomas Schneider, http://www.lecb.ncifcrf.gov/~toms/paper/ev/.

2. This paragraph is drawn from Dennett, 2003c, p. 303.

3. Most of this huge increase in mass relative to "wild" nature is due to our livestock and pets, which now outweigh us in total by more than three to one. It is hard to estimate the ratio of domesticated plants to wild plants, but of course that ratio has changed dramatically as well.

Bibliography

||||\\\||||

Abed, Riadh, 1998, "The Sexual Competition Hypothesis for Eating Disorders." *British Journal of Medical Psychology*, vol. 17, no. 4, pp. 525–47.

Ainslie, George, 2001, *Breakdown of Will*. Cambridge: Cambridge University Press.

———, in press, "*Précis* of *Breakdown of Will*." Target article for *Behavioral and Brain Sciences*.

Anderson, Carl J., and Norman M. Prentice, 1994, "Encounter with Reality: Children's Reactions on Discovering the Santa Claus Myth." *Child Psychiatry & Human Development*, vol. 25, no. 2, pp. 67–84.

Andresen, Jensine, 2001, *Religion in Mind: Cognitive Perspectives on Religious Belief, Ritual, and Experience*. Cambridge: Cambridge University Press.

Armstrong, Ben, 1979, *The Electric Church*. New York: Thomas Nelson.

Armstrong, Karen, 1993, *A History of God: The 4000 Year Quest for Judaism, Christianity and Islam*. New York: Ballantine Books.

Ashbrook, James B., and Carol Rausch Albright, 1997, *The Humanizing Brain*. Cleveland, Ohio: Pilgrim Press.

Atran, Scott, 2002, *In Gods We Trust: The Evolutionary Landscape of Religion*. Oxford: Oxford University Press.

———, 2004, "Hamas May Give Peace a Chance." *New York Times*, December 18.

Atran, S., and A. Norenzayan, 2004, "Religion's Evolutionary Landscape: Counterintuition, Commitment, Compassion, Communion." *Behavioral and Brain Sciences*, vol. 27, pp. 713–70.

Auden, W. H., 1946, "A Reactionary Tract for the Times," Phi Beta Kappa poem, Harvard.

Aunger, Robert, 2002, *The Electric Meme: A New Theory of How We Think and Communicate*. New York: Free Press.

———, ed., 2000, *Darwinizing Culture: The Status of Memetics as a Science*. Oxford: Oxford University Press.

Avital, Eytan, and Eva Jablonka, 2000, *Animal Traditions: Behavioural Inheritance in Evolution*. Cambridge: Cambridge University Press.

Awai, Katsuhito, 2001, "The Evolution of Money." In Antonio Nicita and Ugo Pagano, eds., *The Evolution of Economic Diversity*. London: Routledge, pp. 396–431.

Balaschak, B., K. Blocker, T. Rossiter, and C. T. Perin, 1972, "The influence of race and expressed experience of the hypnotist on hypnotic susceptibility." *International Journal of Clinical & Experimental Hypnosis*, vol. 20, no. 1, pp. 38–45.

Balkin, J. M., 1998, *Cultural Software: A Theory of Ideology*. New Haven: Yale University Press.

Bambrough, Renford, 1980, "Editorial: Subject and Epithet." *Philosophy*, vol. 55, pp. 289–90.

Barna, George, 1999, "Christians Are More Likely to Experience Divorce Than Are Non-Christians." Barna Research Group, 1999-DEC-21. Available at http://www.barna.org/cgi-bin/.

Baron-Cohen, Simon, 1995, *Mindblindness and the Language of the Eyes: An Essay in Evolutionary Psychology.* Cambridge, Mass.: MIT Press.

Barrett, David, George Kurian, and Todd Johnson, 2001, *World Christian Encyclopedia.* 2nd ed. New York: Oxford University Press, 2 vols.

Barrett, Justin, 2000, "Exploring the Natural Foundations of Religion." *Trends in Cognitive Science,* vol. 4, pp. 29–34.

Barth, Fredrik, 1975, *Ritual and Knowledge Among the Baktaman of New Guinea.* New Haven: Yale University Press.

Bering, J. M., 2004, "Natural Selection Is Non-denominational: Why Evolutionary Models of Religion Should Be More Concerned with Behavior Than Concepts." *Evolution and Cognition,* vol. 10, pp. 126–37.

Bierce, Ambrose, 1911, *The Devil's Dictionary.* Copyright expired; available at http://www.alcyone.com/max/lit/devils/.

Blackmore, Susan, 1999, *The Meme Machine.* Oxford: Oxford University Press.

Bonner, John Tyler, 1980, *The Evolution of Culture in Animals.* Princeton: Princeton University Press.

Bowles, Samuel, and Herbert Gintis, 1998, "The Moral Economy of Community: Structured Populations and the Evolution of Prosocial Norms." *Evolution and Human Behavior,* vol. 19, pp. 3–25.

———, 2001, "Community Governance." In Antonio Nicita and Ugo Pagano, eds., *The Evolution of Economic Diversity.* London: Routledge, pp. 344–67.

Boyd, Robert, and Peter Richerson, 1985, *Culture and the Evolutionary Process.* Chicago: University of Chicago Press.

———, 1992, "Punishment Allows the Evolution of Cooperation (or Anything Else) in Sizable Groups." *Ethology and Sociobiology,* vol. 13, pp. 171–95.

Boyer, Pascal, 2001, *Religion Explained: The Evolutionary Origins of Religious Thought.* New York: Basic Books.

Boyer, Peter J., 2003, "The Jesus War." *The New Yorker,* September 15.

Brodie, Richard, 1996, *Virus of the Mind: The New Science of the Meme.* Seattle: Integral Press.

Brown, Dan, 2003, *The Da Vinci Code.* New York: Doubleday.

Brown, David, 2004, "Wildlife Tracking on AVIS Lands." Andover, Mass. March 9.

Bruce, Steve, 1999, *Choice and Religion: A Critique of Rational Choice Theory.* Oxford: Oxford University Press.

Bulbulia, Joseph, 2004, "Religious Costs as Adaptations That Signal Altruistic Intention." *Evolution and Cognition,* vol. 19, pp. 19–42.

Burdett, Kenneth, Alberto Trejos, and Randall Wright, 2001, "Cigarette Money." *Journal of Economic Theory,* vol. 99, pp. 117–42.

Burkert, Walter, 1996, *Creation of the Sacred: Tracks of Biology in Early Religions.* Cambridge, Mass.: Harvard University Press.

Cannon, Walter B., 1957, "'Voodoo' Death." *Psychosomatic Medicine,* vol. 19, pp. 182–90.

Carey, Benedict, 2004, "Can Prayers Heal? Critics Say Studies Go Past Science's Reach." *New York Times,* October 10.

Carstairs-McCarthy, Andrew, 1999, *The Origins of Complex Language: An Inquiry into the Evolutionary Beginnings of Sentences, Syllables, and Truth.* Oxford: Oxford University Press.

Cavalli-Sforza, Luigi Luca, 2001, *Genes, Peoples, and Languages.* Berkeley: University of California Press.

Cavalli-Sforza, Luigi Luca, and Marcus Feldman, 1981, *Cultural Transmission and Evolution: A Quantitative Approach.* Princeton: Princeton University Press.

Chatters, Linda M., 2000, "Religion and Health: Public Health Research and Practice." *Annual Review of Public Health*, vol. 21, pp. 335–67.

Christiansen, Morten H., and Simon Kirby, eds., 2003, *Language Evolution (Studies in the Evolution of Language)*. Oxford: Oxford University Press.

Churchland, Patricia, 2002, *Brain-Wise: Studies in Neurophilosophy*. Cambridge, Mass.: MIT Press.

Cloak, F. T., 1975, "Is a Cultural Ethology Possible?" *Human Ecology*, vol. 3, pp. 161–82.

Coe, William C., John R. Bailey, John C. Hall, Mark L. Howard, Robert L. Janda, Ken Kobayashi, and Michael D. Parker, 1970, "Hypnotism as Role Enactment: The Role-Location Variable." *Proceedings: 78th Annual Convention, American Psychological Association*, pp. 839–40.

Coe, William C., et al., 2001, "Hypnosis as Role Enactment: The Role-Location Variable." *Proceedings of the Annual Convention of the American Psychological Association*.

Colvin, J. Randall, and Jack Block, 1994, "Do Positive Illusions Foster Mental Health? An Examination of the Taylor and Brown Formulation." *Psychological Bulletin*, vol. 116, no. 1, pp. 3–20.

Crick, Francis H. C., 1968, "The Origin of the Genetic Code." *Journal of Molecular Biology*, vol. 38, p. 367.

Cronin, Helena, 1991, *The Ant and the Peacock*. Cambridge: Cambridge University Press.

Cronk, Lee, Napoleon Chagnon, and William Irons, 2000, *Adaptation and Human Behavior: An Anthropological Perspective*. Hawthorne, N.Y.: De Gruyter.

Cupitt, Don, 1997, *After God: The Future of Religion*. New York: Basic Books.

Daaleman, Timothy P., Subashan Perera, and Stephanie A. Studenski, 2004, "Religion, Spirituality, and Health Status in Geriatric Outpatients."*Annals of Family Medicine*, vol. 2, pp. 49–53.

Darwin, Charles, 1859, *On the Origin of Species by Means of Natural Selection*. London: Murray.

——, 1886, *The Descent of Man, and Selection in Relation to Sex*. Princeton: Princeton University Press, 1981 (originally published 1871).

Davies, Paul, 2004, "Undermining Free Will." *Foreign Policy*, September/October, p. 36.

Dawkins, Richard, 1976, *The Selfish Gene*. Oxford: Oxford University Press. Rev. ed. 1989.

——, 1982, *The Extended Phenotype*. Oxford: Oxford University Press, paperback, 1999.

——, 1989 [rev. ed. of Dawkins, 1976]. Oxford: Oxford University Press.

——, 1993, "Viruses of the Mind." In Bo Dahlbom, ed., *Dennett and His Critics*. Oxford: Blackwell, pp. 13–27. Reprinted in Dawkins, 2003a.

——, 1996, *Climbing Mount Improbable*. London: Viking Penguin.

——, 2003a, *A Devil's Chaplain: Reflections on Hope, Lies, Science, and Love*. Boston: Houghton Mifflin.

——, 2003b, "The Future Looks Bright." *Guardian*, June 21.

——, 2004a, *The Ancestor's Tale: A Pilgrimage to the Dawn of Life*. London: Weidenfeld & Nicolson.

——, 2004b, "What Use Is Religion? Part 1." *Free Inquiry*, June/July, pp. 13ff. (not consecutive pages).

Deacon, Terry, 1997, *The Symbolic Species*. New York: Norton.

Debray, Régis, 2004, *God: An Itinerary*. Trans. Jeffrey Mehlman. London: Verso.

Dembski, William, 1998, *The Design Inference: Eliminating Chance Through Small Probabilities*. Cambridge: Cambridge University Press.

———, 2003, "Three Frequently Asked Questions About Intelligent Design." Available at http://www.designinference.com.

Dembski, William, and Jay Wesley Richards, eds., 2001, *Unapologetic Apologetics: Meeting the Challenges of Theological Studies*. Downers Grove, Ill.: InterVarsity.

Dennett, Daniel C., 1971, "Intentional Systems." *Journal of Philosophy*, vol. 68, pp. 87–106.

———, 1978, *Brainstorms*. Cambridge, Mass.: MIT Press/A Bradford Book.

———, 1981, "Three Kinds of Intentional Psychology." In R. Healey, ed., *Reduction, Time and Reality*. Cambridge: Cambridge University Press, pp. 37–61.

———, 1982, "Beyond Belief." In A. Woodfield, ed., *Thought and Object: Essays on Intentionality*. Oxford: Oxford University Press. Reprinted 1987.

———, 1983, "Intentional Systems in Cognitive Ethology: The 'Panglossian Paradigm' Defended." *Behavioral and Brain Sciences*, vol. 6, pp. 343–90.

———, 1986, "Information, Technology, and the Virtues of Ignorance." *Daedalus: Proceedings of the American Academy of Arts and Sciences*, vol. 115, pp. 135–53.

———, 1987, *The Intentional Stance*. Cambridge, Mass.: MIT Press.

———, 1988, "The Moral First Aid Manual." In S. M. McMurrin, ed., *Tanner Lectures on Human Values*, vol. 8. Salt Lake City: University of Utah Press; and Cambridge: Cambridge University Press, pp. 119–47. Reprinted, with revisions, in Dennett, 1995b.

———, 1990, "Abstracting from Mechanism" (reply to de Gelder). *Behavioral and Brain Sciences*, vol. 13, pp. 583–84.

———, 1991a, *Consciousness Explained*. Boston: Little Brown.

———, 1991b, "Real Patterns." *Journal of Philosophy*, vol. 87, pp. 27–51.

———, 1991c, "Two Contrasts: Folk Craft Versus Folk Science, and Belief Versus Opinion." In J. D. Greenwood, ed., *The Future of Folk Psychology: Intentionality and Cognitive Science*. Cambridge: Cambridge University Press.

———, 1995a, "Animal Consciousness: What Matters and Why." *Social Research*, vol. 62, no. 3. Fall (1), pp. 691–710. Reprinted in Dennett, 1998a.

———, 1995b, *Darwin's Dangerous Idea*. New York: Simon & Schuster.

———, 1996, *Kinds of Minds: Towards an Understanding of Consciousness*. New York: Basic Books.

———, 1997. "Appraising Grace: What Evolutionary Good Is God?" (review of Walter Burkert, *Creation of the Sacred: Tracks of Biology in Early Religions*). *Sciences*, January/February, pp. 39–44.

———, 1998a, *Brainchildren: Essays on Designing Minds*. Cambridge, Mass.: MIT Press.

———, 1998b, "The Evolution of Religious Memes: Who—or What—Benefits?" *Method & Theory in the Study of Religion*, vol. 10, pp. 115–28.

———, 1999a, "Faith in the Truth." In W. Williams, ed., *The Values of Science* (The Amnesty Lectures, Oxford, 1997). New York: Basic Books, pp. 95–109. Also in *Free Inquiry*, Spring 2000.

———, 1999b, "Protecting Public Health." In *Predictions: 30 Great Minds on the Future*, published by *Times Higher Education Supplement*, pp. 74–75.

———, 2000, "Making Tools for Thinking." In D. Sperber, ed., *Metarepresentations: A Multidisciplinary Perspective*. New York: Oxford University Press, pp. 17–29.

———, 2001a, "Collision, Detection, Muselot, and Scribble: Some Reflections on Creativity." In David Cope, *Virtual Music: Computer Synthesis of Musical Style*. Cambridge, Mass.: MIT Press, pp. 283–91.

———, 2001b, "The Evolution of Culture." *Monist*, vol. 84, no. 3, pp. 305–24.

———, 2001c, "The Evolution of Evaluators." In Antonio Nicita and Ugo Pagano, eds., *The Evolution of Economic Diversity*. London: Routledge, pp. 66–81.

———, 2002a, "Altruists, Chumps, and Inconstant Pluralists" (commentary on Sober and Wilson, *Unto Others: The Evolution and Psychology of Unselfish Behavior*). *Philosophy and Phenomenological Research*, vol. 65, no. 3 (November), pp. 692–96.

———2002b, "The New Replicators." In Mark Pagel, ed., *Encyclopedia of Evolution*, vol. 1. Oxford: Oxford University Press, pp. E83–E92.

———, 2003a, "The Baldwin Effect: A Crane, Not a Skyhook." In B. H. Weber and D. J. Depew, eds., *Evolution and Learning: The Baldwin Effect Reconsidered*. Cambridge, Mass.: MIT Press/A Bradford Book, pp. 60–79.

———, 2003b, "The Bright Stuff." *New York Times*, July 12.

———, 2003c, *Freedom Evolves*. New York: Viking Penguin.

———, 2003d, "Postscript on the Baldwin Effect and Niche Construction." In B. H. Weber and D. J. Depew, eds., *Evolution and Learning: The Baldwin Effect Reconsidered*. Cambridge, Mass.: MIT Press/A Bradford Book, pp. 108–9.

———, 2003e, "Shame on Rea." Available at http://ase.tufts.edu/cogstud/papers/rearesponse.htm.

———, 2004, "Holding a Mirror Up to Dupré" (commentary on John Dupré, *Human Nature and the Limits of Science*). *Philosophy and Phenomenological Research*, vol. 69, no. 2 (September), pp. 473–83.

———, 2005a, "From Typo to Thinko: When Evolution Graduated to Semantic Norms." In S. Levinson and P. Jaisson, eds., *Culture and Evolution*. Cambridge, Mass.: MIT Press.

———, 2005b, "Geography Lessons." *New York Times Book Review*, February 20, p. 6.

———, 2005c, *Sweet Dreams: Philosophical Obstacles to a Science of Consciousness*. Cambridge, Mass.: MIT Press.

De Vries, Peter, 1958, *The Mackerel Plaza*. Boston: Little Brown.

Diamond, Jared, 1997, *Guns, Germs, and Steel: The Fates of Human Societies*. New York: Norton.

———, 2005, *Collapse: How Societies Choose to Fail or Succeed*. New York: Viking Penguin.

Dulles, Avery Cardinal, 2004, "The Rebirth of Apologetics." *First Things*, vol. 143 (May), pp. 18–23.

Dunbar, Robin, 2004, *The Human Story: A New History of Mankind's Evolution*. London: Faber & Faber.

Dupré, John, 2001, *Human Nature and the Limits of Science*. Oxford: Clarendon Press.

Durham, William, 1992, *Coevolution: Genes, Culture and Human Diversity*. Stanford, Calif.: Stanford University Press.

Durkheim, Emil, 1915, *The Elementary Forms of the Religious Life*. New York: Free Press.

Dusek, J. A., J. B. Sherwood, R. Friedman, P. Myers, C. F. Bethea, S. Levitsky, P. C. Hill, M. K. Jain, S. L. Kopecky, P. S. Mueller, P. Lam, H. Benson, and P. L. Hibberd, 2002, "Study of the Therapeutic Effects of Intercessory Prayer (STEP): Study Design and Research Methods." *American Heart Journal*, vol. 143, no. 4, pp. 577–84.

Duster, Troy, 2005, "Race and Reification in Science." *Science*, vol. 307, pp. 1050–51.

Eagleton, Terry, 1991, *Ideology: An Introduction*. London: Verso.

Easterbrook, Gregg, 2005, "There Goes the Neighborhood" (review of Diamond, 2004). *New York Times Book Review*, January 30.

Eliade, Mircea, 1963, *Myth and Reality*. Trans. W. R. Trask. New York: Harper and Row.

Ellis, Fiona, 2004, review of A. C. Grayling, *What Is Good? The Search for the Best Way to Live. Times Literary Supplement*, March 26, p. 29.

Ellison, C. G., and J. S. Levin, 1998, "The Religion-Health Connection: Evidence, Theory, and Future Directions." *Health Education and Behavior*, vol. 25, no. 6, pp. 700–720.

Evans-Pritchard, Edward, 1937, *Witchcraft, Oracles and Magic Among the Azande.* Oxford: Clarendon Press; 2nd ed., abridged, 1976.

Ewing, J. A., B. A. Rouse, and E. D. Pellizzari, 1974, "Alcohol Sensitivity and Ethnic Background." *American Journal of Psychiatry*, vol. 131, pp. 206–10.

Faber, M. D., 2004, *The Psychological Roots of Religious Belief.* Amherst, N.Y.: Prometheus Books.

Feibleman, James, 1973, *Understanding Philosophy.* New York: Horizon.

Feynman, Richard P., 1985, *QED: The Strange Theory of Light and Matter.* Princeton: Princeton University Press.

Flamm, Bruce, 2004, "The Columbia University 'Miracle' Study: Flawed and Fraud." *Skeptical Inquirer*, September/October, pp. 25–31.

Frank, Robert, 1988, *Passions Within Reason: The Strategic Role of the Emotions.* New York: Norton.

———, 2001, "Cooperation Through Emotional Commitment." In R. Nesse, ed., pp. 57–76.

Freud, Sigmund, 1927, *The Future of an Illusion.* New York: Norton, 1989.

Fry, Christopher, 1950, *The Lady's Not for Burning.* New York: Oxford University Press (play first produced 1948).

Gauchet, Marcel, 1997, *The Disenchantment of the World: A Political History of Religion.* Trans. Oscar Burge. Princeton: Princeton University Press.

Geertz, Clifford, 1973, *The Interpretation of Cultures.* New York: Basic Books.

Glimcher, Paul, 2003, *Decisions, Uncertainty and the Brain.* Cambridge, Mass.: MIT Press.

Gopnik, Alison, and Andy Meltzoff, 1997, *Words, Thoughts and Theories.* Cambridge, Mass.: MIT Press.

Gould, Stephen Jay, 1980, *The Panda's Thumb: More Reflections in Natural History.* New York: Norton.

———, 1999, *Rocks of Ages: Science and Religion in the Fullness of Life.* New York: Ballantine.

Grandin, Temple, 1996, *Thinking in Pictures: And Other Reports from My Life with Autism.* New York: Vintage.

Grandin, Temple, and Margaret M. Scariano, 1996, *Emergence: Labeled Autistic.* New York: Warner Books.

Gray, Russell D., and Fiona M. Jordan, 2000, "Language Trees Support the Express-Train Sequence of Austronesian Expansion." *Nature*, vol. 405 (June 29), pp. 1052–55.

Grice, H. P., 1957, "Meaning." *Philosophical Review*, vol. 66, pp. 377–88.

———, 1969, "Utterer's Meaning and Intentions." *Philosophical Review*, vol. 78, pp. 147–77.

Gross, Paul R., and Norman Levitt, 1998, *Higher Superstition: The Academic Left and Its Quarrels with Science.* Baltimore: Johns Hopkins University Press.

Guthrie, Stuart, 1993, *Faces in the Clouds.* Oxford: Oxford University Press.

Hamer, Dean, 2004, *The God Gene: How Faith Is Hardwired into Our Genes.* New York: Doubleday.

Harris, Marvin, 1993, *Culture, People, Nature: An Introduction to General Anthropology.* New York: HarperCollins.

Harris, Sam, 2004, *The End of Faith: Religion, Terrorism and the Future of Reason*. New York: Norton.

Hauser, Marc, 1996, *The Evolution of Communication*. Cambridge, Mass.: MIT Press.

———, 2000, *Wild Minds: What Animals Really Think*. New York: Henry Holt.

Health Sciences Policy (HSP) Board, 2003, *Unequal Treatment: Confronting Racial and Ethnic Disparities in Health Care*. Washington, D.C.: National Academies Press.

Heath, Chip, Chris Bell, and Emily Sternberg, 2001, "Emotional Selection in Memes: The Case of Urban Legends." *Journal of Personality and Social Psychology*, vol. 81, pp. 1028–41.

Hill, Peter C., and Ralph W. Hood, Jr., 1999, *Measures of Religiosity*. Birmingham, Ala.: Religious Education Press.

Hinde, Robert A., 1999, *Why Gods Persist: A Scientific Approach to Religion*. London: Routledge.

Hooper, Lora V., Lynn Bry, Per G. Falk, and Jeffrey I. Gordon, 1998, "Host-Microbial Symbiosis in the Mammalian Intestine: Exploring an Internal Ecosystem." *Bio-Essays*, vol. 20, no. 4, pp. 336–43.

Hopson, J. A., 1977, "Relative Brain Size and Behavior in Archosaurian Reptiles." *Annual Review of Ecology and Systematics*, vol. 8, pp. 429–48.

Horner, J. R., 1984, "The Nesting Behavior of Dinosaurs." *Scientific American*, vol. 250, no. 4, pp. 30–137.

Hubbard, Lafayette Ronald, 1950, *Dianetics: The Modern Science of Mental Health*. Los Angeles: American Saint Hill Organization.

Hull, David, 1988, *Science as a Process*. Chicago: University of Chicago Press.

Hume, David, 1777, *The Natural History of Religion*. Ed. H. E. Root. Stanford, Calif.: Stanford University Press, 1957 (originally composed 1757, but published posthumously 1777).

Humphrey, Nicholas, 1978, "Nature's Psychologists." *New Scientist*, vol. 29 (June), pp. 900–904.

———, 1995, *Soul Searching: Human Nature and Supernatural Belief*. London: Chatto and Windus. (Published in U.S.A. as *Leaps of Faith: Science, Miracles, and the Search for Supernatural Consolation* [New York: Copernicus, 1999].)

———, 1999, "What Shall We Tell the Children?" In Wes Williams, ed., *The Values of Science: Oxford Amnesty Lectures 1997*. Boulder, Colo.: Westview Press.

———, 2002, "Great Expectations: The Evolutionary Psychology of Faith Healing and the Placebo Effect." In *The Mind Made Flesh: Essays from the Frontiers of Evolution and Psychology*. Oxford: Oxford University Press, pp. 255–85.

———, 2004, contribution to the World Question Center (the 2004 Annual Edge Question: "What's your law?"), http: //www.edge.org/q2004/q04_print1.html.

Iannacone, L., 1992, "Sacrifice and Stigma: Reducing Free-Riding in Cults, Communes, and Other Collectives." *Journal of Political Economy*, vol. 100, pp. 271–91.

———, 1994, "Why Strict Churches Are Strong." *American Journal of Sociology*, vol. 99, pp. 1180–1211.

Irons, William, 2001, "Religion as a Hard-to-Fake Sign of Commitment." In R. Nesse, ed., pp. 292–309.

Iyengar, S., 1987, "Television News and Citizens' Explanations of National Affairs." *American Political Science Review*, vol. 81, pp. 815–31.

Jackendoff, Ray, 2002, *Foundations of Language: Brain, Meaning, Grammar, Evolution*. New York: Oxford University Press.

James, William, 1902, *The Varieties of Religious Experience*. Ed. Martin Marty. New York: Penguin, 1982.

Jansen, Johannes J. G., 1997, *The Dual Nature of Islamic Fundamentalism*. Ithaca, N.Y.: Cornell University Press.

Jaynes, Julian, 1976, *The Origins of Consciousness in the Breakdown of the Bicameral Mind*. Boston: Houghton Mifflin.

John Paul II, 1996, "Truth Cannot Contradict Truth." Address of the Pope to the Pontifical Academy of Sciences, October 22.

Johnson, Philip, 1996, "Daniel Dennett's Dangerous Idea." *New Criterion*, vol. 15, no. 2 (October). Available at http://www.newcriterion.com/archive/14/oct95/dennett.htm.

Kahnemann, Daniel, Ed Diener, and Norbert Schwarz, eds., 2000, *Well-Being: The Foundations of Hedonic Psychology*. New York: Russell Sage Foundation/MIT Press.

Kinsey, Alfred C., 1948, *Sexual Behavior in the Human Male*. Philadelphia: W. B. Saunders.

———, 1953, *Sexual Behavior in the Human Female*. Philadelphia: W. B. Saunders.

Klostermaier, Klaus K., 1994, *A Survey of Hinduism*. 2nd ed. Albany, N.Y.: SUNY Press.

Kluger, Jeffrey, 2004, "Is God in Our Genes?" *Time* magazine, October 25, pp. 62–72.

Koenig, Harold, Michael E. McCullough, and Donald B. Larson, 2000, *Handbook of Religion and Health*. Oxford: Oxford University Press.

Koestler, Arthur, 1959, *The Sleepwalkers*. London: Hutchinson.

Kohn, Marek, 1999, *As We Know It: Coming to Terms with an Evolved Mind*. London: Granta Books.

Kuhn, Thomas, 1962, *The Structure of Scientific Revolutions*. Chicago: University of Chicago Press.

Lakoff, George, 2004, *Don't Think of an Elephant! Know Your Values and Frame the Debate*. White River Junction, Vt.: Chelsea Green Publishing.

Laland, Kevin, and Gillian Brown, 2002, *Sense and Nonsense: Evolutionary Perspectives on Human Behaviour*. Oxford: Oxford University Press.

Lawson, E. Thomas, and Robert N. McCauley, 1990, *Rethinking Religion: Connecting Cognition and Culture*. Cambridge: Cambridge University Press.

———, 2002, *Bringing Ritual to Mind: Psychological Foundations of Cultural Forms*. Cambridge: Cambridge University Press.

Lehner, Philip N., 1978a, "Coyote Communication." In Marc Bekoff, ed., *Coyotes: Biology, Behavior, and Management*. New York: Academic Press.

———, 1978b, "Coyote Vocalizations: A Lexicon and Comparisons with Other Canids." *Animal Behavior*, vol. 26, pp. 712–22.

Leslie, Alan, 1987, "Pretense and Representation: The Origins of 'Theory of Mind.'" *Psychological Review*, vol. 94, pp. 412–26.

Lewis, C. S., 1952, *Mere Christianity*. San Francisco: HarperCollins, 2001.

Lewis, David, 1974, "Radical Interpretation," in *Synthese*, vol. 27, pp. 331–44.

Lewontin, Richard, 2004, "Dishonesty in Science." *New York Review of Books*, November 18, pp. 38–40.

Li, C., K. Malone, and J. Daling, 2003, "Differences in Breast Cancer Stage, Treatment, and Survival by Race and Ethnicity." *Archives of Internal Medicine*, vol. 163, pp. 49–56.

LoBue, Carl P., and Michael A. Bell, 1993, "Phenotypic Manipulation by the Cestode Parasite *Schistocephalus Solidus* of Its Intermediate Host, *Gasterosteus aculeatus*, the Threespine Stickleback." *American Naturalist*, vol. 142, pp. 725–35.

Longman, Robert, 2000, "Intercessory Prayer." Available at http://www.spirithome.com/prayintr.html.

Lorenz, K. Z., 1950, "The Comparative Method in Studying Innate Behavior Patterns."

In J. G. Danielli and R. Brown, eds., *Physiological Mechanisms in Animal Behavior.* Cambridge: Cambridge University Press.

Lovelock, J. E., 1979, *Gaia.* Oxford: Oxford University Press.

Lung, S.-C. C., M.-C. Kao, and S.-C. Hu, 2003, "Contribution of Incense Burning to Indoor PM10 and Particle-Bound Polycyclic Aromatic Hydrocarbons Under Two Ventilation Conditions." *Indoor Air,* vol. 13, p. 194.

Lynch, Aaron, 1996, *Thought Contagion: How Belief Spreads Through Society.* New York: Basic Books.

Maalouf, Amin, 2001, *In the Name of Identity: Violence and the Need to Belong.* New York: Arcade Publishing.

McCleary, Rachel M., 2003, "Salvation, Damnation, and Economic Incentives." Project on Religion, Political Economy and Society, working paper no. 39. Cambridge, Mass.: Weatherhead Center, Harvard University.

McCleary, Rachel M., and Robert J. Barro, 2003, "Religion and Economic Growth." Harvard University working paper, available at http://econ.korea.ac.kr/bk21/notice/uploads/Religion_and_Economic_Growth.pdf.

McClenon, James, 2002, *Wondrous Healing: Shamanism, Human Evolution and the Origin of Religion.* DeKalb: Northern Illinois University Press.

MacCready, Paul, 2004, "The Case for Battery Electric Vehicles." In Daniel Sperling and James Cannon, eds., *The Hydrogen Energy Transition.* New York: Academic Press, pp. 227–33.

Mackie, J. L., 1982, *The Miracle of Theism: Arguments For and Against the Existence of God.* Oxford: Oxford University Press.

Mahadevan, P., and Frits Staal, 2003, "The Turning-Point in a Living Tradition" *Electronic Journal of Vedic Studies,* vol. 10. Available at http://users.primushost.com/~india/ejvs.

Manji, Irshad, 2003, *The Trouble with Islam.* New York: St. Martin's.

Masters, William H., and Virginia Johnson, 1966, *Human Sexual Response.* New York: Lippincott/Williams & Wilkins.

Maynard Smith, John, 1977, "Parental Investment: A Prospective Analysis." *Animal Behaviour,* vol. 25, pp. 1–9.

———, 1978, *The Evolution of Sex.* Cambridge: Cambridge University Press.

Mayr, Ernst, 1982, *The Growth of Biological Thought.* Cambridge, Mass.: Harvard University Press.

———, 2004, *What Makes Biology Unique? Considerations on the Autonomy of a Scientific Discipline.* Cambridge: Cambridge University Press.

Melton, J. Gordon, 1998, *Encyclopedia of American Religions.* 6th ed. Detroit: Gale Group.

Miller, Geoffrey, 2000, *The Mating Mind: How Sexual Choice Shaped the Evolution of Human Nature.* New York: Doubleday.

Mithen, Steven, 1996, *The Prehistory of the Mind: The Cognitive Origins of Art, Religion and Science.* London: Thames and Hudson.

Montague, P. R., and G. Berns, 2002, "Neural Economics and the Biological Substrates of Valuation." *Neuron,* vol. 36, pp. 265–84.

Moore, R. Laurence, 1994, *Selling God: American Religion in the Marketplace of Culture.* New York: Oxford University Press.

Moravec, Hans, 1988, *Mind Children: The Future of Robot and Human Intelligence.* Cambridge, Mass.: Harvard University Press.

MotDoc, 2004, "Cargo Cults." Available at http://www.bbc.co.uk/dna/h2g2/A2267426.

Moya, Andrés, and Enrique Font, 2004, *Evolution: From Molecules to Ecosystems.* Oxford: Oxford University Press.

Moynihan, Daniel Patrick, 1970, Memorandum to President Nixon on the status of Negroes, as reported in *Evening Star*, Washington, D.C., March 2, p. A5.

Nagel, Thomas, 1997, *The Last Word*. Oxford: Oxford University Press.

Nanda, Meera, 2002, *Breaking the Spell of Dharma and Other Essays: A Case for Indian Enlightenment*. Delhi: Three Essays Press.

———, 2003, *Prophets Facing Backwards: Postmodern Critiques of Science and Hindu Nationalism in India*. New Brunswick, N.J.: Rutgers University Press.

National Academy of Sciences, 1999, *Science and Creationism*. 2nd ed. Washington, D.C.: National Academy Press.

Needham, Rodney, 1972, *Belief, Language and Experience*. Chicago: University of Chicago Press.

Nesse, Randolph, ed., 2001, *Evolution and the Capacity for Commitment*. New York: Russell Sage Foundation.

Newberg, Andrew, Eugene D'Aquili, and V. Rause, 2001, *Why God Won't Go Away: Brain Science and the Biology of Belief*. New York: Ballantine.

Nietzsche, Friedrich, 1887, *On the Genealogy of Morals*. Trans. Walter Kaufmann. New York: Vintage, 1967.

Norris, Kathleen, 2000, "Native Evil." *Boston College Magazine*, Winter.

Oliver, Simon, 2003, Review of Denys Turner, *Faith Seeking*. *Times Literary Supplement*, November 14, p. 32.

Pagel, Mark, ed., 2002, *Encyclopedia of Evolution*. 2 vols. Oxford: Oxford University Press.

Pagels, Elaine, 1979, *The Gnostic Gospels*. New York: Random House.

Palmer, Craig T., and Lyle B. Steadman, 2004, "With or Without Belief: A New Approach to the Definition and Explanation of Religion." *Evolution and Cognition*, vol. 10, pp. 138–45.

Panikkar, Raimundo, 1989, *The Silence of God: The Answer of the Buddha*. Maryknoll, N.Y.: Orbis Books.

Pennock, Robert, 1999, *Tower of Babel: The Evidence Against the New Creationism*. Cambridge, Mass.: MIT Press.

Perakh, Mark, 2003, *Unintelligent Design*. Amherst, N.Y.: Prometheus Books.

Persinger, Michael, 1987, *Neurophysiological Bases of God Beliefs*. New York: Praeger.

Pinker, Steven, 1994, *The Language Instinct*. New York: Morrow.

———, 1997, *How the Mind Works*. New York: Norton.

Plantinga, Alvin, 1996, "Darwin, Mind and Meaning." *Books and Culture*, May/June.

Pocklington, Richard, in press, "Memes and Cultural Viruses." In *Encyclopedia of the Social and Behavioral Sciences*.

Pocklington, Richard, and Michael L. Best, 1997, "Cultural Evolution and Units of Selection in Replicating Text." *Journal of Theoretical Biology*, vol. 188, pp. 79–87.

Posner, Richard, 1992, *Sex and Reason*. Cambridge, Mass.: Harvard University Press.

Povinelli, Daniel, 2003, *The Folk Physics of Apes: The Chimpanzee's Theory of How the World Works*. Oxford: Oxford University Press.

Premack, David, and Guy Woodruff, 1978, "Does the Chimpanzee Have a Theory of Mind?" *Behavioral and Brain Sciences*, vol. 1, pp. 515–26.

Purves, William K., David Sadava, Gordon H. Orians, and H. Craig Heller, 2004, *Life: The Science of Biology*. 7th ed. Sunderland, Mass.: Sinauer and W. H. Freeman.

Pyper, Hugh, 1998, "The Selfish Text: The Bible and Memetics." In J. C. Exum and S. D. Moore, eds., *Biblical Studies and Cultural Studies*. Sheffield: Sheffield Academic Press, pp. 70–90.

Pyysièainen, Ilkka, 2001, *How Religion Works: Towards a New Cognitive Science of Religion*. Leiden: Brill.

Quine, W.V.O., 1960, *Word and Object*. Cambridge, Mass.: MIT Press.

———, 1974a, "Comment on Donald Davidson." *Synthese*, vol. 27, pp. 325–30.

———, 1974b, "Comment on Michael Dummett." *Synthese*, vol. 27, pp. 413–17.

Ramachandran, V., and Sandra Blakeslee, 1998, *Phantoms in the Brain: Probing the Mysteries of the Human Mind*. New York: Morrow.

Ramachandran, Vilayanur, W. S. Hirstein, K. C. Armel, E. Tecomka, and V. Iragui, 1997, "The Neural Basis of Religious Experience." Paper delivered to the *Annual Conference of the Society of Neuroscience, October. 1997. Abstract no. 519.1*. Vol. 23, (Washington, D.C.: Society of Neuroscience).

Rappaport, Roy A., 1979, *Ecology, Meaning and Religion*. Richmond, Calif.: North Atlantic Books.

———, 1999, *Ritual and Religion in the Making of Humanity*. Cambridge: Cambridge University Press.

Ratzinger, Cardinal, 2000, "Dominus Iesus: On the Unicity and Salvific Universality of Jesus Christ and the Church." Declaration ratified by Pope John Paul II at a plenary session, June 16, 2000. Available at http://www.vatican.va/roman_curia/congregations/cfaith/documents/.

Ridley, Mark, 1995, *Animal Behaviour*. 2nd ed. Boston: Blackwell Scientific Publications.

Ridley, Matt, 1993, *The Red Queen: Sex and the Evolution of Human Nature*. New York: Macmillan.

Rooney, Andy, 1999, *Sincerely, Andy Rooney*. New York: Public Affairs.

Ross, Don, 2005, *Economic Theory and Cognitive Science: Microexplanation*. Cambridge, Mass.: MIT Press.

Rougement, Denis de, 1944, *Le Part du Diable*. Trans. Haakon Chevalier as *The Devil's Share*. New York: Meridian Books, 1956.

Rubin, D. C., 1995, *Memory in Oral Traditions*. New York: Oxford University Press.

Rue, Loyal, 2005, *Religion Is Not About God*. New Brunswick, N.J.: Rutgers University Press.

Ruhlen, Merrit, 1994, *The Origin of Language*. New York: John Wiley & Sons.

Sacks, Oliver, 1995, *An Anthropologist on Mars*. New York: Knopf.

Saenger, Paul, 2000, *Space Between Words: The Origin of Silent Reading (Figurae Reading Medieval)*. Stanford, Calif.: Stanford University Press.

Sahlins, Marshal, 1972, *Stone Age Economics*. Chicago: Aldine.

Sanneh, Kelefa, 2004, "Pray and Grow Rich." *New Yorker*, October 11, pp. 48–57.

Schönborn, Christoph, 2005, "Finding Design in Nature: The Catholic Church's Official Stance on Evolution," July 7, 2005, http://www.nytimes.com/2005/07/07/opinion/07schonborn.html.

Schumaker, John F., 1990, *The Corruption of Reality: A Unified Theory of Religion, Hypnosis, and Psychopathology*. Amherst, N.Y.: Prometheus Books.

Schwarz, N., and G. L. Clore, 1983, "Mood, Misattribution, and Judgments of Well-Being: Informative and Directive Functions of Affective States." *Journal of Personality and Social Psychology*, vol. 45, pp. 513–23.

Seabright, Paul, 2004, *The Company of Strangers: A Natural History of Economic Life*. Princeton: Princeton University Press.

Sen, Amartya, 1999, *Development as Freedom*. New York: Knopf.

———, 2003, "Democracy and Its Global Roots." *New Republic*, October 6, pp. 28–35.

Shanks, Niall, 2004, *God, the Devil, and Darwin: A Critique of Intelligent Design Theory*. Oxford: Oxford University Press.

Sharlet, Jeffrey, 2003, "Jesus Plus Nothing." *Harper's Magazine*, March, pp. 53–58.

Shea, Christopher, 2005, "Big Picture Guy: Does Megaselling Scientist-Historian Jared Diamond Get the Whole World Right?" *Boston Globe*, January 16, p. C1.

Shehadeh, Raja, 2002, *Strangers in the House: Coming of Age in Occupied Palestine*. South Royalston, Vt.: Steerforth Press.

Shermer, Michael, 2003, *How We Believe: Science, Skepticism and the Search for God*. 2nd ed. New York: A. W. Freeman/Owl Book.

Shermer, Michael, and Frank Sulloway, in press, "Religion and Belief in God." Manuscript, January 2005.

Shriver, Mark D., 1997, "Ethnic Variation as a Key to the Biology of Human Disease." *Annals of Internal Medicine*, vol. 127, pp. 401–3.

Siegel, Lee, 1991, *Net of Magic: Wonders and Deceptions in India*. Chicago: University of Chicago Press.

———, 2003, *Love and Other Games of Chance*. New York: Viking Penguin.

Silver, Mitchell, in press, *An Optional God: Secular Reflections on New Jewish Theology*. New York: Fordham University Press.

Skinner, B. F., 1948, "'Superstition' in the pigeon." *Journal of Experimental Psychology*, vol. 38, pp. 168–72.

Sloan, Richard P., and Emilia Bagiella, 2002, "Claims About Religious Involvement and Health Outcomes." *Annals of Behavioral Medicine*, vol. 24, no. 1, pp. 14–21.

Slone, Jason, 2004, *Theological Incorrectness*. Oxford: Oxford University Press.

Small, Maurice M., and Ernest Kramer, 1969, "Hypnotic Susceptibility as a Function of the Prestige of the Hypnotist." *International Journal of Clinical and Experimental Hypnosis*, vol. 17, pp. 251–56.

Sober, Elliott, and David Sloan Wilson, 1998, *Unto Others: The Evolution and Psychology of Unselfish Behavior*. Cambridge, Mass.: Harvard University Press.

Sperber, Dan, 1975, *Rethinking Symbolism*. Cambridge: Cambridge University Press.

———, 1985, *On Anthropological Knowledge*. Cambridge: Cambridge University Press.

———, 1996, *Explaining Culture: A Naturalistic Approach*. Oxford: Blackwell.

———, 2000, "An Objection to the Memetic Approach to Culture." In Aunger, ed., *Darwinizing Culture*, pp. 163–73.

Sperber, Dan, and Deirdre Wilson, 1986, *Relevance: A Theory of Communication*. Cambridge, Mass.: Harvard University Press.

Spong, John Shelby, 1998, *Why Christianity Must Change or Die*. New York: HarperCollins.

Stark, Rodney, 2001, *One True God: Historical Consequences of Monotheism*. Princeton: Princeton University Press.

Stark, Rodney, and W. S. Bainbridge, 1985, *The Future of Religion: Secularization, Revival and Cult Formation*. Berkeley: University of California Press.

———, 1987, *A Theory of Religion*. New York: David Lang.

Stark, R., W. S. Bainbridge, and D. P. Doyle, 1979, "Cults of America: A Reconnaissance in Space and Time." *Sociological Analysis*, vol. 40, pp. 347–459.

Stark, Rodney, and Roger Finke, 2000, *Acts of Faith: Explaining the Human Side of Religion*. Berkeley: University of California Press.

Sterelny, Kim, 2003, *Thought in a Hostile World: The Evolution of Human Cognition*. Oxford: Blackwell.

Stern, Jessica, 2003, *Terror in the Name of God: Why Religious Militants Kill*. New York: HarperCollins.

Sulloway, Frank J., 1996, *Born to Rebel: Birth Order, Family Dynamics, and Creative Lives.* New York: Pantheon.

Taubes, Gary, 2001, "The Soft Science of Dietary Fat." *Science,* vol. 291 (March 30), pp. 2536–45.

Taylor, S. E., and J. D. Brown, 1988, "Illusion and Well-being: A Social Psychological Perspective on Mental Health." *Psychological Bulletin,* vol. 103, pp. 193–210.

Tetlock, Philip, 1999, "Coping with Trade-Offs: Psychological Constraints and Political Implications." In S. Lupia, M. McCubbins, and S. Popkin, eds., *Political Reasoning and Choice,* Berkeley: University of California Press.

———, 2003, "Thinking the Unthinkable: Sacred Values and Taboo Cognitions." *Trends in Cognitive Science,* vol. 7, pp. 320–24.

Tetlock, Philip, A. Peter McGraw, and Orie Kristel, 2004, "Proscribed Forms of Social Cognition: Taboo Trade-Offs, Forbidden Base Rates, and Heretical Counterfactuals." In N. Haslam, ed., *Relational Models Theory: A Contemporary Overview.* Mahwah, N.J.: Erlbaum, pp. 142–61.

Tinbergen, Niko, 1948, "Social Releasers and the Experimental Method Required for Their Study." *Wilson Bulletin,* vol. 60, pp. 6–52.

———, 1959, "Comparative Studies of the Behaviour of Gulls (*Laridai*): A Progress Report." *Behaviour,* vol. 15, pp. 1–70.

Titon, Jeff Todd, 1988, *Powerhouse for God: Speech, Chant, and Song in an Appalachian Baptist Church.* Austin: University of Texas Press.

Tomasello, Michael, 1999, *The Cultural Origins of Human Cognition.* Cambridge, Mass.: Harvard University Press.

Tomasello, Michael, and Josep Call, 1997, *Primate Cognition.* Oxford: Oxford University Press.

Wall, T. L., L. G. Carr, and C. L. Ehlers, 2003, "Protective Association of Genetic Variation in Alcohol Dehydrogenase with Alcohol Dependence in Native American Mission Indians." *American Journal of Psychiatry,* vol. 160, no. 1 (January 1), pp. 41–46.

Weinberg, Steven, 2003, *Facing Up: Science and Its Cultural Adversaries.* Cambridge, Mass.: Harvard University Press.

Whitehouse, Harvey, 1995, *Inside the Cult: Religious Innovation and Transmission in Papua New Guinea.* Oxford: Clarendon Press.

———, 2000, *Arguments and Icons.* Oxford: Oxford University Press.

Whiten, Andrew, and R. Byrne, eds., 1988, *Machiavellian Intelligence.* Oxford: Oxford University Press.

———, 1997, *Machiavellian Intelligence, II: Extensions and Evaluations.* Cambridge: Cambridge University Press.

Wiesel, Elie, 1966, *The Gates of the Forest.* New York: Holt, Rinehart & Winston.

———, 1972, *Souls on Fire: Portraits and Legends of Hasidic Masters.* N.p.: Gerecor, Ltd. Reprint, New York: Random House, 1982.

Williams, George, 1966, *Adaptation and Natural Selection.* Princeton: Princeton University Press.

———, 1992, *Natural Selection: Domains, Levels, and Challenges.* Oxford: Oxford University Press.

Wilson, David Sloan, 2002, *Darwin's Cathedral: Evolution, Religion, and the Nature of Society.* Chicago: University of Chicago Press.

Wilson, David Sloan, and Elliott Sober, 1994, "Re-Introducing Group Selection to the Human Behavior Sciences." *Behavioral and Brain Sciences,* vol. 17, no. 4, pp. 585–654.

Wittgenstein, Ludwig, 1953, *Philosophical Investigations.* Oxford: Blackwell.

Wolfe, Alan, 2003, *The Transformation of American Religion: How We Actually Live Our Faith*. New York: Free Press.

Wynne, Thomas, 1995, "Handaxe Enigmas." *World Archeology*, vol. 27, pp. 10–23.

Yoder, Don, 1974, "Toward a Definition of Folk Religion." *Western Folklore*, vol. 33, pp. 2–15.

Young, Matt, and Taner Edis, 2004, *Why Intelligent Design Fails: A Scientific Critique of the New Creationism*. New Brunswick, N.J.: Rutgers University Press.

Zahavi, A., 1987, "The Theory of Signal Selection and Some of Its Implications." In V. P. Delfino, ed., *International Symposium on Biological Evolution, Bari, 9–14 April 1985*. Bari: Adriatici Editrici, pp. 305–27.

Zimmer, Carl, 2000, "Parasites Make Scaredy-Rats Foolhardy." *Science*, July 28, pp. 525–26.

Index

||||\\\||||

faith (*cont.*)
  and credal athleticism, 229–30
  mysteries understood through, 228–29
  proving our, 231
  religion distinguished from, 230–31
  unquestioning, 295
  "what everyone knows" as based on, 160–62
faith-based initiatives, 230
Faith Seeking (Turner), 232
"fake it till you make it," 217, 250
falsehoods, benefits from belief in, 273
Family (Fellowship Foundation), 338–39, 411n17
"family values," 279, 326
fanaticism, 13, 35, 56, 256–57, 284,
      299–301, 310
"fan loyalty," 35–36
fantasy, 120, 121–22, 127, 315
fasting, 4
fats, craving for, 72–73
fatwas, 256, 285, 301
fear
  of God, 266
  of heights, 189
Feldman, Marcus, 345
Fellowship Foundation (Family), 338–39,
      411n17
female selection, 87–89
female shapes, 74
feminism, 207, 260
Feynman, Richard, 219–20, 231, 233
Fides et Ratio (John Paul II), 228–29
Fifth Amendment, 360
fifth-order intentionality, 111
Finke, Roger, 69, 179, 183, 185, 189,
      190–91, 194–98, 209, 225, 315, 319,
      400n17, 407n3
Finnish language, 318
firms, religious organizations as, 168
First Amendment, 308, 334
First Cause, 241, 242
fish, schooling of, 90, 179
fitness, 3
  defined, 69–70
  differential, 341
  frequency-dependent, 231
  group, 90–91, 178, 182–85
  happiness contrasted with, 156–57
  of memes, 347, 350
  pearls and, 91
  of religion, 85, 177–78
  religion as fostering, 269
flattery, 113, 114
Fleming, Ian, 238, 285
flipping a coin, 132
flocking of birds, 90, 179
flu season of 2003–2004, 38

flying carpets, 125
folklore, 355
folk medicine, 79, 83
folk music, 79–80, 140, 153, 398n8
folk processes, 79–80
folk psychology, 111–14, 163
folk religion, 140–41, 398n7
  design influences of, 77, 140
  as practical know-how, 156–62
  reflection entering into, 162–63
  as seamless part of life, 160–61
  transition to organized religion, 153,
      167–73, 227, 315
  wild memes of, 170
former religions, 10, 391n4
formulas, believing versus understanding,
      217–25
Foucault, Michel, 373, 405n12
foxes, 114
franchises, 168
Frank, Robert, 255
Franklin, Benjamin, 259, 349
Franklin, C. L., 155, 400n1
freedom, 17, 253, 305, 331
Freedom Evolves (Dennett), 408n4
freedom of religion, 305, 334, 359, 360
free-floating rationale, 82, 91, 109, 170, 182,
      192, 283, 304, 309
  for children's obedience to parents, 129–30
  for culturally transmitted designs, 78
  for curiosity, 366
  for divination, 133
  for liking for sweets, 59, 63
  for money, 67
  represented rationales replacing, 177
  for romantic love, 255
  for witnessing, 364
free love, 48
free riders, 361
freethinkers, 300
free will, 202, 245, 325
frequency-dependent fitness, 231
Freud, Sigmund, 7, 127, 373, 392n6
fruit, 58–59, 66–67
Frum, John, 99–100
Fry, Christopher, 74
full-access agents, 126, 131
functionalism, 181–82
fundamentalism, 190–91, 225, 331
future, anticipating the, 37–38
Future of an Illusion, The (Freud), 7

Gaia hypothesis, 181
gambling, 13
Gauchet, Marcel, 98

7/22